Electromagnetic Interaction with Biological Systems

Electromagnetic Interaction with Biological Systems

Edited by
James C. Lin
University of Illinois
Chicago, Illinois

Plenum Press • **New York and London**

Library of Congress Cataloging in Publication Data

Electromagnetic interaction with biological systems / edited by James C. Lin.
　　p.　　cm.
　　Proceedings of the Joint Symposium on Interactions of Electromagnetic Waves with Biological Systems, held as part of the Twenty-Second General Assembly of the International Union of Radio Science, August 25–September 2, 1987, in Tel Aviv, Israel.
　　Includes bibliographies and index.
　　ISBN 978-1-4684-8061-0　　　　　ISBN 978-1-4684-8059-7 (eBook)
　　DOI　10.1007/978-1-4684-8059-7
　　1. Nonionizing radiation – Physiological effect – Congresses. 2. Nonionizing radiation – Diagnostic use – Congresses. 3. Nonionizing radiation – Therapeutic use – Congresses. 4. Nonionizing radiation – Safety measures – Congresses. I. Lin, James C. II. International Union of Radio Science. General Assembly (22nd: 1987: Tel Aviv, Israel)
QP82.2.N64E44　1989
612′.01448 – dc19　　　　　　　　　　　　　　　　　　　　　　88-38957
　　　　　　　　　　　　　　　　　　　　　　　　　　　　　　　CIP

Proceedings of the Joint Symposium on Interactions of Electromagnetic Waves with Biological Systems, held as part of the Twenty-Second General Assembly of the International Union of Radio Science, August 25–September 2, 1987, in Tel Aviv, Israel

© 1989 Plenum Press, New York
Softcover reprint of the hardcover 1st edition 1989
A Division of Plenum Publishing Corporation
233 Spring Street, New York, N.Y. 10013

PREFACE

     Ever since the early 1940's, electromagnetic energy in the nonionizing
spectrum has contributed to the enhanced quality of life in a variety of
ways. Aside from their well-known roles in communication, entertainment,
industry and science, electromagnetic energy has come into wide spread use
in biology and medicine. In addition to the intended purposes, these
energies produce other effects which have been shown to influence the life
processes of living organisms. It is noteworthy that these energies are
not only harmless in ordinary quantities but are actually necessary for
modern life, indeed without which life as we know it would be impossible.

     The purpose of this book is to present a succinct summary of the
interaction of electromagnetic fields and waves with biological systems as
they are now known. The subject matter is interdisciplinary and is based
primarily on presentations scheduled for a joint symposium at the XXII
General Assembly of the International Union of Radio Science, held in Tel
Aviv, Israel from Tuesday, August 25 to Wednesday, September 2, 1987. The
symposium was jointly sponsored by the Bioelectromagnetics Society in
cooperation with the International Radiation Protection Association.

     The choice of topics was made to facilitate the application and to
stimulate the use of nonionizing electromagnetic energy in biology and
medicine, and to increase the awareness and to promote the consideration of
radiation safety by electrical engineers and experimental physicists.
Therefore, the book is organized into three parts: Part One is devoted to
selected topics of current applications and investigations in diagnostics
and therapeutic uses of nonionizing electromagnetic energy including
noninvasive sensing, radiometry and thermography, diagnostic imaging, and
hyperthermia treatment for cancer. In Part Two, the biological effects are
summaried for stationary, time-varying as well as radio frequency and
microwave fields. The physical properties of and mechanisms for the
interaction of electromagnetic energies with biological molecules, cells
and tissues are also discussed. Part Three surveys available safety
protection guides from North America, Eastern and Western Europe as well as
other parts of the world; a special emphasis is placed on rationales
leading to the establishment of exposure standards and protection guides.

     I take pleasure in expressing here my appreciation to the Naval
Medical Research and Development Command in Bethesda, Maryland, for its
support of the research covered in my own chapter. The Walter Reed Army
Institute of Research in Washington, D.C., and the Office of Chief of Naval
Research, Arlington, Virginia provided generously for the travel of many of
the authors to the URSI General Assembly. The occasion has permitted

engineers and scientists from many countries an opportunity to interact and exchange their latest findings and observations with one another. Lastly, it is a pleasure to recognize the secretarial assistance of Kristine Grzyb and Su Lee, whose skill and diligence have made the tasks of writing and editing less of a drudgery, and to acknowledge the cooperation of the authors, whose intellectual endeavor made the publication of this volume a reality.

James C. Lin

University of Illinois
Chicago, Illinois

CONTENTS

PART I

MEDICAL DIAGNOSTICS AND THERAPY

MICROWAVE NONINVASIVE SENSING OF PHYSIOLOGICAL SIGNATURES

James C. Lin

Department of Bioengineering
University of Illinois
Chicago, IL 60680-4348

INTRODUCTION

Knowledge of the physiologic or pathophysiologic status of the heart and vessels as a transportation system for blood, and lungs as the site of gas exchange are factors that can greatly assist medical practitioners in the management of cardiovascular and pulmonary diseases. In light of prevalence of mortality and morbidity associated with heart, vascular and lung diseases, considerable efforts have been devoted to the development of noninvasive diagnostic techniques which not only are safe, but also offer the possibility of earlier detection as well as quantification of these disease states. Most recently, the advantage afforded by non-contact and remote sensing has engendered a great deal of excitement regarding the use of such technologies for monitoring patients with critical burns and premature developments, and personnel fell prey to such hazardous environments as fire, chemical or nuclear contamination and natural or man-caused disasters.

The application of microwaves for noninvasive sensing of physiological variables may be classified into active and passive modes (Fig. 1). The quantity of interest in both situations is energy transfer between a source and a receiver. This involves the analysis of waves that propagate from emitter to receiver. There are several possible methods for active sensing. Spatially resolved images may be formed through either projection or tomographic reconstruction processes to depict dielectric permittivity changes associated with tissue structural discontinuities (Larsen and Jacobi, 1986). Alternatively, time-varying signatures can be detected to permit active interrogation of cutaneous and subcutaneous tissue movements, even though the spatial resolution that can be attained is somewhat limited. In the case of reflection measurement, the microwave energy transmitted from the source antenna is backscattered by the biological target and received by the detection system. The backscattered wave provides information on the biological target and on factors that govern the propagation to and from the biological target. In contrast, passive measurement involves observation of microwave emissions from subcutaneous tissues, and conversion of those thermally generated microwave emissions to tissue temperatures. This radiometric technique can noninvasively measure subcutaneous tissue temperatures to a depth of several centimeters. Moreover, when medical radiometric measurements are made at several different frequencies or positions, it is possible to retrieve the

3

temperature profiles as a function of depth in tissue. However, a stable and unique solution to this inverse problem is not guaranteed. Nevertheless, some promising approaches have emerged in recent years.

The objective of this paper is to provide an overview of active modes of noninvasive microwave sensing for interrogation of movements attending cardiovascular and pulmonary activities, and to present recent developments in clinical and laboratory experimentation. The discussion will begin with a brief summary of the principal phenomena associated with microwave propagation and scattering in biological tissues.

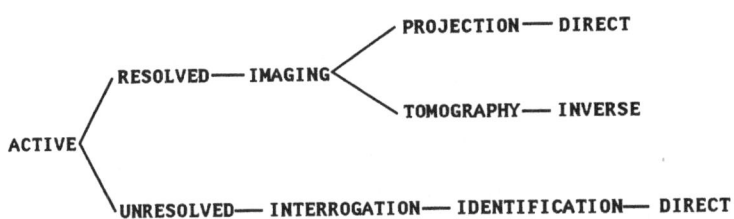

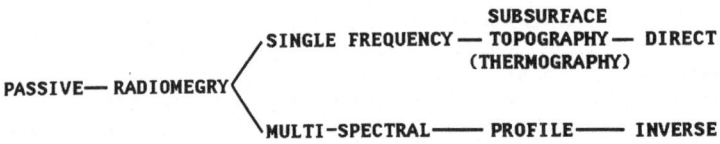

Fig. 1. Active and passive modes for microwave noninvasive sensing of physiological variables.

PROPAGATION AND SCATTERING OF MICROWAVES

Microwaves are refracted, scattered and transmitted at boundaries separating different material media. These phenomena are governed by the source frequency, antenna configuration, dielectric permittivity and geometry of the biological body or tissue. The transmission and backscattering (reflection) are characterized by the transmission coefficient T, and reflection coefficient R, respectively. For a plane wave impinging normally from a medium of permittivity ε_1, on a medium of permittivity ε_2,

$$T = (2 \sqrt{\varepsilon_1})/(\sqrt{\varepsilon_1} + \sqrt{\varepsilon_2}) \qquad (1)$$

where $\varepsilon = \varepsilon_0 (\varepsilon_r - j \sigma/\omega\varepsilon_0)$ with relative dielectric constant ε_r, free-space permittivity ε_0, conductivity σ, and radian frequency $\omega = 2\pi f$ (frequency). The reflection coefficient is given by $R=T-1$. The fraction of incident power reflected by the discontinuity is R^2 and the transmitted fraction is

T^2. As the transmitted wave propagates in the second medium, energy is extracted from the wave and this in a reduction in E, magnitude of the electric field strength, such that

$$E = E_o \, e^{-\alpha z - j\beta z} \qquad\qquad (2)$$

In this case E_o is the maxium value of the electric field at the interface, and and are the attenuation and propagation coefficients, respectively. The energy extraction will result in a progressive reduction of the wave's power density as it advances in the medium. This reduction is quantified by the depth of penetration $\delta = 1/\alpha$ which is the distance in which the power density decreases by a factor of $\exp(-2)$. It should be noted that, and are frequency dependent and take on different values for differing materials.

Dielectric Permittivity

The dielectric permittivity of biological materials in the frequency range of interest is largely determined by cell membranes and tissue water. They appear as lossy dielectrics to microwave radiation, and consequently, have magnetic permability equal to that of free-space and independent of frequency. In contrast, electrical properties (permittivity) of tissues are very frequency dependent in that, dielectric constants decrease and conductivities increase with increasing frequency. Representative values of measured dielectric constant and conductivity at 37°C are given in Table 1 for selected body tissues along the calculated depth of penetration and power transmission coefficient (Michaelson and Lin, 1987).

It is worthy of note that the dielectric constant and conductivity of muscle or tissues with high water content are an order of magnitude greater than the corresponding values for fat or tissues with low water content. This difference yields a depth of penetration for fat nearly ten times greater than for muscle. In general, the depth of penetration into tissue varies inversely with frequency. The transmitted powers at air-tissue interfaces are quite substantial, especially at higher frequencies. It can be seen from Table 1 that transmitted power at air-fat interface is about twice as great as for air-muscle interface.

Table 1. Propagation Characteristics of Microwave Radiation in Biological Tissues at 37°C

Frequency (MHz)	Dielectric Constant		Conductivity (S/m)		Depth of Penetration (cm)		Power Transmission Coefficient from Air	
	Muscle	Fat	Muscle	Fat	Muscle	Fat	Muscle	Fat
915	51	5.6	1.60	.10	2.50	12.8	0.40	0.83
2450	47	5.5	2.21	.16	1.67	8.13	0.43	0.84
5800	43	5.1	4.73	.25	0.80	4.75	0.44	0.85
10000	40	4.5	10.3	.44	0.33	2.59	0.45	0.87

The difference in dielectric permittivities also gives rise to wavelengths for higher water content materials about one-third of the wavelength in tissues with low water content. Furthermore, the wavelength in tissue is nearly ten times shorter than that in air at a given frequency. This fact will help to improve the resolving power of microwaves in medical diagnosis. For example, the wavelength in air at 10 GHz is 30 mm. Table 2 shows a wavelength that is reduced to 5 mm in muscle. This will improve the spatial resolution of 10 GHz radiation in muscle by a factor of six.

Reflection at Planar Tissue Interfaces

The reflection of microwaves at boundaries separating different tissue media is an important element in all noninvasive sensing techniques associated with cardiovascular and pulmonary interrogation. A basic understanding of the phenomenon can be obtained from a consideration of the reflection of plane waves at a planar surface.

The reflection of plane wave at a plane interface depends on the frequency, polarization, and angle of incidence of the wave, and on the dielectric constant and conductivity of the tissue. A wave of general polarization usually is decomposed into its orthogonal linearly polarized components whose electric or magnetic field is parallel to the interface, i.e. E and H polarizations, respectively. For E polarization, there is only a slight variation in magnitude and phase of the reflection coefficient with incidence angle. For H polarization, however, there is a pronounced dependence on incidence angle. The reflection coefficient reaches a minimum magnitude and has a phase angle of 90° at the Brewster angle (Lin, 1986). Thus, the H polarized wave is totally transmitted into the muscle medium at the Brewster angle.

For a normally impinging plane wave Table 3 summarizes the magnitude of reflection coefficient for planar boundaries separating various tissues in the thorax at four frequencies of most interest to noninvasive sensing of cardiovascular and pulmonary variables. The fraction of normally incident power reflected by the discontinuity is obtained from R^2. Clearly, the reflected power at air-tissue interfaces is quite substantial at all frequencies and about one-half of the incident power is reflected at these boundaries. The reflection coefficients for tissue-tissue interfaces generally are smaller than air-tissue interfaces. The values range from a low of five for muscle-blood to a high of 60 for fat-blood interfaces. This suggests that the greater the difference in dielectric properties across the interface, the higher the power reflection.

Table 2. Wavelength Contraction in Biological Tissues at 37^oC.

Frequency (MHz)	Wavelength (mm)			
	Blood	Air	Muscle	Fat
915	41	328	44	137
2450	16	173	18	52
5800	7	52	8	23
1000	4	30	5	14

Table 3. Reflection Coefficients between Biological Tissues at 37°C

	Frequency (MHz)	Air	Fat (Bone)	Lung	Muscle (Skin)	Blood
Air	915	0	43	73	78	79
	2450	0	41	71	76	77
	5800	0	39	70	75	76
	10000	0	37	70	74	76
Fat (Bone)	915		0	43	52	54
	2450		0	42	50	53
	5800		0	42	50	53
	10000		0	45	52	54
Lung	915			0	12	14
	2450			0	10	15
	5800			0	10	14
	10000			0	10	13
Muscle (Skin)	915				0	4
	2450				0	5
	5800				0	4
	10000				0	3
Blood	915					0
	2450					0
	5800					0
	10000					0

When there are several layers of different tissues, the reflection and transmission characteristics become more complicated. Multiple reflections can occur between the skin and subcutaneous tissue boundaries, with a resulting modification of the reflection and transmission coefficients (Michaelson and Lin, 1987). In general, the transmitted wave will combine with the reflected wave to form standing-waves in each layer. Figure 2 shows the distribution of electric field strength in a semi-infinite layer of heart muscle beneath finite layers of fat, muscle and bone for two frequencies. It is seen that in addition to frequency dependence, the electric fields exhibit considerable fluctuation within each tissue layer. While the standing-wave oscillations become bigger at 2450 MHz than 915 MHz, microwave energy at both frequencies can penetrate into more deeply situated tissues. This implies that at these frequencies sufficient energy may be transmitted and reflected to allow interrogation of the cardiovascular and respiratory organs.

It should be noted that reflection from an air-skin interface will give rise to similar standing-wave pattern with peak and valley that vary as function of distance. Hence, the electric field strength will oscillate as a function of location away from the biological target. This suggests the risk of very little power at the standing-wave minima. This difficulty may be alleviated by using slightly different viewing aspects (spatial diversity) or with different frequencies (frequency diversity) or quadrature signal processing that takes into account the sine and cosine temporal variations.

Although reflection and transmission characteristics, and depth of penetration in planar tissue structures provide considerable physical insight into coupling, distribution and scattering of microwave radiation. Biological bodies have complex geometries and exhibit substantial curvature that can modify microwave transmission and reflection. For bodies and organs with complex shape, the propagation characteristics depend critically on the polarization and orientation of the incident wave with respect to the target and the ratio of target size to wavelength. These complications place severe limitations on transmission and reflection calculations for bodies of arbitary shape and complex permittivity. Some

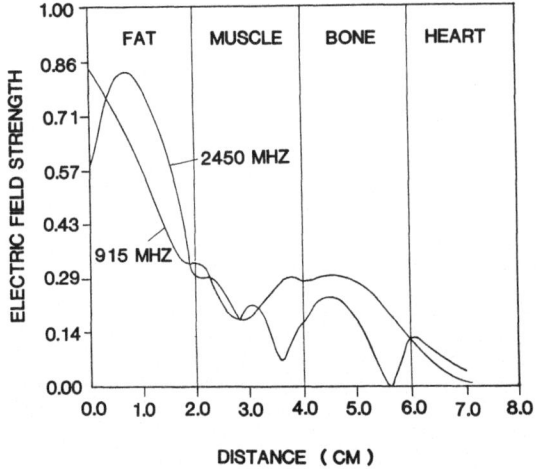

Fig. 2.   Distribution of electric field strengths in a planar model of fat, muscle, bone, and heart muscle.

results that have been obtained from prolate spheroidal models may serve as a pattern that can be used to estimate reflected energy for other cases (Michaelson and Lin, 1987). While the scattering coefficients for the E and H polarization usually differ from each other except for the forward- and back-scattered (reflected) components. The scattered energy varies widely with angle of observation. In general, the smaller the ratio of body size to wavelength the more uniform the distribution of scattering coefficient as a function of observation angle. Furthermore, measurement in the E-plane (plane parallel to electric field vector) usually would be more advantageous compared to the H-plane.

Doppler Phenomena from Target Motion

When microwave is scattered from a biological target moving relative to a receiver, the received energy undergoes an apparent frequency change, generally referred to as the Doppler shift. Using the scheme illustrated in Fig.3 a relation can be obtained (Lin, 1986) between the Doppler frequency change f_d and the target velocity u as

$$f_d = -(1/2\pi)(\underline{k}_r - \underline{k}_t)\cdot\underline{u} \tag{3}$$

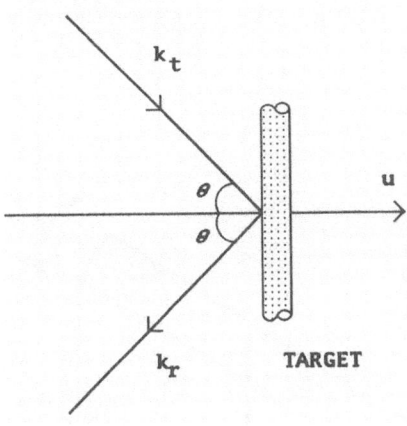

TRANSMITTER

k_t

θ

u

θ

k_r

TARGET

RECEIVER

Fig. 3.   Plane wave scatter   by a moving biological target.

where $\underline{k}_r$ and $\underline{k}_t$ are propagation vectors associated with the receiver and transmitter, respectively. If the receiving and transmitting antennas are located in close proximity of each other or are the same, i.e., $\underline{k}_r = -\underline{k}_t$, Eq. (3) then reduces to

$$f_d = (1/2\pi) \underline{k}_r \cdot \underline{u} = -2f(u/v)\cos\theta \tag{4}$$

where f is the source frequency, v is the velocity of microwave and θ is the angle between the target velocity vector and the direction of wave propagation. It is seen from Eq. (4) that f_d is directly proportional to the target velocity and takes on the largest value when $\theta = 0$ or 180° such that

$$f_d = \pm 2fu/v \tag{5}$$

9

where the plus and minus signs account for movements toward and away from the transmitter, respectively.

In noninvasive sensing of physiological signatures, various parts of the target fill all or an appreciable portion of the incident beam, the target velocity varies over the beam so that the Doppler component has a spectrum of frequencies. For example, during contraction the heart rotates anteriorly by about 4°. If we consider the situation depicted in Figure 4 where a rotation imparts an angular velocity ω of the target about its center of gravity, two fixed points on the target a distance s apart will have a relative radial velocity toward the transmitter of

$$\Delta u = u_1 - u_2 = \omega s \cos \zeta \qquad (6)$$

Hence from Equation (5) the Doppler frequency difference between these two points is

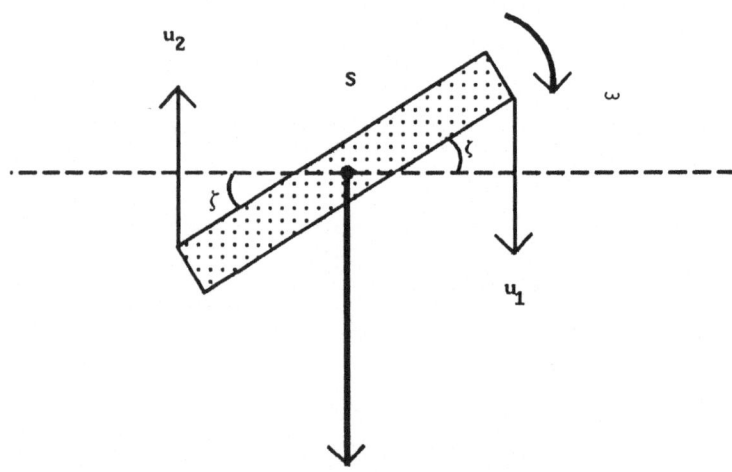

DIRECTION OF TRANSCEIVER

Fig. 4.   Doppler shift produced by the rotation of a moving biological target.

$$\Delta f_d = 2f\Delta u/v = 2f\omega s \cos \zeta /v \qquad (7)$$

Thus, the Doppler spectrum will be proportional to the angular velocity of the target and the gross aspect of the target. In a Doppler system the spectrum will be detected as frequency shifts relative to the transmitter frequency. This is usually accomplished through mixing the back-scattered wave with the transmitted wave and then measure the difference frequency by using a digital counter or by passing the demodulated signal through a set of bandpass filters.

If Equation (5) is multiplied and integrated over time, while neglecting the constant term of integration, then

$$\phi(t) = 2\pi \int f_d(t)dt = 4\pi f\, x(t)/v \qquad (8)$$

where $\phi(t)$ is the instantaneous phase variation corresponding to the distance $x(t)$ traveled by the target. Thus, the Doppler phase shift is related to target displacement, while Doppler frequency is related to target velocity. If the demodulated waveform is fed through a low-pass filter, the output $g(t)$ will be a signal that varies with time and target motion such that

$$g(t) = G \sin \phi(t) = G \sin [4\pi f\, x(t)/v] \qquad (9)$$

where G is the amplitude related to the reflected wave. Since the instantaneous displacement of the target involved in cardiovascular and pulmonary interrogation is small compared to the wavelength in tissue (see Table 2) for microwave frequencies of most interest, an approximate relation for $g(t)$ can be obtained as

$$g(t) = 4\pi Gf\, x(t)/v = 4\pi G\, x(t)/\lambda_t \qquad (10)$$

where $f\lambda = v$. Under these conditions, it may be concluded that there is a linear relationship between the output of the receiver and the displacement of the biological target.

Thus far the propagation phenomenon has been described in terms of plane waves impinging on parallel layers of tissues and simple surfaces isolated in free space. In many practical situations however interaction occurs in the near-field rather than the far-field where plane wave predominates. Furthermore, they involve applicators or antennas that are comparable in dimensions to the wavelength. This combination, together with the fact that most biological targets consist of complex surfaces of irregular shape makes the propagation characteristics too complicated to permit accurate description.

Because of the fundamental nature of plane wave interaction the understanding obtained serves a useful purpose when used with proper precautions, although to be used effectively they must be supplemented by further detailed information on the nature of the problem. For example, the data given in Table 1 shows that electromagnetic energy at 915 MHz can penetrate three times as deep into the tissues as energy at 5800-MHz. This implied advantage of lower frequencies is somewhat misleading in that as the frequency is decreased, the wavelength and the applicator size become increasingly large until it is no longer possible to direct the wave to a desired biological target with reasonable applicator dimensions. If the applicator is not increased in size as frequency is lowered, the radiated energy will rapidly diverge and be scattered by the target; only a small fraction of the available energy will penetrate the tissues.

Other Factors

In addition to above, there are some other considerations that have relevance in microwave noninvasive sensing of physiological signatures. The filtration scheme alluded to in the previous section offers a simple technique for detection of the Doppler signal under continous-wave (CW) operations. However, this CW approach does not provide target range determination and is sensitive to motions from all ranges. An alternative approach is to employ a train of narrow pulses transmitted at a rate lower than the expected Doppler frequency. This approach is referred to as pulse-Doppler or range gating and will be described at greater length in the next section. A brief discussion on range and velocity resolution is presently in order.

Resolution defines the ability of a system to differentiate as separate two closely spaced targets. This ability depends partially on the signal processing scheme employed. In the case of range, resolution is achieved mainly through proper selection of bandwidth of the transmitted waveform. It has been found that two targets can be resolved in range quite simply if they are separated by at least 0.8 pulse-width (Skolnik, 1980). When a number of targets with different velocities are present, the received signal is a composite of corresponding number of Doppler shifted signals. In order to adequately separate multiple targets in motion, it will be necessary to apply the signal for a sufficient duration. Thus, resolution in Doppler frequency depends on the signal duration. It is possible to isolate multiple targets by the use of either high range resolution or narrow band Doppler filtration. However, large signal-to-noise ratios are necessary if good resolution is desired. A rule-of-thumb suggests that the signal-to-noise ratio must be greater than 20 dB in order to obtain reasonable results.

The pulse Doppler technique combines the range discrimination ability of pulse systems with the frequency discrimination ability of CW systems by using a coherent pulse train, i.e., a train of pulses which are samples of a single unmodulated sine wave. When the pulse train is reflected by a moving target, the signal is Doppler frequency shifted in proportion to the target's velocity. In the presence of multiple targets with different velocities, the reflected signal is a superposition of a corresponding number of pulse trains, each with its own Doppler shift. A range gate is employed to select only those pulse reflected from the target. A narrow-hand filter following the range gate passes only the frequencies corresponding to a particular Doppler component, thus blocking all those pulse trains which pass the range gate but do not have the proper Doppler shift. The pulse Doppler technique offers the capability of measuring range and velocity unambiguously over a predetermined plane of the ambiguity domain in the presence of multiple targets, albeit at the expense of considerably more system complexity.

DETECTION CIRCUITS AND PROCESSING SYSTEMS

Over the years, a number of systems have been reported for detection of cardiovascular and respiratory signatures with microwave radiation. These developments generally make use of amplitude and phase information contained in the transmitted and reflected waves and operate under CW conditions. A simplified functional block diagram is shown in Figure 5. The system consists of a microwave signal source, a sampling device, a pair of transmitting and receiving antennas, an amplitude or phase detector, a set of filters, a visual display unit and/other output devices.

The microwave signal source in the desired frequency range provides signal strength required for detection. The selection of a microwave frequency depends on serveral factors. At higher frequencies where spatial resolution would be best, the penetration depth is very short (see Tables

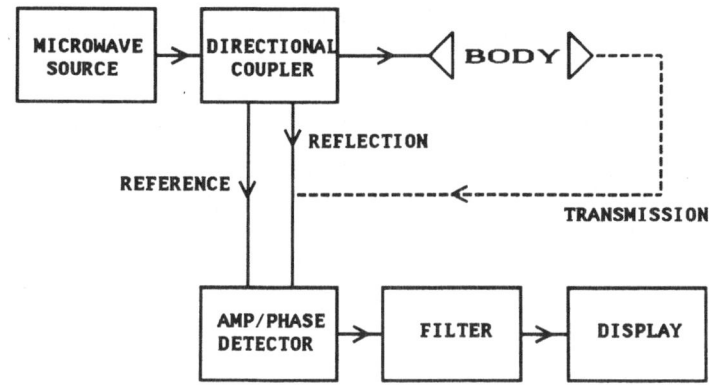

Fig. 5. A functional block diagram of microwave noninvasive sensing system.

1,2,3). The detection of energy transmitted through and reflected from deep-lying tissue interfaces becomes impractical. As the frequency is lowered, the penetration depth increases. However, the antenna aperture needed to efficiently deliver the electromagentic energy also increases. This along with the increase in wavelength would degrade the spatial resolution of the system to the point where it becomes useless as a noninvasive sensing device. The choice of operating frequency thus depends on a compromise between signal intensity that decreases and spatial resolution that increases with frequency. Fortunately, wavelength contraction that naturally occurs in biological tissues, and by use of antennas loaded with high dielectric constant materials permit, microwave interrogation of deep-lying organs at frequencies between 2 and 8 GHz with manageable attenuation loss and practical spatial resolution in tissue media (Lin, 1985). It should be noted that frequencies as high as 25 GHz have been used successfully for sensing subcutaneous arterial wall motion (Lee and Lin, 1986; Papp et al, 1987). Similarly, sources that operate at 35GHz have been shown to provide reliable signatures of heart rate and respiration at distances as far as 3.0 meters (Seals et al, 1983) . The sampling device or directional coupler diverts a small portion (20-50db) of the forward power to serve as a continuous reference for the detector. It also samples the back-scattered signal for use in reflection measuring systems. The antennas couple the microwave energy into and from the biological targets.

Microwave antennas developed for diagnostic applications utilize two basic schemes for coupling microwave energy into tissue media: non-contact and direct-contact methods. Non-contact antennas are normally spaced at distances of a few cm ten's of meters from the subject and make use of the scattered field. Energy reaching a receiving antenna under these circumstances may follow multitudinous pathways in and around the body. This multipath propagation imposes a severe limitation on experiments involving transmission measurements. For applications in which extraction of fine structures from the physiological signal is important, direct-contact methods that minimize scattered radiation are preferred over non-contact schemes. In contrast, back-scattered radiation can be efficiently employed in many diagnostic and noninvasive sensing applications associated with reflection measurement, since multipath propagation contributes less significantly to back-scattered radiation. Hence both non-contact and direct-contact methods are useful.

The radiation pattern of an antenna determines the distribution of radiated energy. This pattern is the same for transmitting and receiving operations. Antennas developed for biomedical applications usually have radiation patterns that give maximal intensity in the forward direction. However, the radiation pattern in the near field is quite different from that in the far field although there is a smooth transition from one to the other. Near the antenna, the radiation pattern adheres to the same width as the width of the antennas aperture; that is; the beam is confined to within a cylindrical column in the region. The finite beam-width permits electromagnetic energy to be directed to the biological target of interest and provides better spatial resolution. Also power density is not uniquely defined. However, it may be estimated from dividing the total power delivered by the anenna aperture. Using this relationship, the average radiated electromagnetic power density from present systems ranges from approximately 0.001 to one mW/cm^2. The antenna aperture ranges from 0.25 to 200 cm^2.

The distance from the antenna for which the beam dimension matches the aperture dimension is a function of aperture size and wavelength. This distance is equal to square of the antenna width or aperture dimension D, divided by twice the wavelength, such that

$$D^2/2\lambda \qquad\qquad (11)$$

At greater distances, the m-axis power density varies inversely as the square of the distance but in the near field oscillates about the constant value indicated above.

The Doppler shifted signal may be detected by sensing the instantaneous phase difference between the received signal and the reference source signal. The output from the detector after low-pass or band-pass filtration is a signal whose amplitude reflects the instantaneous displacement of the moving target and whose envelope variation corresponds to the target velocity. It should be mentioned that a matched filter or its equivalent will be most appropriate for optimal detection.

The remainder of this section will discuss several practical microwave systems for noninvasive sensing of physiological signatures. It will include some succinct descriptions of signal processing techniques and feature extraction schemes which have been used to treat the detected Doppler signals.

Microwave Circuitry

The microwave hardware can vary from very simple devices to reasonably complex designs intended to minimize temporal drift and spurioas noise since the system employed in practice, almost always operates with reflected energy, the transmissions and reception functions usually share the same antenna.

For contact applications, a microwave oscillator may be combined with a small ribbon antenna to serve as a simple yet reliable pulse motion sensor (RCA,1987). The microwave circuit consists of a bipolar transistor oscillator with a lightly coupled antenna (Fig. 6). A moving tissue interface in close proximity of the sensor will affect the loading (antenna coupling) on the oscillator current. This changing current can be detected

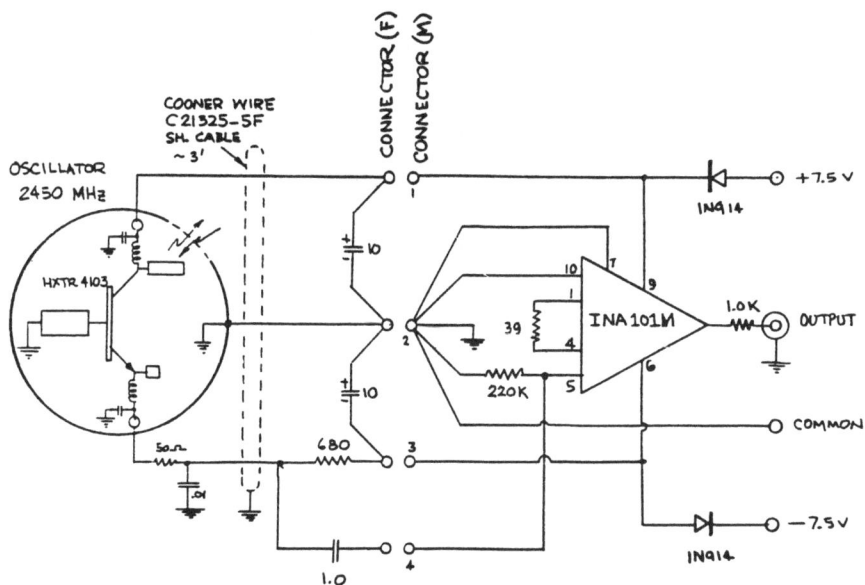

Fig. 6. The schematic diagram of a miniaturized superficial temporal artery monitor (RCA, 1987).

using an AC coupled amplifier. To minimize sparious signals and extraneous motion artifacts, the sensor must be well shielded small, and having all microwave elements located within the sensing head. This may be accomplished by constructing the sensor as a two-piece assembly and by housing the amplifier and power supply away from the sensing head in a separate package. This scheme alleviates microwave transmission through a cable which is prone to motion and interference generated artifacts, since the change in current flow is essentially DC, the cable length is not restructed. Such a device may be operated with a DC power supply of 6 to 10-V at a nominal current drain of 10 to 15-mA. An output power of approximately 0.5 mW at 2.45GHz could be obtained using two 9-V transistor batteries.

There are several Doppler transceivers designed for commercial applications, e.g., speed measurement and motion sensing. The transceiver not only transmit microwave energy, it is also able to receive the back scattered signal. The output power is typically 5-10 mW in the frequency range of 10 to 25 GHz. The transceiver comprises a fixed tuned CW Gunn diode oscillator and a Schottky barrier mixer diode, assembled into a compact waveguide package (Fig. 7). The mixer diode combines the reflected signal with a small portion of the microwave signal generated by the Gunn oscillator a reference to produce a Doppler shifted signal that is proportional to velocity of the moving target.

For contact applications the output waveguide flange may be used to couple microwave energy directly into biological tissue. In contrast, for measurement at long distance a directional antenna attached to the waveguide output flange is preferred for gain and sensitivity enhancement. The transceiver is simple to operate and requires only the supply of a bias voltage to energize the device. The operating voltage and current are typically + 10V and 150 mA. This will give rise to an output signal of approximately 10-20 mV peak to peak, which is sufficient for direct display on most oscilloscopes and chart recorders.

15

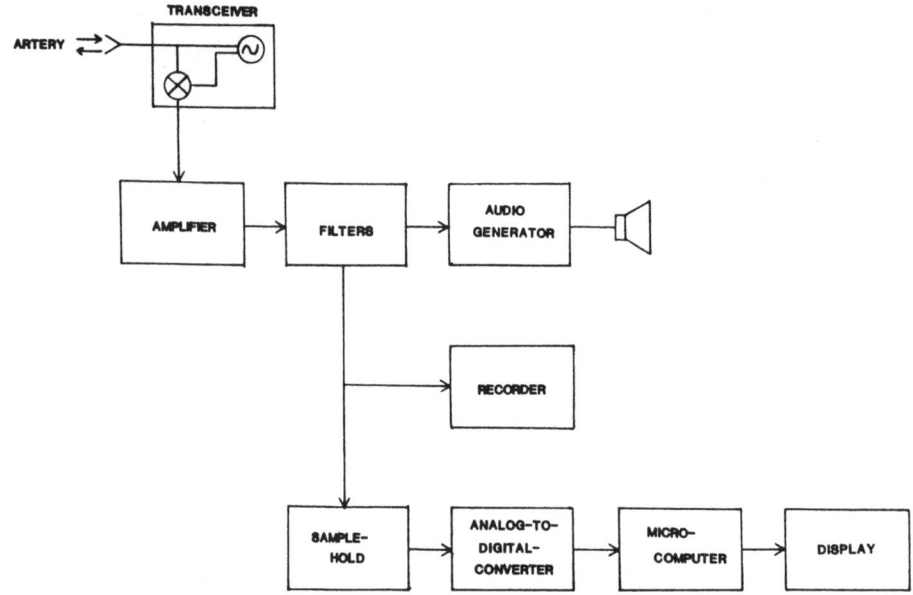

Fig. 7. A simplified diagram of a CW Doppler microwave transceiver.

    A limitation that can severely degrade the performance of these
simpler devices relates to the intrinsic channel noise and extrinsic motion
artifacts associated with extraneous body movements and time-varying
clutter in the environment. This problem may be partially overcome by
using the above mentioned pulse Doppler technique or appropriate signal
processing schemes to extract the signal from noisy measurement. It may
also be somewhat mitigated through the use of stable phase-locked microwave
oscillators and clutter cancellation circuitry (Chen et al., 1986).
Another common difficulty often encountered in these simpler systems is the
dead spot associated with nulls of the standing wave pattern. Slight
variation in positioning the detector with respect to the subject will
remove this ambiguity. Alternatively, a quadrature detection scheme which
splits the received signal into two orthogonal components could be utilized
to eliminate the dead spot (Sharpe et al., 1986). Examples of the
respiratory motion obtained using quadrature detection are shown in Fig 8.
If can be seen that at a given range the signal provided by one channel is
considerably stronger than the other and vice versa.

    As shown in Tables 1-3 the differences in propagation characteristics
of microwave energy give very different reflection and transmission
coefficients. While the reflection coefficient remains fairly constant
over the frequency range of interest, the depth of penetration varies
inversely with frequency. In fact, the depth of penetration at 2450 MHz is
about 3-5 times greater in tissues than 10 GHz. Such considerations have
led to the development of dual frequency systems that would allow efficient
detection of chest movements of a person with either light and dry or thick
and wet clothing (Popovic et al., 1984).

Signal Processing

    Several simple techniques have been applied to extract the heart and
respiration rates from the microwave signals including peak detection and
autocorrelation (Lin et al., 1979; Hoshal et al., 1984). However, in the

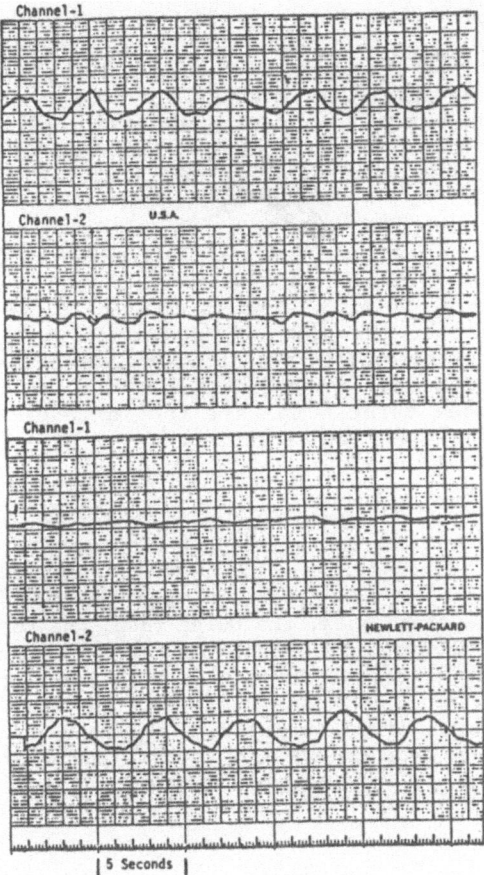

Fig. 8.   Quandrature detection of respiratory motion: The sensitivity of
channels 1 and 2 are adjusted such that one is maximized over the
other at a given distance (Sharpe et al., 1987).

presence of time-vary clutters and extraneous body movements, greater
sophistication is required of the signal processing schemes to reliably
detect these variables. For example, adaptive filtration techniques such
as least squares lattice filter, recursive least squares estimation and
first-order linear prediction have been evaluated for microwave heart
monitors. Of these, the normalized least squares lattice filter was found
to give the best performance in real-time operations (Byrne et al., 1986).
The adaptive filter technique does not produce beat per minute estimates of
the heart rate in the usual senses, which is the average of a large number
of heart beats over a considerable period of time. Instead, this technique
estimates the time interval between heart beats which is inversely
proportional to the instantaneous heart rate. These valves vary
considerably from beat-to-beat, so that farther processing is needed to
produce a smooth estimate of heart rate. The advantage of this approach is
that the detection of heart rate is separated from the final estimate of
the heart rate. Moreover, performance is not degraded by aperiodicity in
the signal, a severe drawback of autocorrelation techniques. The
normalized least squares algorithm is simple to implement and can reliably
detect heart beat in noisy environments given between 3-10 s of data,
sampled at a 120 Hz rate. This may permit fast identification of heart
activities such as arrhythmias.

Further, a pattern recognition technique have been used to extract the heart and respiration rates (Chan and Lin, 1985; 1987). Since the waveforms of heart and breathing signals differ only slightly from each other, the same algorithm is useful for both. Specifically, the recognition algorithm start by locating all the local maximum points. A slope test is performed by comparing the slopes approaching the data point under test and leaving it. Points at which the signal changes from upslope to downslope are the local maxima. The search for the foot of each cycle is accomplished by proceeding backwards from the remaining local maxima to a point where the slope becomes negative. This point is defined as the foot or beginning of the cycle, if its amplitude is less than a preset value. The first maximum point from the foot is defined as the peak of the cycle (either systolic peak or the peak of inspiration). After all the peaks have been located, the time difference between two succeeding peaks is then calculated. A median of all the periods calculated is used as a reference point to delete those values that are too large or too small, and an average is calculated from those remaining.

Feature Extration

A variety of techniques has been applied to extract from the microwave signal signatures of diagnostic quality. One feature extraction algorithm finds three peaks, two dicrotic notches, and two foots for the arterial pulse waveform. From these data, the pulse rate, the foot-to-systolic peak time, and the foot-to-dicrotic notch time are calculated (Lee amd Lin, 1985). The algorithm first locates three positive slopes (Fig. 9). The beginning of each slope (i.e., potential foot) is designated as the point where the slope changes from a negative slope to a positive slope. Once the potential foot is located, the positive slope is tracked until a peak

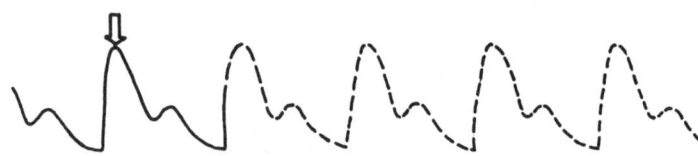

SEARCH FOR A MAXIMUM POINT.

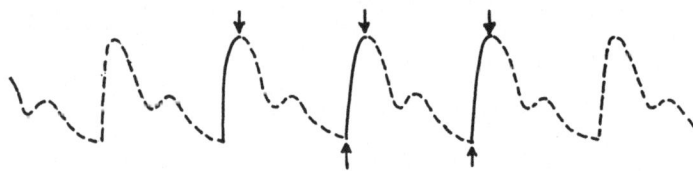

SEARCH FOR FOOTS AND SYSTOLIC PEAKS.

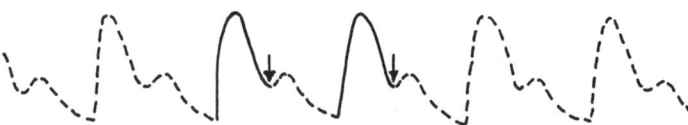

SEARCH FOR DICROTIC NOTCHES.

Fig. 9. Pulse feature extraction algorithm for the analysis of pressure pulse waveform (Lee and Lin, 1985).

18

is found. The peak must be above a threshold in order to avoid recognizing an anacrotic notch as the systolic peak. If the peak found is greater than the threshold, then the foot and systolic peak are declared found and are stored for later use. If the systolic peak is less than threshold, the search proceeds to seek another potential foot. In locating the systolic peak, the anacrotic shoulder is seen as a change in the ascending limb to a different upslope.

To find the dicrotic notch, two edge detectors or operators are used. The first operator performs a global search for the general area of the notch while the second operator performs a more localized search. For the global search, the algorithm begins by setting a search range to avoid erroneous fluctuations. The search range begins 21 samples (34 ms) after the systolic peak and ends either 70 samples (280 ms) after the systolic peak or one-half the distance between the systolic peak and the next foot whichever occurs first.

To begin the global search, a pointer is placed at the beginning of the search range. Then the 11 samples centered at the pointer are weighted using the following operator:

$$1 \quad 0 \quad 1 \quad 0 \quad 1 \quad -6 \quad 1 \quad 0 \quad 1 \quad 0 \quad 1.$$

A summation is made of the weighted samples and the sum is saved. The pointer is advanced and the weighting and summation is repeated until the end of the search range is reached. The sample point with the maximum sum is declared the potential dicrotic notch.

The localized search for the dicrotic notch is performed over ten sample points surrounding the potential dicrotic notch. The procedure used during the global search is repeated here, but the weighting only involves three sample points at a time. The three sample points are weighted with the following sequence of numbers:

$$1 \quad -2 \quad 1.$$

The summation performed in this instance produced the second difference. The sample point which has the maximum second difference is the dicrotic.

MEASUREMENT OF PHYSIOLOGICAL SIGNATURES

Sensing of body movements associated with the expansion and contraction of the circulatory and respiratory systems provides a noninvasive technique for measureing such vital physiological variables as respiration, heart and pulse rates and circulation. Microwaves provide a simple approach to detect the heart, arterial wall, and respiratory movements without compromising the integrity of these physiological phenomena. In this case, microwave energy is directed to the target and the reflected signal is processed to yield information on the organ of interest or the subject under interrrogation. This nonivasive technique provides a capability for continuous monitoring as well as quantification of time-dependent changes in the cardiopulmonary system. The advantage afforded by remote sensing suggests the use of this technology for monitoring patients with critical burns and premature developments, and personnel fell prey to such hazardous situations as fire, chemical or nuclear contamination and natural or man-made disaster. In the section to follow, recent results on microwave sensing of physiological signs and signatures will be briefly discussed to review current efforts and to indicate potential applications.

Noninvasive Sensing

Currently, there are three major areas in which a noninvasive microwave contact or noncontact, close range approach holds promise. These include ventricular movement, pressure pulse sensing and monitoring of superficial arterial circulation.

Low-frequency displacements of the precordium overlying the apex of the heart is related to movements in the left ventricle and echoes the hemodynamic events within the left ventricle. Microwave apexcardiograms obtained using 2.45 GHz showed close correlation to the hemodynamic events occuring within the left ventrical (Lin et al., 1979). It involves detecting the reflected Doppler signal using an antenna located a few cm over the apex of the heart. An example of microwave sensed ventricular movement in a healthy young male who held his breath throughout the measurement is shown in Fig. 10, together with simultaneously recorded electrocardiographic and phoncardiographic tracings. It is seen that toward the end of systole, a rapid rising wave occurs due to ventricular filling, which is completed by atrial contraction occuring between the P-wave and QRS-complex in the electrocardiogram. A rapid downward deflection represents maximal ventricular ejection following a period of isometric contraction, just after the QRS-complex. The ventricular movement reaches a plateau at the level of midsystole, and it is then followed by

MICROWAVE APEXCARDIOGRAM

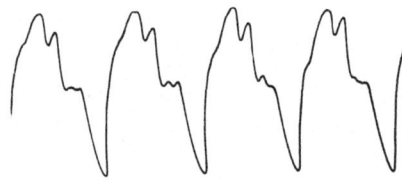

PHONOCARDIOGRAM

ELECTROCARDIOGRAM

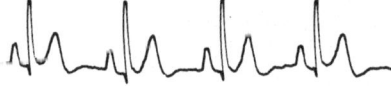

TIME (sec.)

Fig. 10. The microwave apexcardiogram - microwave sensed ventricular movement (Lin et al., 1979).

20

another downward delection which coincides with the aortic valve opening and completes the cardiac cycle. The method may therefore be used to delineate fine structures in ventricular activity.

Doppler microwaves have been employed to interrogate the wall properties and pressure pulse characteristics at a variety of arterial sites, including the carotid, brachial, radial and femoral arteries (Stuchly et al., 1980; Lee and Lin 1985; Papp et al., 1987). An example of microwave-sensed carotid pulse waveform in a patient using contact application of 25 GHz energy is shown in Fig. 11, along with simultaneously recorded intra-aortic pressure waves. Note the characteristic resemblance of the microwave-sensed arterial pulse and the invasively recorded pressure wave. Because of its basis on motion detection, this continuous wave device will detect other localized movements as well. For example, the microwave tracings Fig. 11 have a characteristic positive wave (beginning at arrows) before each carotid upstroke, which result from jugular venous expansion and are not present in the corresponding intra-aortic pressure waves. The consistency of the microwave pulse waves with respect to the intra-aortic pressure measurements was evaluated using five sequential pulse waves from each patient. The recordings were used to calculate left ventricular ejection time (LVET) and time required for a pressure to reach one half of its maximum amplitude in ms ($T_{1/2}$). Table 4 lists both the means and standard deviations for these two quantities. The results confirm that a noninvasive Doppler microwave sensor can successfully and reproducibly detect pressure pulse waveforms of diagnostic quality. With minimal training, a technician using this microwave sensor can usually obtain a good quality pulse waveform within 30 to 45 seconds as compared to an acoustic system which when used, often requires much longer detection times.

Table 4. Comparison of Intra-Aortic and Noninvasive Pulse Wave Data

	Patient	Measurement	T1/2 Mean	T1/2 S.D.	LVET Mean	LVET S.D.
A)	E. W.	(Intra-Aortic)	48.4	2.87	336.0	5.96
		(Microwave)	45.8	5.38	359.6	9.20
B)	M. O.	(Intra-Aortic)	49.6	0.80	265.0	5.47
		(Microwave)	42.0	1.89	261.0	3.74
C)	M. B.	(Intra-Aortic)	46.8	2.92	286.0	5.83
		(Microwave)	39.6	1.49	275.6	3.38
D)	A. D.	(Intra-Aortic)	38.2	3.06	322.0	6.78
		(Microwave)	36.0	2.00	330.0	7.07
E)	J. V.	(Intra-Aortic)	49.4	3.38	406.0	6.63
		(Microwave)	41.2	1.47	402.0	5.10

LVET = Left Ventricular Ejection Time in msec.
T1/2 = Time required for a pressure pulse to reach one half of its maximum amplitude in msec.

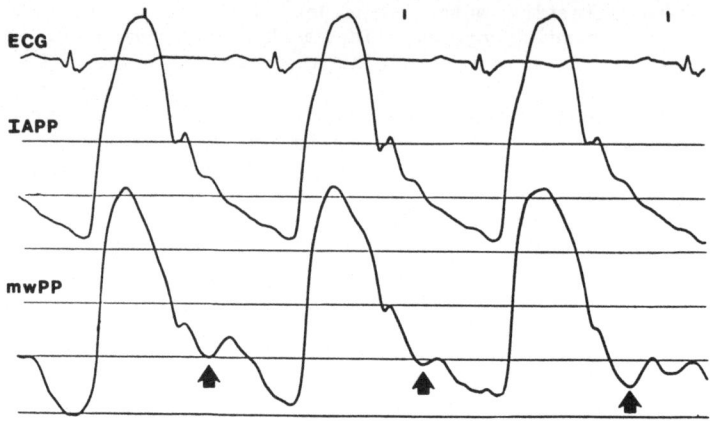

Fig. 11.   Comparison of Doppler microwave pulse (mWPP) and intra-aortic
           pressure pulse (IAPP) waveform. Note the jugular venous
           expansion artifact (beginning of arrow) in the microwave tracing
           (Papp et al., 1987).

The pulse waves obtained by this microwave technique can not be
interpreted on an absolute scale since they are derived from the relative
expansion of the sensed artery. Therefore, if used alone, microwave
Doppler can not provide numeric values for systolic and diastolic blood
pressure. However, this device could be used in conjunction with an
automatic blood pressure cuff which could provide the necessary calibration
for the pressure waves (Papp et al., 1987).

The miniaturized, mixer-less microwave sensor previously described has
been used to monitor pulsations at a number of superficial cranial artery
sites (RCA, 1987). It was found that while extraneous noise associated
with speech and head movement would occasionally produce debilitating
artifacts, positioning the sensor over the occipital branch of the
superficial temporal artery gave nearly noise-free recordings. It was
suggested that by placing the miniature sensor within a flight helmet may
allow monitoring of changes in the superficial temporal artery pulsations
during simulation or actual flight test.

Remote Sensing

The ability to remotely detect such vital signs as heart beat and
respiration is particularly useful in situations where direct contact with
the subject is either impossible or undersirable. Indeed, heart beat and
respiration have been detected at distances of a few to tens of meters,
with or without intervening physical barriers. The potential applications
include neonatal monitoring, burn management, and rescue operations
necessitated by fire, chemical or nuclear accidents, and such natural
disasters as earthquake and avalanche of snow.

The use of microwave in measuring respiratory movements of men and
animals have been demonstrated in several reports (Lin et al., 1973, 1975;
Lin and Salinger, 1975). The microwave-sensed respiratory history of a cat
subjected to differential heating of the head region is shown in Fig. 12.
A 10 GHz standard gain horn is directed toward the upper torso of the
subject at a distance of about 2 meters. It can be seen that the sensor is
capable of registering instantaneous changes in respiration. The

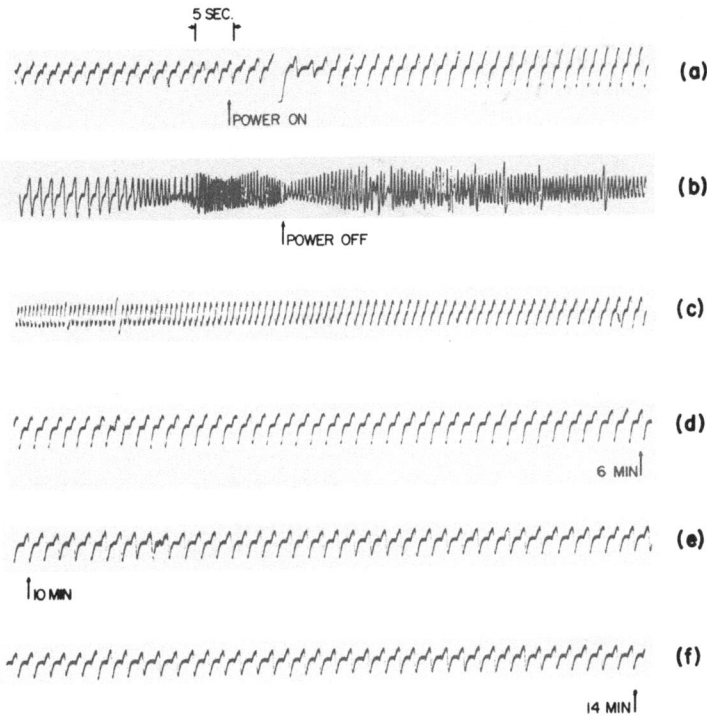

Fig. 12. Microwave tracing of the respiratory activity of a cat subjected
to a brief period of differential heating of the brain (Lin et
al., 1983).

respiration rate increased simultaneously with heating. A period of
hyperventilation was followed by an intense tachypnea. The rapid panting
ended about 14 minutes after brain heating. This approach has several
advantages over more conventional techniques because it does not require
any physical contact with the subject. Problems such as skin irritation,
restriction of breathing and electrode connections are easily eliminated.
The method appears to be well suited for monitoring of burn patients and
infants exhibiting apneic spells. It should be mentioned that Doppler
microwaves have been used to detect chest movement at a distance of a few
centimeters from the subject to extract both heart beat and respiration
rate (Lin et al., 1977; 1979; Chen et al., 1986; Chan and Lin, 1987).

Remote sensing of heart rate and respiration has been reported at
distances over 30 meters. The most frequently used frequencies are 2 and
10 GHz microwaves (Popovic et al., 1984; Chen et al., 1986; Sharpe et al.,
1987). Aside from the more directive and higher gain antennas, designs
that minimized various noise sources and achieve high sensitivity with low
levels of radiated power are some of the most important considerations
necessary for success. As mentioned earlier, the extended range and
sensitivity compared to the simpler Doppler systems are attained at the
expense of considerably greater system complexity. However, the resulting
capability is impressive. These systems are capable of detecting heart
beat and respiration rate of human subjects lying on the ground at a
distance of 30 m or more (Chen et al., 1986; Sharpe et al., 1987).
Moreover, a 2 GHz system was able to detect heart rate and respiration of

subject behind a 1-m thick, dry brick wall (Fig. 13). It is conceivable that similar devices may find uses in a variety of rescue related operations where direct physical contact with the subject is either impossible or undesirable.

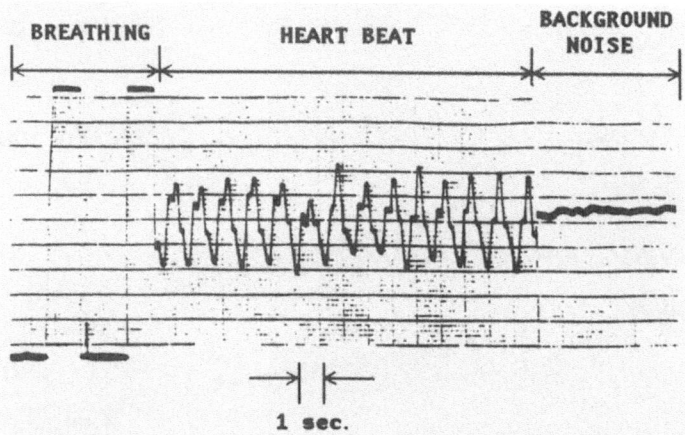

Fig. 13. Microwave (2 GHz) tracing of the heart beat and respiration of a human subject lying face-up under six layers (50cm) of dry bricks (Chen et al., 1986).

SUMMARY

The last decade has witnessed a growing interest in the use of microwave energy for noninvasive detection and monitoring of vital physiological signs and signatures. Some of the leading applications have been delineated in this paper, together with the basic techniques and operating principles. While it is unlikely that all the microwave pathways into the body have been enumerated, the afore-mentioned techniques are providing a capability for noninvasive detection and monitoring of physiological signs and signatures with unprecedented opportunities. Indeed, the large variety of potential uses of noninvasive measurement using microwaves, whether with direct contact or remote sensing, are just beginning to be realized.

It is also clear that each techniques has its own set of potential and critical limitations that preclude an overall solution for all the situations mentioned. Much work remains to be done in enhancing and extending the capabilities of these modalities. It is especially noteworthy that the photon energy of microwaves is sufficiently low to prevent any ionization occuring in biological tissue at ordinary intensities. Moreover, the average power emitted from these sensors are less than 20 mW, giviing rise to average power densities that are well below the American National Standards Institute recommendations for safe personnel exposure.

REFERENCES

Byrne, W., Flynn, R., Zapp, R. and Siegel, M., 1986, Adaptive filter processing in microwave remote heart monitors, IEEE Trans. on Biomed. Engg., 33: 717-722.

Chan, K.H. and Lin, J.C., 1985, An algorithm for extracting cardiopulmonary rates from chest movements, Proc. IEEE Engg. Medicine Biology Conf., 466-469.

Chan, K.H. and Lin, J.C., 1987, Microprocessor based cardiopulmonary rate monitor, Med. Biol. Engg. and Computing, 25: 41-44.

Chen, K.M., Misra, D., Wang, H., Chuang, H.R. and Postow, E., 1986, An X-Band microwave life-detection system, IEEE Trans. on Biomed. Engg., 33: 697-701.

Hoschal, G. Siegel, M., and Zapp, R., 1984, A microwave heart monitor and life detection system. In: IEEE Frontiers Engg and Computing in Health Care, 331-333.

Larsen, L.E. and Jacobi, J.H., 1986, "Medical Applications of Microwave Imaging," IEEE Press, New York.

Lee, J.Y., Lin J.C. and Popovic M.A., 1983, Microprocessor-based arterial pulse wave analyzer, Ann. Conf. Engg. Med. Biol., 79.

Lee, J.Y. and Lin, J.C., 1985, A microprocessor based non-invasive pulse wave analyzer, IEEE Trans. Biomed. Engg., 32: 451-455.

Lin J.C., Guy A.W. and Kraft G.H., 1973, Microwave selective brain heating, Journal of Microwave Power, 8: 275-286.

Lin, J.C. and Salinger, J., 1975, Microwave measurement of respiration, International Microwave Symposium, Palo Alto, Calif., 285-287.

Lin, J.C., 1975, Noninvasive microwave measurement of respiration, Proc. IEEE, 63: 1530.

Lin, J.C., Dawe, E. and Majcherek, J., 1977, A noninvasive microwave apnea detector, Proc. San Diego Biomed. Symp., Academic Press, 441-443.

Lin, J.C., Kiernicki, J., Kiernicki, M. and Wollschlaeger, P.B., 1979, Microwave apexcardiography, IEEE Trans. MTT, 27: 618-620.

Lin, J.C., 1985, Frequency optimization for microwave imaging of biological tissue, Proceedings of IEEE, 72: 374-375.

Lin, J.C., 1986, Microwave Propagation in Biological Dielectrics with Application to Cardiopulmonary Interrogation, In: Medical Applications of Microwave Imaging, L.E. Larsen and J.H. Jacobi, Eds., New York IEEE Press, 47-58.

Michaelson, S.M. and Lin, J.C., 1987, "Biological Effects and Health Implications of Radiofrequency Radiation," Plenum, New York.

Papp, M.A., Hughes, C., Lin, J.C. and Pouget, J., (1987) Doppler Microwave, A clinical assessment of its efficacy as an arterial pulse sensing technique, Invest Radiol, 22: 569-573.

Popovic, M.A., Chan, K.H. and Lin, J.C., 1984, Microprocessor-based noncontact heart rate/respiration monitor, IEEE Engg. Medicine Biol. Conf., Los Angeles, 754-757.

RCA Laboratories, 1987, "Miniature Superficial Temporal Artery Monitor," Final Report, Princeton, New Jersey.

Rozzell, T. and Lin, J.C. 1987, Biomedical application of electromagnetic energy, IEEE Engineer in Med. and Biol. Mag., 6: 52-56.

Seal, J., Sharpe, S.M., Schaefer, D.J. and Studwell, M.L., 1983, A 35 GHz FM-CW system for long-range detection of respiration in battlefield casualties. Abstract of the Bioelectromagnetic Society Meeting, 35.

Sharpe, S.M., MacDonald, A., Seals, J. and Crowgey, S.R., 1986, An electromagnetic-based non-contact vital signs monitor," Georgia Tech. Res. Inst., Biomed. Div., Atlanta.

Skolnik, M.E., 1980, "Introduction to Radar Systems," 2nd ed., McGraw-Hill, New York.

Stuchly, S.S., Smith, A., Goldberg, M., Thansandote, A., and Menard, A., 1980, A microwave device for arterial wall motion analysis. Proc. 33rd Annual Conf. Engg. Med. Biol., 22: 47.

MICROWAVE RADIOMETRY AND THERMOGRAPHY

Y. Leroy, A. Mamouni, J. C. Van de Velde, and B. Bocquet

Centre Hyperfréquences et Semiconducteurs
U.A. C.N.R.S. n°287 - Bât. P4
Universite des Sciences et Techniques de Lille Flandres Artois
59655 Villeneuve D'Ascq Cedex - France

INTRODUCTION

Based on the measurement of the electromagnetic thermal noise generated by living tissues, microwave radiometry is a starting point for a non invasive thermometric process in a depth of the subcutaneous tissues of up to several centimeters.

Thermological investigations are carried out in order to find ways of defining new processes of interest to medicine and especially for diagnostic methods and for a non invasive thermometric control in hyperthermia.

After a presentation of the physical bases of this process, we give a review of the most relevant research carried out in this field. We also present the results obtained recently by our research group.

* a multiprobe radiometric system has been constructed, and a high resolution imaging process has been settled. This process is now being submitted to clinical evaluation.

* a canonical inversion process fitted to the case of compact thermal structures has been defined, and is based on the data processing of the above mentionned images and of radiometric measurements obtained at two frequencies.

THE PRINCIPLE OF MICROWAVE RADIOMETRY

A probe or antenna, placed flush on a lossy material (in medical applications, the living tissues) sums up the thermal noise powers generated by the subvolumes of the tissues facing the probe. The power transmitted from one subvolume to the probe is proportionnal to

* the coupling parameter subvolume-probe

* the absolute temperature in the subvolume.

The coupling parameter (or weighting factor) is defined by the application of the reciprocity theorem to the couple of antennas constituted by the probe and the subvolume. This weighting factor is proportionnal to the conductivity in the subvolume multiplied by the square of the electrical field radiated at the same point when the probe is fed by a microwave generator at the same frequency (Robillard et al, 1982). In this way, the knowledge of the field radiated by the probe in the lossy material is needed ; note that this near field radiation is generally quite different to the T.E.M. mode and can be conveniently described only if a lot of propagation modes are taken into account (Robillard et al, 1982, Mamouni et al, 1987). For a homogeneous lossy material, this coupling parameter decreases with the distance to the probe, and for a given distance, it decreases with the frequency. In living tissues, these weighting factors have significant values in a depth of several centimeters at frequencies between 1 and 5 GHz.

The factor temperature found in the expression of the noise power results from Planck's law in terms of the Raleigh-Jeans law (an approximation available in microwaves for the considered temperatures).

The electronic system connected to the probe reduces to a quadratic receiver. In practice, several functions are encountered in the radiometer, for example an amplifier is needed due to the very low level of the thermal noise power. The construction of such a radiometer has to take into account the noise factor and the bandwidth of the receiver (G. Evans and C.W. Mc Leish, 1977) and also the possible mismatch at the interface between the probe and the material (A. Mamouni et al, 1977).

A REVIEW OF THE RESEARCH IN MICROWAVE RADIOMETRY

The pionneers in this field are Barret and Myers (1977) who constructed radiometers at 1.3 and 3.3 GHz and experimented a new process of breast cancer screening. This method is based on the combination of microwave and infrared thermographic investigations.

We began our work in radiometry at the end of the 70 s. After the construction of radiometric systems (Mamouni et al 1977, Leroy 1982) several series of thermological investigations were carried out (Gautherie et al 1980 ; Robert et al 1982). We also published the first paper explaining how such radiometric signals can be computed (Robillard et al 1982) and also to demonstrate the possibility of combining heating by microwaves and microwave radiometry in the same system (N'guyen et al 1979), a function which is well suited to carrying out a non invasive control of temperature in hyperthermia. After series of experiments on models and clinical evaluations, our studies led to the development of Hylcar systems (Hyperthermia with non invasive control by radiometry) devoted to the treatment of moderately deep seated tumors (Odam-Bruker Corporation).

Note also our work into correlation radiometry (Mamouni et al 1981-1983) based on the achievement of the correlation fonction of the noise signals received by two probes.

J. Edrich (1980) developed several radiometric systems working at 34 and 10 GHz in which the microwave emission produced by the tissues is focused through a large elliptic reflector into a horn connected to the radiometer. (Note that Edrich is the only author who is using remote antennas for medical applications). Edrich (1987) carried out different kinds of thermological investigations and has continued studying cancer screening such as that initiated by Barrett and Myers. Some of these studies are also devoted to the measurement of the subcutaneous heating produced by electrosurgical dispersive electrodes.

K. Carr (1981) developed a radiometer at 4.7 GHz. This system is experimented mainly on animals and on patients. Note that this radiometric system is associated to a transmitter (915 MHz) which can heat the tissues in front of the probe.

D. Land (1983) made several radiometers which are being experimented at the present time. These investigations are devoted to breast cancer detection rheumatoïd arthritis in the knee and inflammation of the appendix.

R. Paglione (1986) has constructed a radiometer working at 4 GHz which is used for the detection of appendicitis

Mizushina (1986) has made a three band radiometer (1.5-2.5 and 3.5 GHz) which is used for experiments on physical models and animals, in order to reconstruct the temperature profile in the tissues.

The literature also mentions several other radiometric systems (Iskander et al 1983).

A MULTIPROBE RADIOMETER FOR AN IMAGING PROCESS

Now that radiometric systems are available the next problem consists in the processing of the radiometric data suited to medical investigations. If we reconstruct an image with the radiometric intensities measured at the different points of the surface of the living tissues, we get a qualitative information about the temperature distribution which is present in the depth of the tissues accessible to the radiometric measurements. The examination of such an image makes possible the location of thermogen volumes, which can be associated to a pathology or have a biological meaning.

The synthesis of such images needs a lot of radiometric data. For this reason, we have first constructed a multiprobe radiometer (central frequency 3 GHz) (Enel et al, 1984) in order to make easier the acquisition of the data and their localization.

The multiprobe consists in six probes, joined close to each other, (the geometrical configuration of their apertures is shown in figure 1) sequentially connected to the radiometer by a multiport switch.

A microcomputer manages the commutation of the probes, stores the radiometric data, carries out their processing leading to the synthesis of a radiometric image, and also works in the temperature profile reconstruction which is described in the last section of this paper.

Prior to the synthesis of the image, it is necessary to know the volume of tissues the radiometric measurement is concerned with. For this reason a modal method has been applied and a numerical computation based on the integral equation at the discontinuity probe-material has been devised, which gives the electrical field radiated in the material (Mamouni et al 1987) and makes possible the computation of the radiometric signals for a defined situation. From these data, the weighting which has to be attributed to the different areas located in the aperture of the probe in terms of the evaluation of a radiometric sensitivity can be determined.

Next, we have to define the sequence of positioning of the multiprobe in the acquisition of the radiometric data.

Due to the fact that the foot print of the probes is mainly associated with the central area of their aperture, we have to carry out small displacements of the multiprobe leading to an overlapping of the surfaces which are covered successively by the multiprobe.

In the process shown in figure 2, 72 radiometric data, resulting from 12 positionings of the multiprobe, are obtained on an area of 7 cm x 7 cm (Leroy et al, 1987).

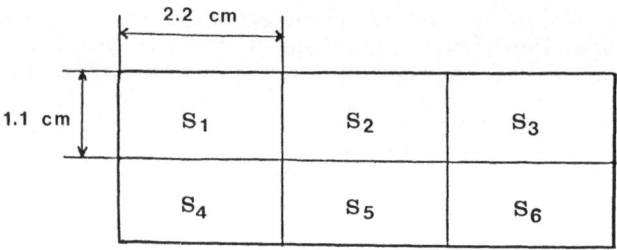

Fig. 1 . Geometry of the apertures of the multiprobe.

The synthesis of the radiometric image is then carried out : interpolations are made with respect to the radiometric data and to the weighting alloted to the different parts of the apertures of the probes. In figure 2, we show a so-called "longitudinal mode" of positioning in which the large sides of the probes are vertical ; a "transverse mode" of positioning can also be carried out in which the large sides of the probes are horizontal. It is also possible to combine the data of the "longitudinal mode" and "transverse mode" for the same area in order to get a 144 point radiometric image.

In practice, radiometric measurements are first carried out for a 72 point image. If the difference between the maximal and the minimal radiometric temperatures (considering the 72 radiometric data) is greater than 1.5°, a

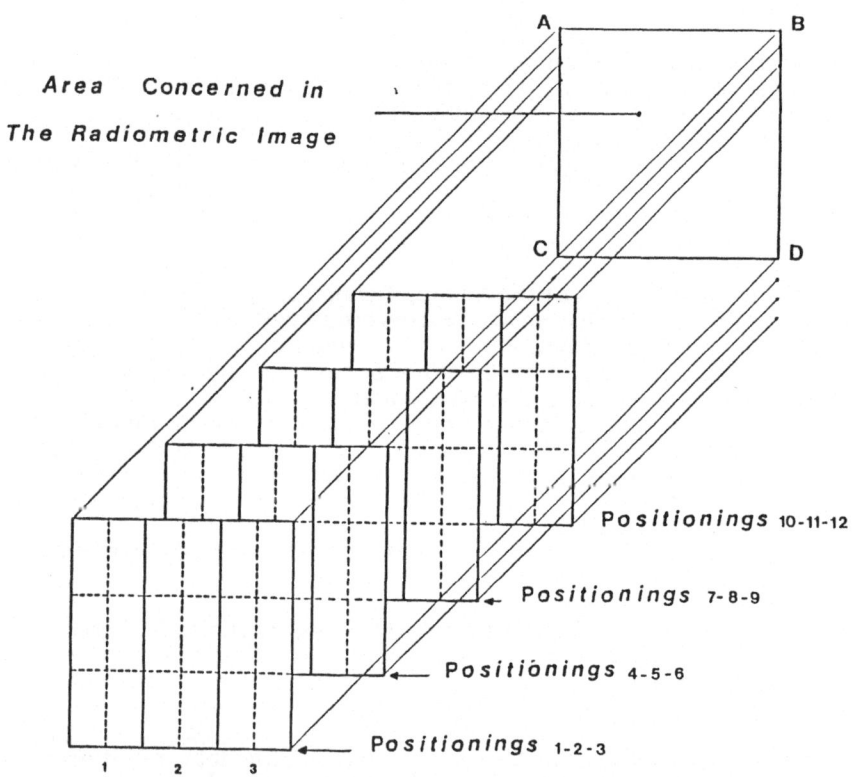

Fig. 2 . Mode of positionnings of the multiprobe in the realization of a 72 point radiometric image.

significant radiometric image can be synthesized. However if this temperature range is smaller than 1.5°, experience has shown that 72 data are not sufficient ; then we continue the data acquisition until we get 144 radiometric measurements. Note that in this process we observe an important progress in the spatial resolution because for a 144 point image the average distance between two near centres of the probe apertures is only 5 mm.

The clinical evaluations made with this process are based on the comparison of two radiometric images. The first one is obtained on an area of tissues (generally pathological tissues) in order to define new diagnostic processes based for example on the observation of hot or cold spots on the radiometric image. A second image, which corresponds to a similar area chosen on healthy tissues (generally on the symetrical part of the body), is considered as a reference. These images use 16 pseudocolours defined from the temperature range covered in the series of the radiometric measurements.

This imaging process is being used at this time in the Oncology Center of Lille (Pr. Giaux, Dr. Delannoy, Dr. Delvallée) for cancer investigations.

It has also been evaluated in the Atomic Energy Centre - Lab. of Radiobiology - in Jouy en Josas on pigs in order do observe the evolution of erythemas following an acute local irradiation by a radioactive source (Daburon et al 1983, 1987). The conclusion of these studies have shown microwave thermography to be a very efficient process for the follow-up of such diseases and for the test of drugs.

INTERPRETATION OF THE RADIOMETRIC MEASUREMENTS

The problem of the physical meaning of the radiometric signals in terms of a quantitative temperature profile retrieval has existed for as long as radiometric systems have been made ; attempts have been made in this field, but the problem is very difficult and has not yet been completely solved. Note that a complicated temperature profile can lead to ill-posed problems, and that their solution in a general case can be reached only by a lot of radiometric measurements at the different points of the area under investigation and at numerous frequencies.

Edenhoffer (1981) and Schaller (1984) have treated the case of the inversion of radiometric data at three frequencies (T.E.M. mode) on a multilayered material (skin, fat, muscle) by an optimization method.

Chive et al (1984) are interested in the temperature profite on the axis of a radiometric probe during a hyperthermia process, from the knowledge of the surface temperature, of the radiometric temperatures at two frequencies and, by application of the bio-heat equation.

Mizushina et al (1986) obtain the temperature versus depth profile on physical models and also on a living rabbit from the radiometric data measured by their system which works at three frequencies. They introduce a functional form for the unknown temperature distribution in depth (an exponential law). In this way they can estimate the temperature distribution over a depth of up to 5 cm with an accuracy of ± 0.5° C.

Bardatti et al (1985-1987) are tackling an inverse problem modeled as the solution of a first kind Fredholm integral equation. For example a retrieval algorithm, which consists of a "windowing" of the singular function expansion of the solution is applied to a bidimentionnal situation.

We are also interested in this kind of problem ; we are considering first the case of compact thermal structures, which seems to be a situation often encountered in practice. Our method starts from experiments on physical models.

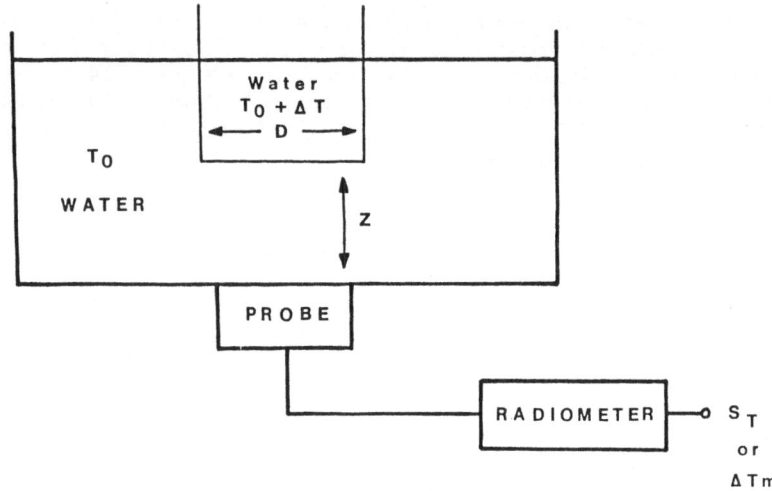

Fig. 3 . The experimental set up used for the radiometric measurements on the compact thermal structures.

The first step of this method is based on experimental data obtained with our radiometric imaging process (3 GHz) and with another radiometer working at 1.5 GHz. A lot of radiometric measurements have been carried out on items, generally cylinders, made of a thin material transparent to microwaves, filled up with water ; these items are put in a container which is also full of water (Fig. 3). The temperature T of the item is slightly different to the temperature T_0 of the water of the container (generally $\Delta T = T-T_o = 5°C$). Different diameters D of the item, and different depths Z -with respect to the surface on which are put the radiometric probes- are considered. The radiometric signal is defined as an excess of temperature ΔTm ; then we can define the ratio $\Delta Tm/\Delta T$ as a function of the depth Z for different values of the diameter D at 1.5 GHz (fig. 4) and at 3 GHz (Fig. 5).

This kind of experiment first leads to defining the conditions of visibility of an item (the limit of visibility corresponds to a signal to noise ratio of 0 dB) taking into account the sensitivity of the radiometers (0.1°C) as a function of its geometry (D), localization (z) and excess of temperature ΔT (Bocquet et al, 1986). In this way we have shown that in water, the visibility criterion is such as for $\Delta T = 1°C$ the depth of an item must be smaller than 4.5 cm at 1.5 GHz and 3 cm at 3 GHz ; its diameter must be greater than 1 and 2 cm at the frequencies 3 and 1.5 GHz.

We have also shown that data such as the values of $\Delta Tm/\Delta T$ as a function of D and z can be used in order to define a canonical method of inversion of the radiometric data. This method starts from the following property : the threshold of the radiometric image at 3 GHz is directly related to D ; then graphs $\Delta Tm/\Delta T$ (D, Z) at 1.5 and 3 GHz (with the probe just in front of the item) give directly z and ΔT. Note here the usefullness of the radiometric imaging, which is able to provide the size D of the structure : it is obvious, by examination of results such as fig. 4 and fig. 5, that without the knowledge of D, the radiometric data have no quantitative meaning.

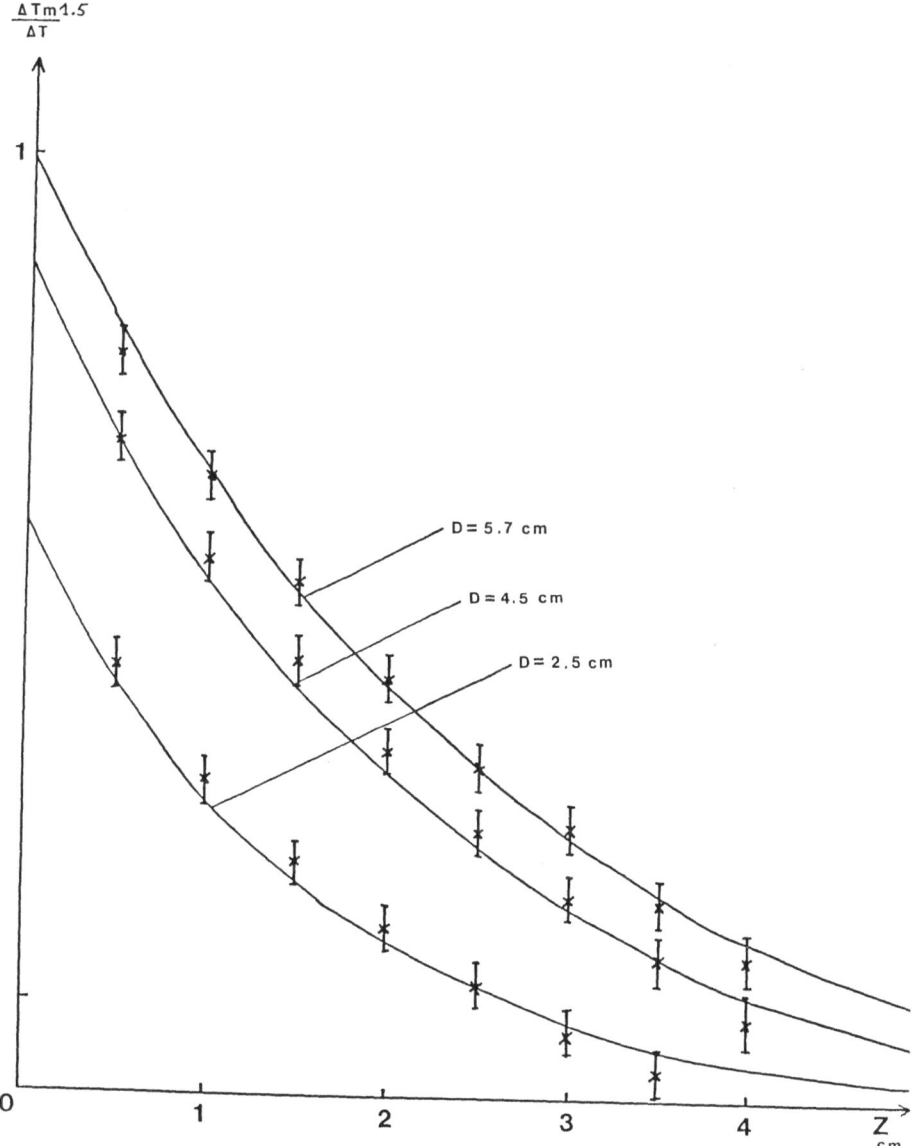

Fig. 4 . Radiometric signals for the situation defined in Fig. 3 on water
(To = 33°C, ΔT = 5°C) f = 1.5 GHz.

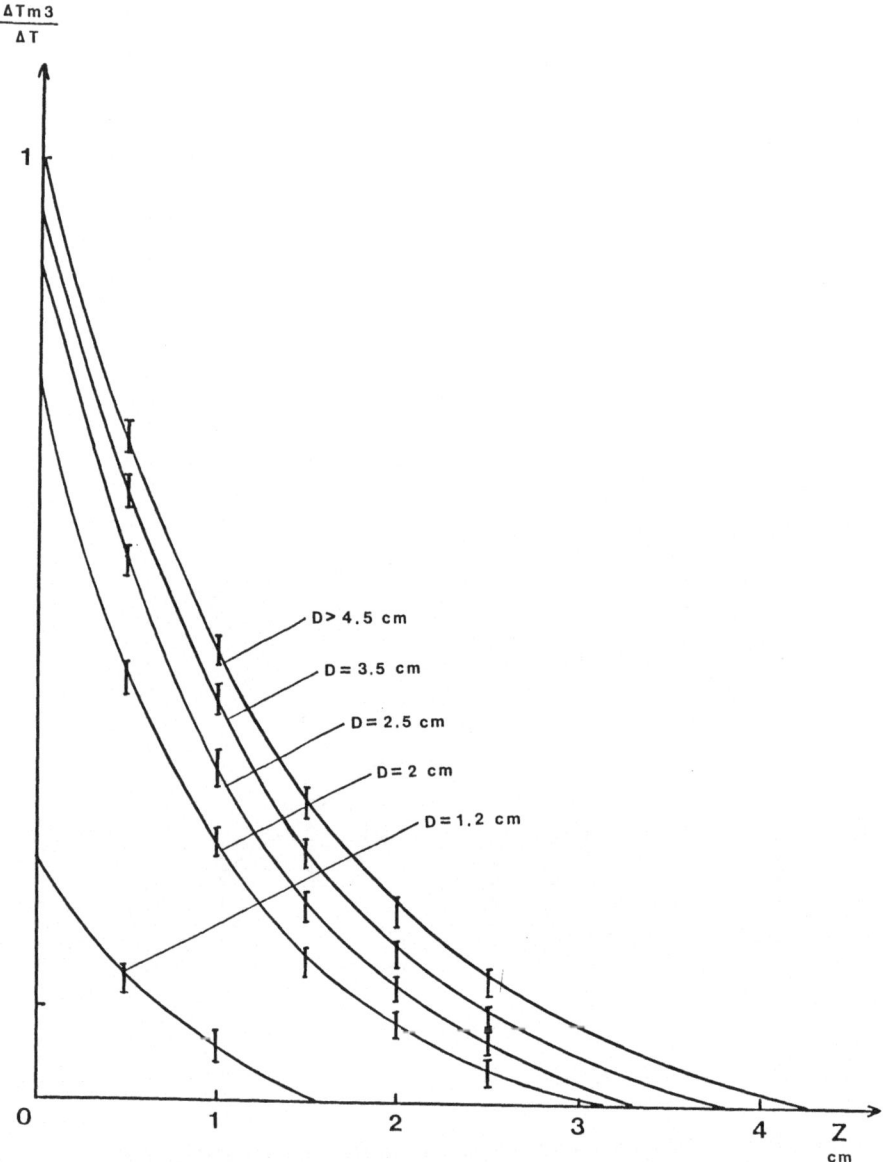

Fig. 5 . Radiometric signals for the situation defined in Fig. 3 on water (To = 33°C, ΔT = 5°C) f = 3 GHz

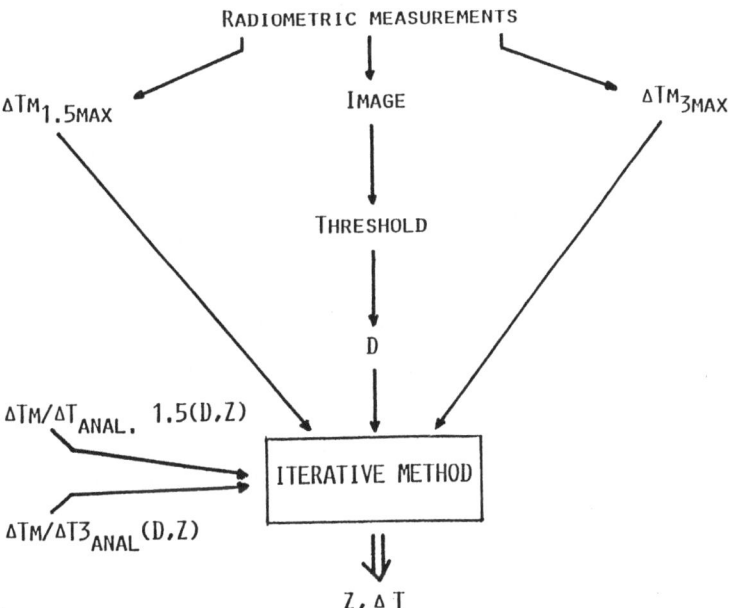

Fig. 6 . Logigram of the computation which gives the characteristics (Z, D, ΔT) of a compact thermal structure.

This method of interpreation is now perfected in the estimation of the inaccuracies and in its application to the case of living tissues. These three directions of research are explained now.

We have first established analytical expressions of $\Delta Tm/\Delta T$ as a function of D and z at 1.5 and 3 GHz. In this way, the experimental data (ΔTm at 1.5 and 3 GHz and D) are introduced in the resolution of an equation using the Newton method and so can define ΔT and z. The logigram of this computation is represented in fig. 6.

The knowledge of the inaccuracies introduced in the different steps is considered in order to determine the accuracy on the final parameters D, z and ΔT.

Experiments on models made of water are useful in order to test the feasibility of our interpretation. The absorbtion of water, at the considered frequencies, is not very different to the absorbtion of living tissues ; then the present results give an estimation which is quite interesting. However, if we have to improve our method, we have to know the data such as $\Delta Tm/\Delta T$ (D, z) for a permittivity which is strictly the permittivity of the tissues. Then we can compute the radiometric data by means of our modal method.

CONCLUSION

In this paper, we present an overview of the present state of the art in microwave radiometry applied to medical applications. This process is now available for non invasive thermometric investigations for a depth in the tissues which reaches 4 cm. This technique has perhaps not yet come of age, however, we observe an increasing number of laboratories working in this topics : construction and improvement of systems, clinical evaluations, and research into the inversion of the radiometric data in terms of temperature retrieval.

This paper also sets out our recent results in radiometric thermal imaging, which are still being evaluated by hospital doctors. We are also developing a quite simple and streight-forward method for temperature retrieval applied to the case of compact thermal structures, based on radiometric measurements at two frequencies.

ACKNOWLEDGMENTS

This research is supported by te "Agence Nationale pour la valorisation de la Recherche", the "Ministère de la Recherche et de la Technologie" and the "Etablissement Public Régional Nord-Pas-de-Calais".

REFERENCES

Bardatti F., Mongiardo M. and Solimini D. Inversion of microwave thermographic data by the singular function method. IEEE MTTS Digest, 75-77 (1985).

Bardatti F., Calamai G., Montgiardo M., Solimini D. Multispectral radiometric system. Proc. 17 European Microwave Conference, Roma, 1987.

Barret A.H., Myers P.C. and Sadowski N.L. Detection of breast cancer by microwave radiometry, *Radiosciences*, **12**, 167-171 (1977).

Bocquet B., Mamouni M., Hochedez M., Van de Velde J.C., Leroy Y. Visibility of local thermal structures and temperature retrieval by microwave radiometry, *Electronics Letters,* n° 3, **22**, 120-122 (1986).

Carr K., Morsi El-Mahdi and Shaeffer J. Dual-mode microwave system to enhance early detection of cancer, *I.E.E.E. Transactions on Microwave Theory and Techniques, MTT,* **29,** n° 3 (March 1981).

Chivé M., Plancot M., Giaux G., Prévost B. Thermal dosimetry in microwave hyperthermia. 4th Int. Symp. Hyperthermic Oncology, Aarhus, pp. 863-866, 1984.

Daburon F., Lefaix J.L., Remy J. and Fayart G. Microwave radiometry of subcutaneous temperature evolution after acute local irradiation in pigs, *Proc. 7th Int. Cong. Radiation Research,* Amsterdam (July 1983).

Daburon F., Lefaix J.L., Remy J. Méthodes thermographiques après irradiation localisée chez le porc. To be published in I.T.B.M., 1987.

Edenhofer E., Electromagnetic remote sensing of the temperature profile by stochastic inversion, *Radio Science,* **16,** 1065-1069 (1981).

Edrich J., Jobe W.E., Hendee W.R., Cacak H.K., Gautherie M. Imaging thermograms at cm and mm wavelength. Ann. NY Acad. Sc. 335 (1980) 443-455.

Edrich J. Microwaves in breast cancer detection. To be pubished in Europ. J. of Radiology, 1987.

Enel L., Leroy Y., Van de Velde J.C., Mamouni A. Improved recognition of thermal structures by microwave radiometry, *Electronics Letters,* **20,** (March 1984).

Evans G. and McLeish C.W., *R.F. Radiometer Handbook,* Artech House, 1977.

Gautherie M., Mamouni A., Samsel M., Guerquin-Kern J.L., Leroy Y. and Gros Ch. Microwave radiothermometry (9 GHz) applied to breast cancer, *J. Optic and Photonics applied to Medicine,* **2,** 154-160 (1980).

Iskander M.F. and Durney C.H. Microwave methods of measuring changes in lung water, *J. Microwave Power,* **18,** 265 (1983).

Land D.V., Radiometer receivers for microwave thermography, *Microwave J.,* **26,** n° 5 (1983).

Leroy Y. Microwave radiometry and thermography. Present and prospective, in *Biomedical Thermology,* A.R. Lyss Inc., New York, (1982), pp. 485-499.

Leroy Y., Mamouni A., Van de Velde, J.C., Bocquet B., Dujardin B. Microwave radiometry for non invasive thermometry. Automedica-Gordon and Breach Science Published Inc. Special Issue "Non invasive thermometry" 1987.

Mamouni A., Van de Velde J.C. and Leroy Y. New correlation radiometer for microwave thermography, *Electronics Letters,* **17,** 554-555 (1981).

Mamouni A., Bliot F., Leroy Y. and Moschetto Y. A modified radiometer for temperature and microwave properties measurements of Biological substances, Seventh E.M.C. Microw. Exhib. and Publ. Ltd (1977), pp. 703-707.

Mamouni A., Leroy Y., Van de Velde J.C. and Bellarbi L. Introduction to correlation microwave thermography, *J. Microwave Power,* **18**(3), 285-293 (1983).

Mamouni A. and Van de Velde J.C. (unpublished results) 1987.

Mizushina S., Yamamura Y. and Siguera T. Three band microwave radiometer system for non invasive measurement of the temperature at various depth, *Proc. I.E.E.E. MTT.S Digest,* 759-762 (1986).

Nguyen D.D., Mamouni A., Leroy Y. and Constant E. Simultaneous microwave local heating and microwave thermography. Possible clinical applications, *J. Microwave Power*, **14**, 135-137 (1979).

Paglione R. Portable diagnostic radiometer. R.C.A. Review, Vol. 47, Déc. 1986, p. 635-643.

Robert J., Edrich J., Mamouni A., Escanye J.M. and Itty C. Microwave thermometry in intracranial Pathology, *Biomedical Thermology,* Alan Lyss, New York (1982), pp. 501-508.

Robillard M., Chivé M., Leroy Y., Audet J., Pichot Ch. and Bolomey J.C. Microwave thermography. Characteristics of waveguide applicators and signatures of thermal structures, *J. Microwave Power*, **17**, n° 2 (June 1982).

Schaller G. Inversion of radiometric data from biological tissue by an optimisation method. Elect. Lett. 26 April 1984, Vol. 20, n° 9, pp. 380-382.

PROGRESS IN MAGNETIC RESONANCE IMAGING FOR MEDICAL DIAGNOSIS

Hermann Weiss

Philips GmbH Forschungslaboratorium Hamburg
Vogt Koellnstrasse 30
D 2000 Hamburg 54, West Germany

1. INTRODUCTION

Although Magnetic Resonance Imaging (MRI) established itself in the past five years to a well accepted diagnostic tool, it still is in the state of extensive research. This is due to the availability of large bore superconductive magnets with ever increasing field-strengths. Thus not only MR-imaging is possible but also MR-in vivo spectroscopy becomes clinically relevant: the spatial separation of spectral lines being proportional to the magnetic field the trend to higher magnetic field strengths is evident. However, with increasing magnetic fields some problems arise in the high frequency excitation of the nuclei as well as in detection of the irradiated MR-signal. In this paper results obtained with the world's first 2 Tesla MR-system are discussed and some considerations on a 4 Tesla system are presented.

In addition a third paragraph is dealing with field independent image processing and application of Artifial Intelligence is proposed for a kind of computer-aided diagnosis using MR-images as input.

2. RESULTS WITH A 2 TESLA-SYSTEM

Beginning in 1983 an MRI-center at the Philips Research Laboratory Hamburg has been built up now consisting of two wooden houses (Fig. 1) in which two systems have been established for research in nearly all relevant fields of MR for medical diagnosis. Fig. 2 gives a view on the 2 Tesla system during installation in February 1984 followed by the world's first 2 Tesla images in March 1984 [1] (Fig. 3). They exhibit an excellent quality which really was not expected because of the theoretical considerations on the penetration depth of the rf-field used for excitation of the hydrogen atoms in the body. However, it turned out that there is no principal limit for imaging at 2 Tesla main magnetic field. This very soon has been confirmed by the first 2 Tesla clinical images (Fig. 4) obtained in cooperation with physicians of the University Hospital Hamburg-Eppendorf which investigated patients at our laboratory before they had their own 1.5 Tesla system. The four slices of Fig. 4 are images of an i d e n t i c cut through the brain. It demonstrates possibilities and difficulties of MR imaging at the same time: The MR-image is a mixture of three tissue parameters: the proton

Fig. 1   MR-research center at Philips Research Labs Hamburg showing the transfer of a cold 2 Tesla magnet (weight 8 tons) from one house to the other

Fig. 2   2 Tesla MR-system during installation; insertion of a phantom into the Faraday rf-shielded bore

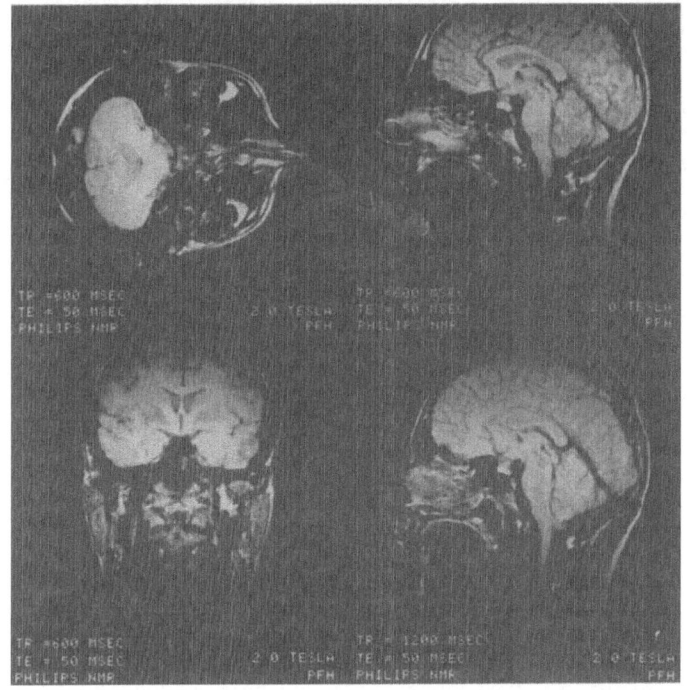

Fig. 3   World's first 2 Tesla MR images

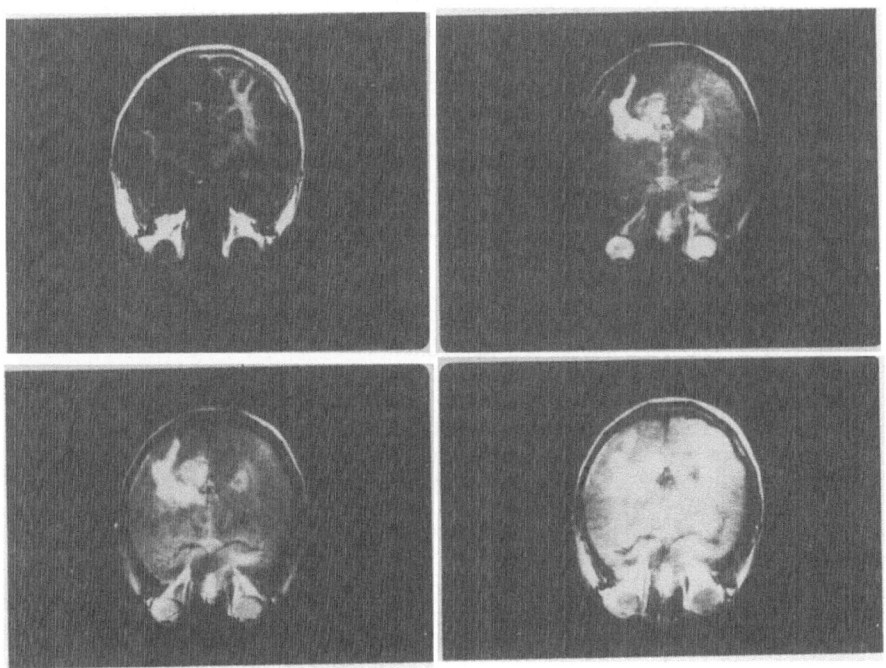

Fig. 4   Head-slice of a patient case showing a brain tumour surrounded
by an edema

density rho (i.e. mainly water content), the longitudinal relaxation time T1, the transverse relaxation time T2 characterizing the molecular binding of the hydrogen in the tissue. Depending on the physical parameter setting during the imaging cycle the mixture of rho, T1, T2 is different, resulting in different contrast in the image. It is the experience of the radiologist which defines then the image quality! The complete MR-system is a quite complicated set up of main magnet, gradient coils, rf-coil, rf-spectrometer and computer system for controlling the data acquisition and performing the reconstuction of the image (Fig. 5). The reader, who is not familiar with the principles of MR-images may refer to [2]. One very critical component of the MR-system is the r f - c o i l which irrdadiates the high frequency power (at 2 Tesla hydrogen imaging 85 MHz) into the body for excitation of the nuclei, which then precess with the Larmour frequency

$$\omega = \gamma B_0,$$

where γ is the gyromagnetic factor and B_0 the strength of the main magnetic field. Parameters to be optimized are the sensitivity of the coil and the homogeneity of the magnetic vector B_1 penetrating the body. The sensitivity directly influences the signal-to-noise ratio in the images, whereas the homogeneity of the B_1-field characterizes the brightness in the image. It could be shown, that at high field strength B_τ, the conventional saddle-coil arrangement is not sufficient, but that resonators are the best solution [3]. Fig. 6 shows the larger total body r f-r e s o n a-t o r and the smaller head rf-resonator constructed for the 2 Tesla system. The field of view obtained with this resonator type coils is about 50 cm x 50 cm as demonstrated in the very "homogeneously illuminated" images of Fig. 7, showing cross sections through the abdomen in transverse (Fig. 7a) and longitudinal mode (Fig. 7b).

The excitation of a total cross section already demands an r f-power of some kW. This considerable high power consumption, the limit of which is given by patient absorption - a heating has to be avoided - can be drastically reduced using so-called surface coils [4]. They are applicated only to a region of interest at the surface of organs like eye, spine, female breast, joints etc. A set of different kinds are shown in Fig. 8a. The B_1-field distribution then of course is inhomogeneous (Fig. 8b) and only surface-near points of the body can be imaged. However, two advantages sometimes are used, i.e. the possibility of zooming an organ (Fig. 9) and the overview-mode with low power consumption (Fig. 10).

CHEMICAL SHIFT AND FAT/WATER SEPARATION

The advantage of higher field MR-imaging, namely the increased signal-to-noise ratio seemed to be threatened by the effect of chemical shift. This unwanted effect occures due to the fact that proton bound in water have a slightly different resonance frequency then those bound in fat. This frequency difference results in the image in a shift of the water protons against the protons of the fat, smearing out the image. By applying high gradient fields these two images can again be forced to lay upon one another, but higher gradient fields on the other side decrease the signal-to-noise level. A possible solution of the dilemma is to image the fat protons separately from the water protons. This indeed can be achieved by the CHESS-method [5], suppressing the water resonance while exciting the fat resonance and vice versa. The resulting images are to be seen in Fig. 11. This fat/water separation technique not only solves the chemical shift artifact problem but also offers new better diagnostic possibilites as shown in Figs. 12 and 13. In Fig. 12

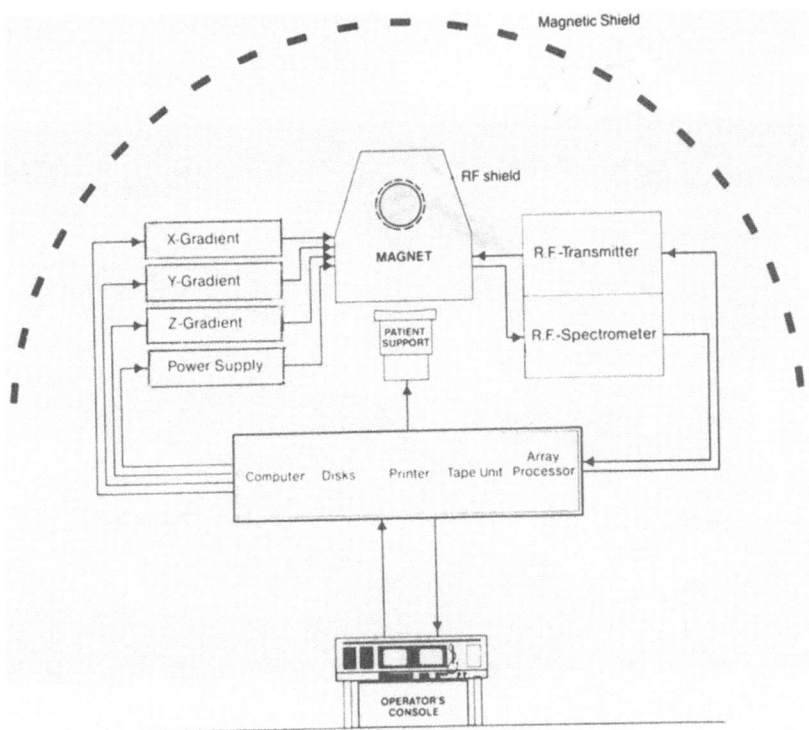

Fig. 5   Schematic picture of an MR-imaging system

Fig. 6   Rf total body and head resonators
          for the 2 Tesla system

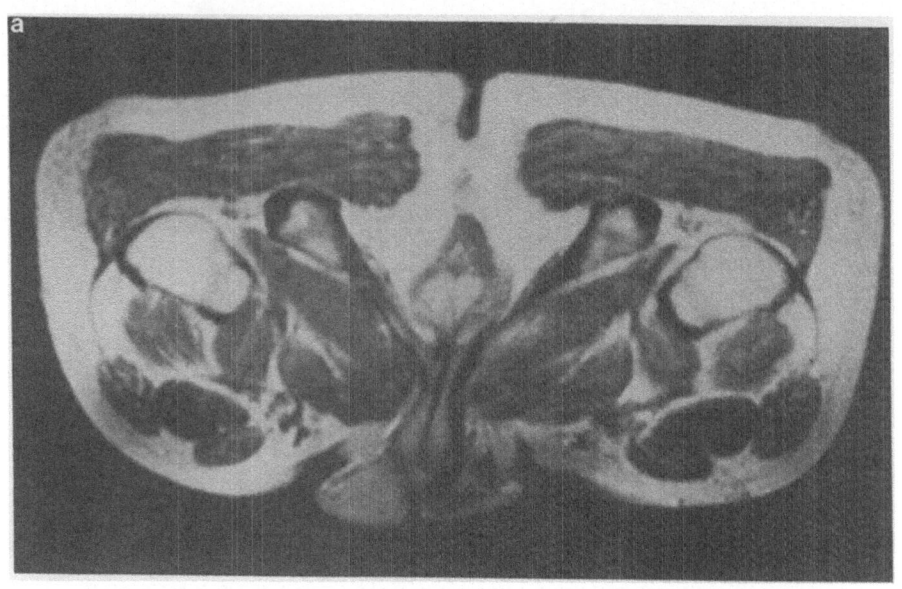

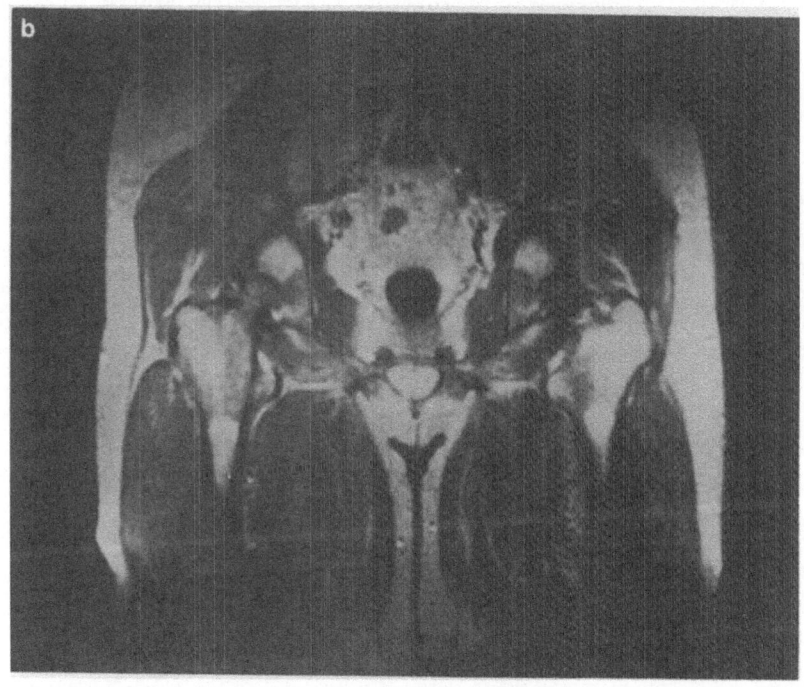

Fig. 7   Cross section of the abdomen
          a) transverse to the main magnet field B_0
          b) parallel to the main magnet field B·

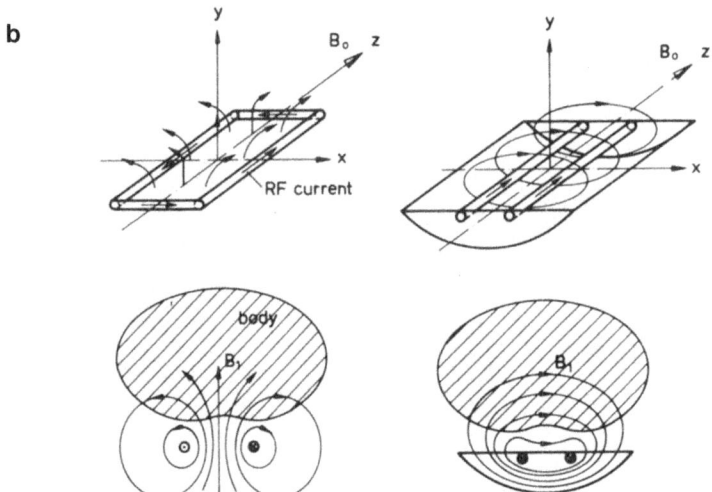

Fig. 8   MR-surface coils
        a) set of different surface coils for thorax and spine (upper two),
           eye and joints (lower three)
        b) inhomogeneous B_1 field distribution of the thorax, spine coils

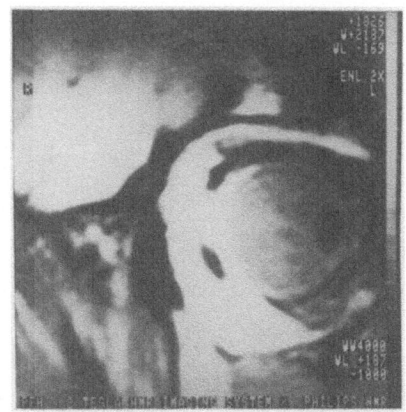

Fig. 9   Zooming the eye with a surface coil:
          spatial resolution 0.5 m x 0.5 m

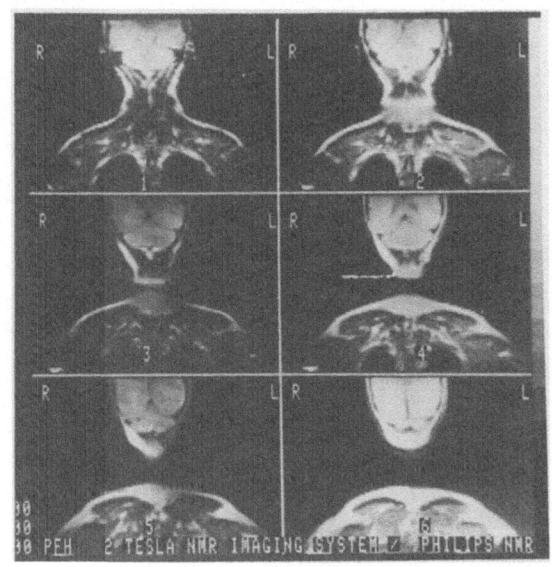

Fig. 10   6 slices of an overview of a patient:
           field of view 50 cm x 50 cm

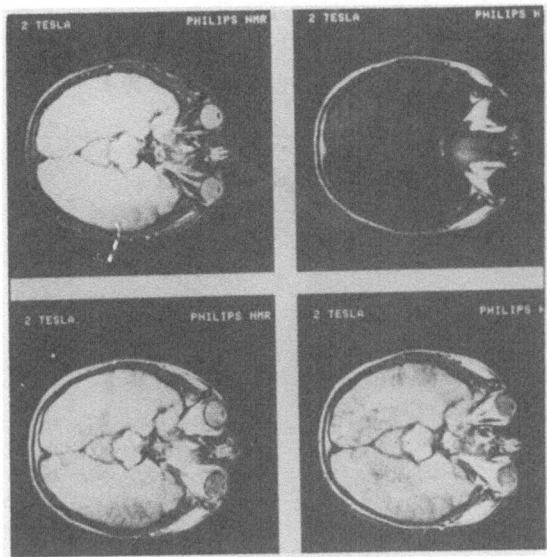

Fig. 11   Fat/water separation in a head slice:
           bottom right: conventional MR-image fat and water smeared
           top right: fat image only
           top left: water image only
           bottom left: computer composed in focus image by
                            adding water and fat image

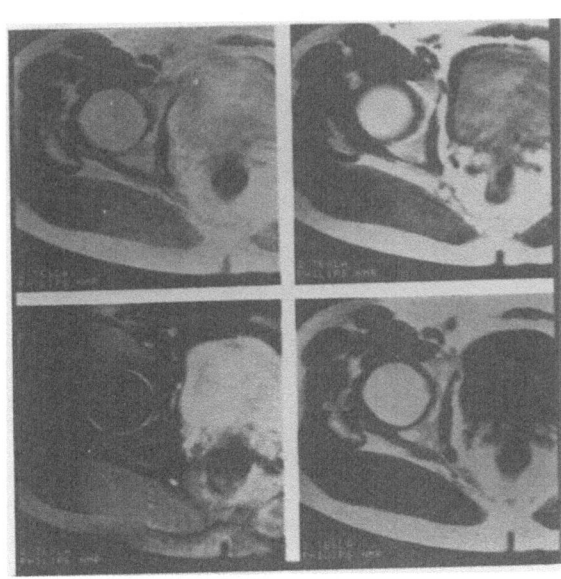

Fig. 12   Clinical case fat/water separation:
           top right: conventional MR-image
           bottom right: fat-image
           bottom left: water image
           top left: composed image

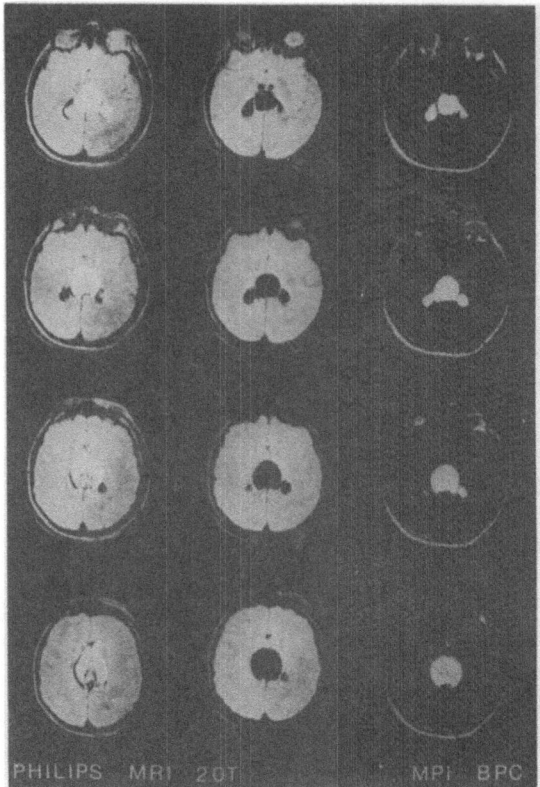

Fig. 13

Series of simultaneously
taken brain slices
left column:
    conventional series
middle column:
    "water series"
right column:
    "fat series"

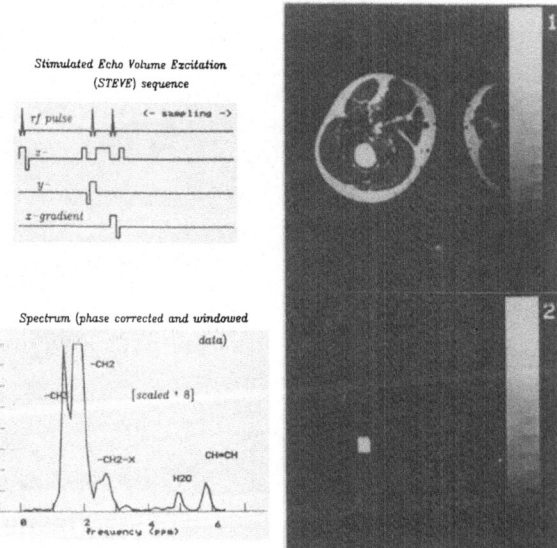

Fig. 14   Localized MR spectroscopy using the STEVE method [6]
           top left: the pulse sequence of rf-power and gradient
                            switching
           top right: slice through a femur
           bottom right: selected volume in the marrow
           bottom left: H-spectrum of the marrow

the "water-image" shows the relevant diagnosis, the water content in the hip joint, in Fig. 13 the "fat-image" delivers the relevant diagnosis by discovering the tumour in the brain to be a "fatty" one, a lipoma. The fat/water separation technique is the first step towards spectroscopy in vivo by MR-techniques: a small volume only in the body is selected to be excited and within this volume a spectrum of all the hydrogen-components in tissue can be obtained by carefully measuring the contributions of different frequencies with the spectrometer. The volume size typically is several cm^3 for proton spectroscopy. The medical goal is the visualization of the metabolism, thus for going beyond the pure morphological imaging of tissue. Fig. 14 shows a proton spectrum of the fatty consistency of bone marrow where the individual hydrogen peaks are well separated [6]. In Fig. 15 spectroscopic imaging technique is shown, where the physician tips a very small volume in between the grid via a cursor on the screen, and at the same time the corresponding H-spectrum is displayed thus having a clear correspondence between image and spectrum.

Since the separation for the individual lines is proportional to the main field strength spectroscopy is only possible from about 1.5 Tesla onwards and is increased in performance by higher fields. Thus the demand for a Super-High-Field system arose and resulted in plans to build a 4-Tesla system.

3. THE 4-TESLA SYSTEM

Constructing a 4-Tesla superconducting total body magnet seemed to be feasible. Is it possible to build a complete MR-system around the magnet? First question was, would it be possible at all to have MR-images at the proton Larmour frequency, i.e. can the rf-high frequency of 170 MHz penetrate the body in order to excite the nuclei.

This question is answered by combining experiments and theoretical considerations taking into regard the fact, that the penetrating magnetic field B_1 of the rf-coil is partly displaced by the conducting parts of the body, i.e. muscle tissue, resulting in a frequency shift of the exciting rf-power [7]. Measuring this frequency shift experimental values of the rf-penetration depths could be obtained (Fig. 16). It demonstrates that at 170 MHz a penetration depth of about 10 cm is to be expected. This means that head-images should be possible, whereas total body proton images are hardly to be achieved. Nevertheless it seemed worthwhile to build a 4 Tesla-system to verify (or not) these results.

A second problem is the homogeneity of the main field $B\tau$. The linewidth of the spectra is proportional to the deviation of the homogeneity of the field and it only makes sense to try to separate the lines better if at the same time the linewidth is not increased. So a very carefully shimming of the magnet is a precondition for spectroscopy. Unfortunately the fine shimming is destroyed by inserting the patient due to its conductivity. Hence the field has to be shimmed spatially variable with the patient in the system. This is a very tedious and time consuming procedure. We developed automatic shim procedures in order to make spectroscopy faster: A selected volume (Fig. 17a) is excited and the irradiated pulse Fourier-transformed to give the line within the excited volume. The linewidth automatically is measured and via a computer program the shim coils of the main magnet are geared in order to minimize this linewidth resulting in an optimal spatially varying shimming of the magnet. The result of this auto-shim is shown in Fig. 17b, where the homogeneity of the magnetic main field has been increased by a factor of 10 within the selected volume [8].

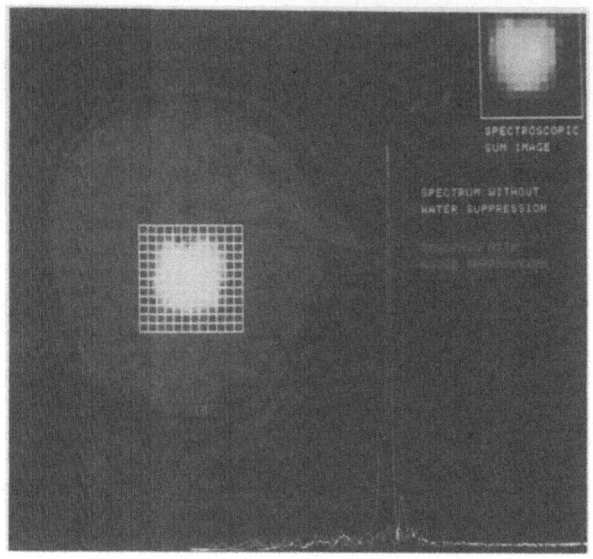

Fig. 15　Spectroscopic Imaging:
　　　　left:　　　　　cut through the brain and marked grid,
　　　　　　　　　　　where spectroscopic data have been obtained
　　　　right top:　　spectroscopic data of the region of interest
　　　　bottom:　　　 H-spectrum of "one pixel"

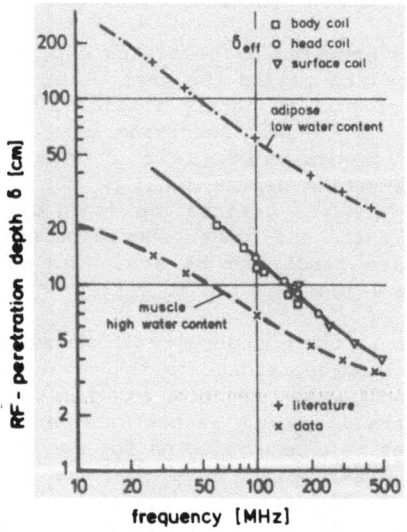

Fig. 16　Penetration depth experiment:
　　　　upper and lower curves are values from literature for in vitro
　　　　measurements,
　　　　the middle curve are results obtained from in vivo
　　　　measurements of the frequency shift within the body

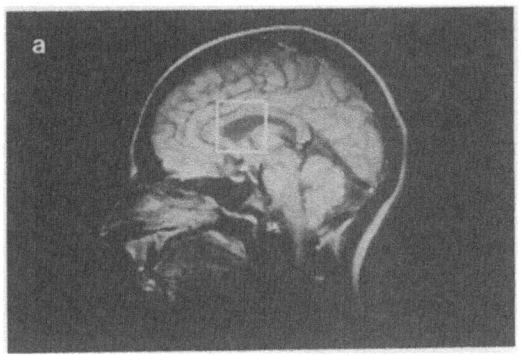

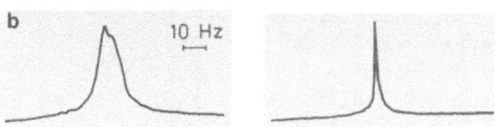

CSF - line width

basic shim   after autoshim

13 Hz   1.2 Hz

Fig. 17   Automatic shim of the main magnet
           a) selected volume of excitation within the ventricle filled
              with CSF
           b) linewidth without extra shim (left) and with auto shim
              (right)

Fig. 18   Gradient coils for the 4 Tesla system

Another problem are the gradient coils of the MR-system which localize the pixels of the image. The very high current (100 A/gradient)
have to be switched in some milliseconds during scanning the body re-
peatedly causing very strong Lorentz forces in the copper wire of the
coil. An X-, Y-, Z-gradient coil system has been designed and constructed (Fig. 18) which now is used in our 2 Tesla system and in future
serves the 4 Tesla-system. In Fig. 19 the gradient coil mounted within a
4 m long support is seen in the 2-Tesla system. One clearly can imagine
the very long 4 Tesla magnet to be installed, when the gradient support
will completely disappear. A further challenge is the spectrometer
capable of imaging up to 170 MHz. Such flexible, broadband spectrometer
was made and operates at the moment successfully in the 2 Tesla system.
Thus the complete 4 Tesla surrounding is ready for the magnet to be delivered in due time.

4. MR-IMAGE PROCESSING

The large amount of digital data arising in MR-imaging of course
suggest to use all the possibilities of modern computerized image processing. A few examples improving diagnostic relevance here are presented. Fig. 20 shows 3 - dimensional representa-
tions of a set of 50 MR-slices. The contours of the slices have
been interconnected by triangles, and after a smoothing procedure
displayed in a manner that all the organes have transparent surfaces in
order to give to the physician an overview on the diagnostic situation.

As already mentioned the MR-image is composed from the three parameter rho, T1 and T2 the mixture of which is very much dependent on the
parameter setting during MR-data acquisition. The Synthetic
Imaging method [9] is a tool to optimize a posteriori the image
contrast. To this aim a set of two special MR-images is made, then from
this the pure values of rho, T1 and T2 are calculated and mixed in order
to obtain all possible images, which could be obtained by corresponding
parameter setting during imaging. But now all possible images can be obtained by simple using a cursor on the computer display thus finally
giving the "best image" which in the case of Fig. 21 is the image on top
right, where the tumour within the brain clearly is visible on top of
the ventricle. This synthetic imaging procedure drastically reduces the
time of investigation of the patient.

Finally the use of the method of Artifial Intelligence is demonstrated by automatic interpretation of MR-images in the head-cut shown
in Fig. 22. The procedure starts by segmentation of the MR-image in the
three-dimensional rho-, T1-, T2-mathematical space. After this separation a set of rules and pixel operations like histogram transformation
more and more isolates the disease. (Fig. 23). An intermediate step is
shown in Fig. 24, the final result in Fig. 25. The brain slice completely is interpreted and the tumour (top) recognized via computer-aided
diagnosis, a first step in supporting the physician to handle the amount
of MR-data [10].

ACKNOWLEDGEMENTS

This work is funded by the Minister of Science and Technology of the
FRG.

52

Fig. 19   The 4 Tesla gradient support mounted in the 2 Tesla system

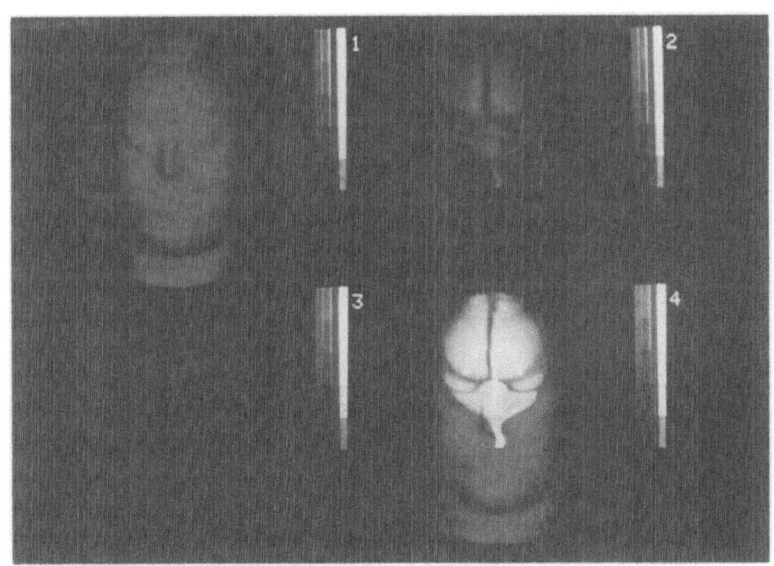

Fig. 20   Three-dimensional display of MR-images
           top left: skin of the head
           top right: brain
           bottom left: ventricle
           bottom right: all together

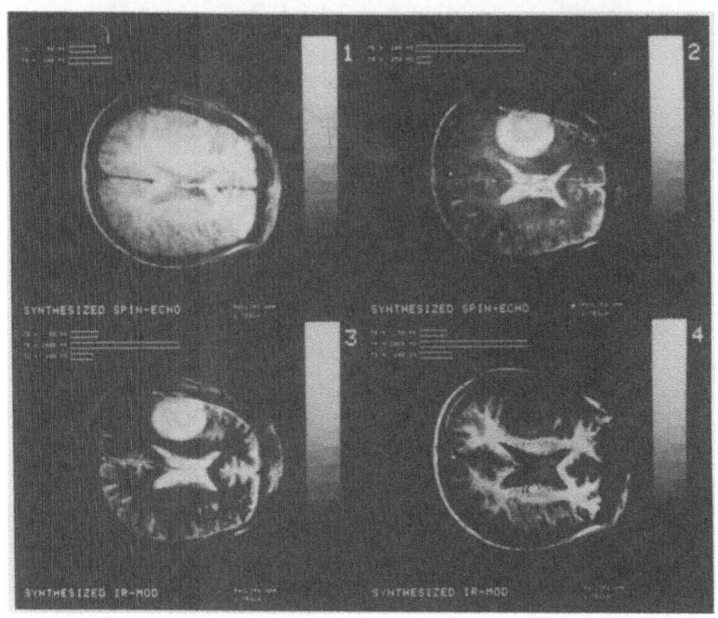

Fig. 21   Synthetic Imaging
          The same slice a posteriori obtained by different parameter
          setting

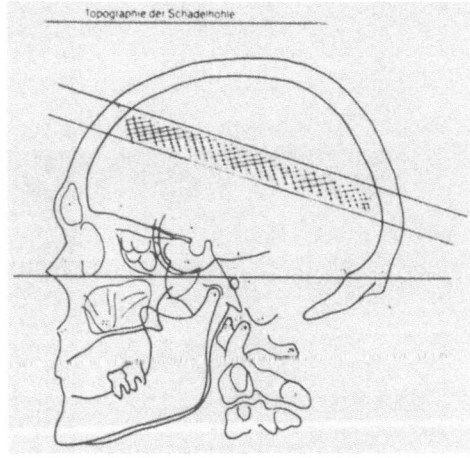

Fig. 22   Automatic image interpretation:
          selection of slice within the head

```
                         +---------------------+
                         | ALL_POSSIBLE_TISSUES |
                         +---------------------+
                                    |
                              RHO-HISTOGRAM
                             /              \
       +---------------------+        +-------------+
       | BACKGROUND_OR_BONE  |        | SOME_TISSUE |
       +---------------------+        +-------------+
                  |                          :
             RULE-SET 1                      :
             /        \                       |
    +------------+  +------+            RULE-SET 2
    | BACKGROUND |  | BONE |            /        \
    +------------+  +------+    +------------------+  +------+
                                | NON_SKIN_TISSUES |  | SKIN |
                                +------------------+  +------+
                                          |
                                    T2-HISTOGRAM
                                   /            \
       +-----------------------+        +---------------+
       | GRAY_OR_WHITE_MATTER  |        | FLUID_OR_TUMOR |
       +-----------------------+        +---------------+
                  |                            |
             T1-HISTOGRAM                      |
             /          \               RULE-SET 3
    +-------------+  +--------------+    /        \
    | GRAY_MATTER |  | WHITE_MATTER |  +-------------+  +-----------+
    +-------------+  +--------------+  | CSF_OR_TUMOR |  | VENTRICLE |
                                       +-------------+  +-----------+
                                              |
                                         RULE-SET 4
                                         /        \
                                    +-----+    +-------+
                                    | CSF |    | TUMOR |
                                    +-----+    +-------+
```

Fig. 23 Rule Set for isolating the relevent disease: tumour

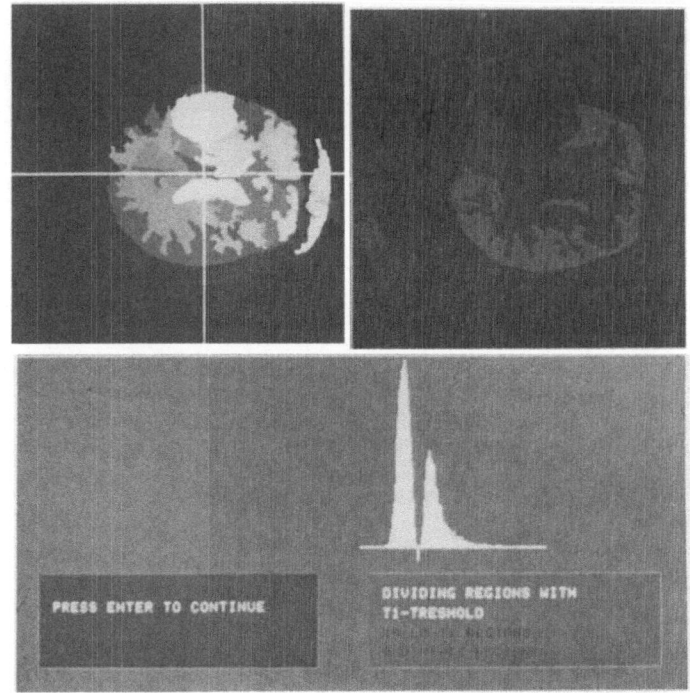

Fig. 24   Intermediate step using a histogram of the paramete

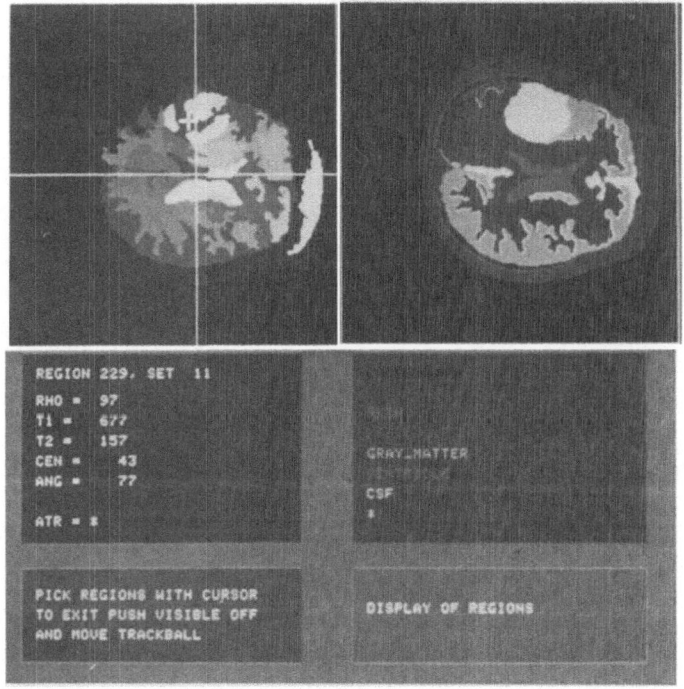

Fig. 25   Final result of the automatic interpretation

REFERENCES

[1] Bomsdorf, H., Buikman, D., Helzel, T., Kuhn, M., Kunz, D., Luedeke, K.M., Meyer, W., Roeschmann, P., Tischler, R., Vollmann, W., Weiss, H., 1984, First MR Body Images at 2 Tesla, 3rd Annual Meeting of the Society of Magnetic Resonance in Medicine, New York

[2] Mansfield, P., Morris, P.G., 1982, NMR Imaging in Biomedicine, Academic Press

[3] Roeschmann, P., Wetzel, Ch., 1986, The Importance of Rf-Coils for MR-Tomography (German), Biomedizinische Technik 31:178

[4] Roeschmann, P., Tischler, R., 1986, Proton Imaging with Novel Surface Coil Resonators at 2 Tesla, Lecture for 71st RSNA, Chicago 1985, published in Radiology 161:251

[5] Frahm, J., Haase, A., Heinicke, W., Matthaei, D., Bomsdorf, H., Helzel, T., 1985, Chemical Shift Selection (CHESS) MR Imaging Using a 2.0 Tesla Whole Body Magnet, Radiology 152:79

[6] Tschendel. O., Helzel, T., Kunz, D., Bomsdorf, H., Wieland, J., Rathke, S., 1987, A New Highly Flexible Broadband Frequency Data Acquisition System for Whole Body Imaging and Spectroscopy Studies at Highest Field Strength, SMRM 6th Annual Meeting, New York 1987, Book of Abstracts p. 371

[7] Roeschmann, P., 1987, Radiofrequency Penetration and Absorption in the Human Body: Simulations to High Field Whole Body NMR Imaging. To be published in Medical Physics

[8] Tochtrop, M., Vollmann, W., Holz, D., Leussler, C., 1987, Automatic Shimming of Selected Volume in Patients, Proc. 6th Annual Meeting SMRM New York 1987, p. 818

[9] Kuhn, M.H., Menhardt, W., 1987, Interactive MR Image Synthesis, Philips Technical Review 43:95

[10] Menhardt, W., Schmidt, K.H., 1987, Automated Interpretation of Transaxial MR Images of the Brain, Proc. of CAR 87, p. 386

TECHNICAL AND CLINICAL ADVANCES IN HYPERTHERMIA TREATMENT OF CANCER

J.W. Hand

Medical Research Council Cyclotron Unit
Hammersmith Hospital
Ducane Road
London W12 0HS

INTRODUCTION

Although significant improvements have been achieved in the treatment of cancer by surgery, chemotherapy and radiotherapy, control of local or regional disease is a common problem. Since approximately one third of patients who die of cancer have uncontrolled local or regional disease, it has been argued that worthwhile improvements in survival rates could be expected if the efficacy of the local therapy could be increased. During the past few years there has been increasing interest in investigating the role that hyperthermia may have in cancer therapy. Hyperthermia is therapy in which tissue temperature is elevated to $41^{\circ}C$ or higher by external means in opposition to the thermoregulatory processes which control body temperature around the normal set-point. The biological rationale for using hyperthermia is related to the ability of heat to destroy, in a virtually selective manner, malignant cells situated in a physiologically deprived environment and to act as a direct sensitising agent to the effects of ionising radiation. Clinically, the combination of hyperthermia and radiotherapy has led to improvement of irradiation treatment of some advanced tumours in an ever increasing number of studies. However, the true potential of hyperthermia remains to be determined by the outcome of controlled clinical trials. A prerequisite for such trials is availability of equipment which is capable of delivering and monitoring treatments in a safe, effective and predictable manner. This is a major challenge particularly when dealing with tumours deep in the body. However, significant improvements to the technology associated with hyperthermia have been made in recent years and some clinical trials involving the treatment of patients with superficial tumours are now underway.

Clinical applications of hyperthermia fall into three broad categories- systemic, regional and local. The methods of inducing systemic hyperthermia in which the patient's core temperature is maintained at $41-42^{\circ}C$ for several hours and the clinical results achieved have been reviewed elsewhere (eg Milligan, 1984; Van der Zee et al, 1987) and will not be addressed further in this paper. When hyperthermia is administered regionally or locally, the aim is to induce temperatures in the range $42-45^{\circ}C$ within tumour for a period of about 1 hour. During regional hyperthermia energy is deposited throughout a large volume of tissue and differences in

59

energy absorption and blood flow between tumour and normal tissue are relied upon to produce higher temperatures in the tumour than in normal tissue. This approach is the basis of all non-invasive electromagnetic techniques that have been investigated for heating tumours deep in the body. During local hyperthermia, the tumour and some surrounding normal tissue are heated. When non-invasive techniques are used, treatment is generally limited to superficial tumours within approximately 3cm of the skin. The limitations and difficulties encountered when using non-invasive techniques have led to an increasing popularity in the use of invasive techniques which may be applied to both superficial and deep seated tumours. The purpose of this paper is to present brief reviews of various electromagnetic methods which are either in current use or about to undergo evaluation in hyperthermia clinics and of the clinical results achieved.

NON-INVASIVE TECHNIQUES FOR LOCAL HYPERTHERMIA

Non-invasive techniques may be classed as being capacitive, inductive or radiative according to how power is transferred from the applicator to the tissues. Two physical parameters, the effective penetration depth, D_{eff}, (the distance in an homogeneous medium over which the absorbed power density (W m^{-3}) is reduced to e^{-2} (-13%) of its initial value) and the wavelength in the tissue, λ, can give considerable insight into the performance of an applicator. Penetration from an applicator depends upon the thickness of tissue layers, curvature of tissues and, in particular, upon the dimensions of the applicator (all in relation to λ). For example, if an applicator has dimensions which are small compared with λ, it will be a very poor radiator and the fields close to it will decrease more rapidly with distance into the tissue than the exponential decrease seen with plane waves. Table 1 shows D_{eff} and λ for an homogeneous medium with dielectric properties representative of tissues with high water content. In low water content tissues the values are varied since the electrical properties of the tissue are more sensitive to variations in water content. Typically, penetration depths are an order of magnitude greater and wavelengths are about three times greater than those shown in Table 1.

Radiofrequency applicators can be classified into two groups - capacitive or inductive devices - and are usually operated at a frequency within the range from about 10-30MHz. A simple capacitive applicator consists of a pair of electrodes placed about the body. For local treatment an electrode with dimensions of a few centimetres is placed over the tumour whilst the second, larger electrode is often placed beneath the patient. Since a large electric field is present near the edges of the electrodes, a bolus is used between electrode and the skin to prevent burning superficial tissues [Brezovich et al, 1981]. A drawback to the technique is that the electric field is predominantly normal to the boundary between subcutaneous fat and other tissues. This produces a specific absorption rate (SAR, W kg^{-1}) in the subcutaneous fat which is an order of magnitude greater than that across the boundary in the underlying higher water content tissues and limits effective use of the technique to cases in which the thickness of the subcutaneous fat layer is less than 1.5 to 2cm.

A simple inductive applicator for superficial treatments is a planar coil of 3 or 4 turns placed parallel to and about 3cm from the skin. Currents induced in the tissues flow predominantly parallel to boundaries of the subcutaneous fat layer, thus avoiding the problem of overheating this layer encountered with capacitive electrodes. A drawback is that the absorbed power distribution exhibits large peaks beneath the coil windings and a null on the axis of the coil. Recently, other applicators which can be described in terms of a current carrying loop with its plane perpendicular to the tissue surface have been developed. For example, Kato

60

Table 1. Effective penetration depth D_{eff} for various applicators and
frequency dependence of wavelength λ and plane wave penetration
depth D. The medium is assumed to be homogeneous with dielectric
properties representative of tissues of high water content
(Johnson and Guy, 1972)

f(MHz)	D_{eff}(cm)				λ(cm)	D(cm)
915	3.0*		3.0**		4.5	3.1
433	3.1*		3.3**		8.8	3.6
300	3.1*		3.8**		11.8	3.9
200	3.1*		4.3**		16.6	4.8
100	3.2*		5.4**		27.0	6.7
27	3.0+	3.0π	5.9φ	6.7ψ	68.1	14.3

* Radiating aperture 10cm x 10cm (Hand, 1987)
** Radiating aperture 20cm x 20cm (Hand, 1987)
+ Capacitive electrode, diameter 5cm (Hand and Johnson, 1986)
π Inductive 'pancake' coil, 3 turns, diameter 12cm (Hand and Johnson, 1986)
φ Water-filled ridged waveguide, effective aperture 29cm x 13.7cm (van Rhoon et al, 1984)
ψ Inductive 'current sheet' applicator, aperture 20cm x 22cm (Johnson et al, 1987)

and Ishida (1983) discuss the use of a one-turn, square column-like coil
driven at 6MHz. The lateral dimensions of the applicator described were
20cm x 20cm and its depth was 60cm but since the resonant frequency of the
device is determined by its inductance and capacitance, its size can be
chosen to meet specific requirements. Bach Andersen et al (1984) developed
a distributed current magnetic dipole applicator in which current carrying
conductors, tuned to resonance by a capacitor, are placed by a metallic
reflector. Dimensions are typically 6-10cm x 4-9cm and operating frequency
is around 150MHz. In order to achieve greater penetration, Franconi et al
(1986) have developed a twin dipole applicator for operation at 27MHz. Two
parallel dipoles of narrow width are connected axially at a suitable
distance to behave as a pair of loosely coupled rectangular loops carrying
currents of the same phase and amplitude. A range of applicator dimensions
was discussed and penetration depth was up to 7cm in muscle-like tissue.
Johnson et al (1987) described an alternative applicator in which the
resonator is formed from a high conductivity current carrying plate folded
back to form a 'U' shape. The basic design can be used to construct
applicators operating at a frequency as low as 10MHz or as high as 1000MHz.
The advantage of all of these devices is that their absorbed power
distributions do not have a null and are considerably more uniform that
that of the planar coil.

Applicators for use at frequencies above about 200MHz are better
radiators since their dimensions are comparable with λ. Open-ended wave-
guides or horns, either placed in direct contact with the tissues or used
with a bolus, have been used as applicators and high permittivity material
has often been used to match their dimensions to specific requirements. By

using devices with various cross-sections and/or polarisation the uniformity of the absorbed power distribution across the aperture can be improved in relation to that associated with an applicator of rectangular cross-section operated in its fundamental mode. A review of these applicators has recently been given by Hand and Hind (1986). Newer designs feature antennas based on microstrip or related techniques. These offer a number of advantages including small size, low weight and the ability to conform to tissue surfaces. The operating frequency, bandwidth and heating characteristics of microstrip applicators can depend critically on the load presented to them (Bahl and Stuchly, 1980; Sandhu and Kolozsvary, 1984) although predictable performance can be achieved if suitably thick bolus is used. A related type of applicator which is relatively insensitive to loading consists of a resonant metallic patch sandwiched between dielectric slabs (Johnson et al, 1984). Microstrip spiral antennas which offer good coupling, broad bandwidth and a circularly symmetrical field have also been reported (Tanabe et al, 1983).

NON-INVASIVE TECHNIQUES FOR DEEP HYPERTHERMIA

There have been frequent attempts in the past to use capacitive electrodes to induce deep body hyperthermia. However, the relatively small electrodes used produce predominantly superficial heating (Armitage et al, 1983). Recently, work in Japan has led to the development of radiofrequency machines which operate at 8 or 13.56MHz and use large electrodes with diameters of 20 to 25cm. When such electrodes are placed about a patient their separation is comparable with their diamater and so significant absorbed power can be expected deep in the body (Song et al, 1986). The technique is limited in application to cases where subcutaneous fat layers are less than 1.5 to 2 cm thick for the reasons outlined in the previous section. A further difficulty is that the electrically inhomogeneous tissues can lead to regions of high current density and to hot spots which may cause patient discomfort (Oleson, 1982; Armitage et al, 1983).

There has also been interest in inductive applicators intended for heating deep seated tumours. One type consists of a single cylindrical electrode formed from copper sheet and placed around the patient, concentrically with his longitudinal axis. This applicator, referred to generically as a concentric coil device, is driven at 13.56MHz. Theoretical calculations and phantom measurements suggest that the SAR distribution is not conducive for deep heating (Oleson 1982; Armitage et al, 1983; Halac et al, 1983; Strohbehn et al, 1986).

A useful arrangement for inducing deep hyperthermia is one in which a linearly polarised electric field at a frequency below 100MHz is generated parallel to the patient's longitudinal axis (Brezovich et al, 1982). One device which is capable of achieving this field distribution is the helical coil applicator. Ruggera and Kantor (1984) investigated the fields produced by self-resonant helices in which the coil winding was chosen to be either λ or $\lambda/2$ (full wave or half wave operation) at frequencies of 13.56, 27.12 or 40.68MHz. The axially directed electric field showed little dependence on radial position but varied along the axial direction such that its maximum could be located in the central region of the coil. The uniformity of field across the transverse plane could be optimised for half wave operation by choosing the length of the coil to be twice its diameter. In the case of full wave operation, the length should be four times the diameter. Hagmann and Levin (1984) also analysed the helical coil applicator and suggested that improved characteristics might be achieved if the coils were externally tuned to a lower frequency (rather than being self-resonant) or were operated in a travelling wave mode.

62

The desirable field distribution referred to above can also be achieved using aperture type applicators. One approach has been to use a pair of water filled, single ridged waveguide applicators designed to operate at 27MHz such as those described by Paglione et al (1981). The effective aperture of these applicators is the area defined by the ridge. A bolus is incorporated to protect tissues from fringing fields close to the aperture and to provide skin cooling. An 'annular array' device which consists of 16 radiating apertures arranged in pairs within an octagonal array has been developed by Turner (1984a; 1984b). In its basic form, power from a single 2kW, 50-110MHz amplifier is split by a 4-way power divider and fed with equal amplitude and phase to four 'quadrants'. Each quadrant consists of four apertures located in two adjacent faces of the octagonal array. Each aperture directs its energy toward the centre of the array where the patient is positioned. The space between the aperture sources and the patient is filled with a liquid bolus to improve coupling to the patient, to reduce stray field levels and to control skin temperature. Distributions of SAR and temperature produced by this device have been modelled using information from CT scans of patients (Paulsen et al, 1984; Strohbehn et al, 1986). Other applicators under development include a coaxial TEM applicator (Lagendijk and de Leeuw, 1986) and a segmented cylindrical phased array system (Bach Andersen and Raskmark, 1985).

ARRAY APPLICATORS FOR LOCAL HYPERTHERMIA

The treatment field associated with a single stationary applicator of any design is inadequate for tumours of large area (eg recurrences on the chest wall). In addition, there is a need to adjust and control the distribution of absorbed power in the tissues during treatment to allow optimization of the temperature distribution. One approach to overcoming these problems is to scan a single applicator over the tissues by means of a programmable robot arm (Sterzer et al, 1986; Lee et al, 1986). Another approach under investigation is to use arrays of microstrip applicators (Sandhu and Kolozsvary, 1984; Tanabe et al, 1983) or inductive devices (Hand et al, 1987) in which relative amplitudes are controlled.

Arrays with amplitude and phase control may offer moderate focusing resulting in a small increase in penetration. Microwave heating of cylinders of lossy material has been studied to gain insight into methods which may be applicable to treatments of tumours in the neck or in limbs. An analytical solution to focussed heating in cylindrical targets due to circumferential arrays of 915MHz horn antennas has been described (Wait, 1985; Wait and Lumori, 1986). It was calculated that a local maximum in SAR could be achieved at the primary focus of the array within a 12cm diameter muscle-like cylinder when either 4, 8, or 16 aperture sources were assumed. However, the global maximum in SAR was produced at the periphery of the cylinder. A practical system consisting of an array of 8 tapered dipoles in the form of strip radiators located within a plastic cylinder and spaced from the tissue by a bolus has been designed for heating limb tumours (Turner, 1986a). The normal range of operating frequencies for this applicator is 105-180MHz and phantom studies have suggested that significant heating can be achieved on the central axis of 20cm diameter cylindrical phantoms. In another theoretical study, the use of phase and amplitude control of a 3 x 3 array of 1.225GHz applicators to improve power deposition in neck tumours was investigated (Jouvie et al, 1986). The results showed that such an array could produce a more desirable SAR distribution within an anatomically real model of the neck than those associated with a single 433MHz aperture source or a pair of 13.56 MHz capacitive electrodes. The effects of changing relative amplitudes and phases and operating frequency within the range 300-450MHz of an array of

4 aperture sources on the SAR distribution in an homogeneous lung model have been described by Bach Andersen (1987).

Clearly, if multiple applicators are to be placed around a particular region of the body, the size of each applicator must be decreased if the number of applicators is increased. A trade-off must be made between the effective penetration from each applicator (favouring large applicators) and the gain due to constructive interference of the fields from the applicators in the array (favouring a larger number of applicators). If the elements in an array are electrically small, an adequate bolus must be an integral part of the array to prevent excessive heating of superficial tissues.

INVASIVE TECHNIQUES FOR LOCAL HYPERTHERMIA

Doss and McCabe (1976) described the technique of 'localised current fields' (LCF) in which a low frequency current (200-1000kHz) is passed through a volume of tissue defined by electrodes which can be either interstitial or a combination of interstitial and superficial in direct contact with the skin. The temperature distribution produced is strongly dependent upon the spacing between electrodes and blood flow rate in the tissue (Strohbehn and Mechling, 1986). Experience has shown that electrodes should be spaced at intervals of 10 to 15mm and that they should be parallel since hot spots are produced in regions where electrodes converge. It is desirable that a means of switching the RF currents between electrodes in a suitable temporal and spatial sequence be incorporated in a LCF system to optimise temperature distributions.

Taylor (1978) described an implantable microwave antenna which consisted of miniature coaxial cable terminated by extending the central conductor approximately $\lambda/4$ beyond the outer conductor. Other designs for implantable antennas have been reported subsequently (Strohbehn and Mechling, 1986; Samaras, 1984). A drawback of some of the earlier antennas is that very little heating is produced towards their tips. Recent designs have addressed this problem. In some cases, greater capacitance at the tip of the antenna is achieved by increasing the diameter of the central conductor in this region (Roos and Hamnerius, 1985; Turner, 1986b). In another design Lin and Wang (1987) use 2 radiating slots separated by $\lambda/8$ over a ground plane in the form of a sleeve. Most clinical applications require an array of these antennas because the heating pattern associated with a single antenna extends no more than 1 or 2cm in the radial direction. Choice of frequency (300-2450MHz) is largely determined by the length of the desired heating pattern along the axis of the antenna. The number of antennas required is typically 1 per $2cm^2$ to 1 per cm^2, depending upon the volume of tissue to be heated (Strohbehn and Mechling, 1986). Investigations into the effects of driving frequency, antenna length and relative phase and amplitude control have shown that a coherently driven antenna system should lead to much better SAR distributions than can be achieved with an incoherent system (Trembly, 1985; Stein and Trembly, 1985; Wong et al, 1986).

A third technique is to implant ferromagnetic 'seeds' and heat them inductively in an RF magnetic field (Atkinson et al, 1984; Stauffer et al, 1984). For long cylindrical ferromagnetic seeds implanted parallel to the magnetic field, the absorbed power density (W m^{-3}) per unit length of seed is proportional to ($H^2f^{05}\mu^{05}a$) where H and f are the amplitude and frequency of the magnetic field and μ and a are the permeability and radius of the seed. On the other hand the absorbed power density in tissue at a radial distance r from the axis of the induction coil is proportional to ($H^2r^2f^2$). The ratio of relative power absorbed from the magnetic field by seeds and

Table 2. Response following treatments by electromagnetically induced
local hyperthermia alone (after Overgaard, 1987).

Reference	Method	No of tumours	Complete Responses
Abe et al, 1982	RF(capacitive)	6	0/6 (0%)
Dubois et al, 1983	Microwave	27	1/27 (4%)
Dunlop et al, 1986	Microwave	9	1/9 (11%)
Hiraoka et al, 1984a	RF(capacitive)	9	0/9 (0%)
Kim & Hahn, 1979	RF(inductive)	19	4/19 (21%)
Luk et al, 1981	Microwave	11	1/11 (9%)
Manning et al, 1982	LCF	11	2/11 (18%)
Overgaard, 1981	RF(capacitive)	13	0/13 (0%)
Perez et al, 1981	Microwave	5	2/5 (40%)
U et al, 1980	Microwave	6	0/6 (0%)

tissue is therefore frequency dependent. Virtually selective heating of the ferromagnetic seeds can be achieved when the frequency of the magnetic field is in the range 200-2000kHz. The optimum frequency is determined by several factors including location of the implant relative to the axis of the induction coil and the permeability and radius of the seeds used. The number of implanted seeds required is about 1 per cm^2. The fact that the heating rate of the implant is particularly dependent upon the permeability of the seed material can be used to advantage in designing 'constant temperature seeds' made from material with a Curie temperature close to the maximum temperature desired in the tissue (Brezovich et al, 1984; Lilly et al, 1985).

CLINICAL STUDIES

To produce local control of a tumour using hyperthermia alone, a sufficiently high temperature would have to be achieved throughout the tumour volume and maintained for a sufficient period of time. Even if such ideal temperature distributions could be achieved within the bulk of the tumour, physical arguments based on heat transfer processes (Lagendijk, 1987) and biological arguments based on studies which suggest that the sensitivities to the direct cytotoxic effects of heat of normal and tumour cells in a well vascularised area with normal physiological conditions are similar (Hahn, 1982) lead to the conclusion that if adequate hyperthermic treatment is to be administered to the peripheral regions of the tumour, there will be inevitable significant damage to the surrounding normal tissue. Indeed, clinical experience to date has shown that the use of hyperthermia alone is ineffective for local control of tumours. Despite the variety of techniques used (electromagnetic fields, ultrasound, hot water, etc), the overall response rate (partial and complete responses) to hyperthermia alone has been approximately 50% with a complete response rate of 10 to 15% (Overgaard, 1987). Table 2 summarises the complete responses reported in several studies for treatments which involved only electromagnetically induced hyperthermia.

To date the combined use of chemotherapeutic agents and hyperthermia has usually involved systemic hyperthermia or the treatment of limbs by hyperthermic perfusion. These topics have been reviewed recently elsewhere (eg Dahl, 1986; Engelhardt, 1987). Treatments using drugs and localised hyperthermia have received relatively little attention and detailed knowledge regarding the distribution and metabolism of drugs in heated tumours and normal tissues is sparse. Most clinical experience with local or regional hyperthermia has been in combination with either external beam

Table 3. Effect of local hyperthermia induced by electromagnetic techniques on radiation response (after Overgaard, 1987).

Reference	Method	No of patients or tumours	Complete Responses	
			Radiation alone	Radiation + heat
Arcangeli et al,1987	Microwave	192	38%	76%
Dunlop et al,1986	Microwave	86	50%	60%
Gonzalez et al,1986	Mwave & RF(cap)	46	33%	50%
Hiraoka et al,1984a	RF(capacitive)	33	25%	71%
Kim et al,1982a,b,1984	RF(ind & cap)	238	39%	72%
Li et al,1984	Microwave	124	29%	54%
Lindholm et al,1987	Microwave	85	25%	46%
Overgaard,1981,1984	RF(cap & ind)	101	39%	62%
Perez et al,1986	Microwave	164	41%	73%
Scott et al,1984	Microwave	62	39%	87%
U et al,1980	Microwave	14	14%	86%
Valdagni et al,1986	Microwave	91	35%	63%
van der Zee et al,1984	RF(cap) & Mwave	71	5%	27%

radiation or brachytherapy (Meyer, 1984; Perez and Meyer, 1985). Table 3 summarises the complete responses reported in several studies in which radiotherapy and electromagnetically induced hyperthermia were compared with radiotherapy alone.

In combined therapy a question which must be addressed concerns the sequencing of hyperthermia and radiotherapy. In an attempt to determine the optimum sequence and time interval between the two modalities, Kim et al (1984) found no differences in tumour control or tumour clearance rates following treatments in which hyperthermia was given either prior to or after radiotherapy (see Table 4). In general, there was no increase in normal tissue response following treatments involving hyperthermia compared with those following irradiation alone. However, when the temperature of normal tissue reached 42.5°C or more and the dose per fraction was 6.6Gy, there was a tendency for enhanced skin reactions to occur and these were more pronounced when hyperthermia was given prior to radiotherapy. If hyperthermia is given before irradiation, tumour vasculature could be compromised leading to an increase in radioresistance of tumour cells through hypoxia. On the other hand, blood flow in normal tissue is likely to be increased, thus rendering normal cells radiosensitive. For these reasons Song et al (1984) recommended the sequence 'hyperthermia + irradiation' for increased therapeutic gain. To increase further the response in tumour compared with that in normal tissue when hyperthermia is given only a few minutes after radiotherapy, higher temperatures should be induced in the tumour compared with those in normal tissues. If such a temperature differential cannot be assured (the usual case in practice), the treatments should be separated in time, preferably by a period of 3 or 4 hours, to avoid increasing the radiation response in normal tissue (Overgaard, 1987).

NON-INVASIVE TREATMENTS OF SUPERFICIAL TUMOURS

Most of the clinical studies reported to date involving radiotherapy and hyperthermia induced by non-invasive methods have been restricted to superficial tumours due to the obvious limitations of the hyperthermia techniques. Nevertheless, complete response rates markedly higher than those for irradiation alone have been a common feature of these studies (see Table 3) (Overgaard, 1987). Furthermore, the ability to assess tumour responses quantitatively, together with the large number of such lesions available for treatment has proved useful in the development of clinical hyperthermia. For example, Perez et al (1986) compared two non-randomised groups of patients treated for recurrences from carcinoma of the breast. Most of the lesions (95%) were in the chest wall. Treatments and fractionation used are summarised in Table 4. Hyperthermia was induced by 2450MHz or 915MHz aperture sources and heating commenced 15 to 30 minutes after radiotherapy sessions. A minimum of two thermistor probes were inserted into the tumour. A second group of patients with similar tumours were treated with irradiation alone (20-60Gy in 2-3Gy daily fractions). For small tumours (1-3cm diameter) in patients receiving 30-40 Gy, the complete response rate within 3 months of treatment was 80% for combined treatment, a significant difference when compared with 33% for irradiation alone. In the case of tumours with diameters larger than 3cm, the corresponding complete response rates were 65% and 42% but these were not significantly different for the number of patients involved. The complete response rates at a minimum of 6 months after treatment for all patients are summarised in Table 4. There was significantly poorer control of those tumours in which the average temperature was less than 41°C compared with those in which higher temperatures were achieved.

In another study Arcangeli et al (1987) have discussed their clinical experiences using hyperthermia and radiotherapy in the treatment of a variety of human tumours. Ninety patients with a total of 200 multiple and superficial lesions from miscellaneous primary tumours were included. Each patient in this study had at least 2 lesions, one of which received hyperthermia, and a comparison of the responses following radiation alone and the combined therapy was made. There were two main patient populations, namely 38 patients with a total of 81 multiple neck node metastases from squamous cell carcinoma of the head and neck (Arcangeli et al, 1985), and 17 patients with a total of 39 cutaneous and nodal metastases from malignant melanoma of the skin. Treatments and fractionation are summarised in Table 4. Microwave heating was achieved using 500MHz water filled horn antennas (Lovisolo et al, 1984). In the early part of this study tumour temperature was monitored only at a single point and at 5 minute intervals throughout the treatment by means of a copper/constantan thermocouple. Analysis showed that the combination of heat and radiation resulted in increased complete response rates compared with those due to radiotherapy alone for both head and neck and melanoma lesions (see Table 4). The study also demonstrated that it was easier to achieve a prescribed hyperthermic treatment in small lesions than in larger ones. However, when the prescribed treatment could be achieved in the larger lesions, the advantage of giving the combined treatment compared with giving radiation alone was greater than that for small lesions. Relationships between thermal isoeffect dose and both response and duration of response were demonstrated and these suggested that a treatment of approximately 1 hour at 42.5°C was necessary to achieve significant improvement in the complete response rate and its duration.

Marked inhomogeneities in temperature distributions measured within tumours have been a common feature of all studies in which multiple temperature measurements were made. In the study of Perez et al (1986) 74% of tumours smaller than 2cm reached the prescribed temperature in contrast with 60% of larger tumours. No appreciable effect of hyperthermia was seen in those tumours which did not achieve an average temperature of at least 41°C. Of those tumours which did achieve this level, response and control was improved in those reaching higher temperatures. The association between response and tumour volume has been reported frequently with higher complete response rates being observed in small tumours than in larger ones. It is likely that this is a consequence of the increased difficulty of achieving adequate temperature uniformity within larger tumours using the techniques currently available. However, when relatively uniform

Table 4. Details of treatments involving radiotherapy (RT) and hyperthermia (H).

Ref	Tumour/ Site	No of tumours	Treatment	Fraction- ation	No of fractions	Complete Responses		
						H+RT	RT+H	RT
A	Melanoma	97	4.0Gy + 42-43.5°C/ 30min	2/week 2/week	10 10	9/16 (57%)	7/11 (63%)	6/21 (30%)
			6.6Gy + 42-43.5°C/ 30min	1/week 1/week	6 6	9/12 (75%)	8/11 (72%)	15/26 (59%)
							RT+H	RT
B	Breast carcinoma (1-3cm) chest wall recurrence (> 3cm)	101 93	4.0Gy + 41-43°C/ 30-60min	1 /3 days 1 /3 days	5-10 5-10		22/28 (79%) 13/20 (65%)	35/73 (48%) 12/43 (28%)
							RT+H	RT
C	Squamous cell carcinoma head and neck	81	1.5-2.0Gy + 42.5°C/ 45min	3/day 3/week	36 7		30/38 (79%)	18/43 (42%)
			5.0Gy + 42.5°C/ 45min	2/week 2/week	8 8		10/13 (77%)	5/9 (55%)
C	Melanoma	39	6.0Gy + 45°C/ 30min	2/week 2/week	5 5		6/9 (75%)	4/8 (50%)
D	Breast carcinoma chest wall recurrence	112	2-6Gy + 60min/ max tolerated temperature	2-5/week ?/week	Total dose of 13-70Gy 2-13			

References are: A - Kim et al (1984); B - Perez et al (1986); C - Arcangeli et al (1987); D - van der Zee et al (1986).

heating is achieved, the highest therapeutic advantage is achieved with the larger lesions as shown by Arcangeli et al (1985) and others.

The definition of a dose for hyperthermia is currently a topic of controversy (Gerner, 1985). Any definition must take into account the temperature reached and the time at which the tissue is maintained at that temperature. An empirical isoeffect relationship which is often used to compare various time-temperature relationships has been discussed by Field and Morris (1983) and by Sapareto and Dewey (1984). This relationship takes the form:

$$\frac{t_2}{t_1} = R^{(T_1-T_2)}$$

where t_1, t_2 are the times to produce the same biological effects at temperatures T_1, T_2 respectively. According to Field and Morris (1983), R = 2 for temperatures above 42.5°C and R = 6 at lower temperatures. This approach is usually used to relate treatments to the 'equivalent' time at 43°C. When the temperature varies with time, as is the case in clinical treatments, the isoeffect relationship becomes (Dunlop et al, 1986):

$$\Delta t_{eq} = \sum \Delta t . R^{(T-43)}$$

where Δt_{eq} is the equivalent time and Δt is the time at temperature T. The limitations of such a simple approach have been discussed elsewhere (Field, 1987). The problems associated with dosimetry in hyperthermia are increased by the fact that temperatures are monitored at only a few points within the tissues. Nevertheless, it is clear from a number of studies that the efficacy of a hyperthermic treatment is dependent upon the minimum 'dose' achieved within tumour. Dunlop et al (1986) in their study of 28 patients with 116 small recurrent or metastatic tumours, observed that treatments involving radiation plus at least two hyperthermic treatments in which the lowest isoeffect dose monitored within tumour was equal to or greater than 20 minutes at 43°C produced a complete response rate of 86% compared with 50% for radiation alone. No significant differences were seen in responses following two, three or four 'effective' hyperthermic fractions. Kapp et al (1986) carried out a prospective randomised trial in which 31 patients with 88 superficial metastases were given either two or six hyperthermic treatments (43 to 45°C intratumoural temperatures for 45 minutes) as an adjuvant to irradiation. No significant difference was observed in the response rates for lesions receiving two versus six hyperthermic treatments.

Van der Zee et al (1986) analysed the response of 112 tumours (mostly breast carcinoma) in 112 patients treated with a combination of radiotherapy and local hyperthermia. Details of the treatments are summarised in Table 4. Hyperthermia was induced using either microwave dipole or horn antennas (at 433 or 2450MHz) or capacitively coupled electrodes at 27MHz. Temperatures were measured at up to 15 locations within tumours. A typical hyperthermia treatment was of 60 minutes duration and consisted of inducing the maximum temperature possible within the constraints imposed by patient tolerance or the fact that a temperature monitored within normal tissue exceeded 43°C. In 87% of patients there was at least a 50% decrease in tumour area or volume after this combined treatment. The authors attributed this high response rate to the addition of hyperthermia to radiotherapy since they expected the patients included in this study to respond poorly to radiotherapy alone.

In their analysis, van der Zee et al (1986) investigated 4 parameters which could be used to express the heat dose. These were (i) the mean of all temperatures recorded within the tumour during a treatment session, (ii) the maximum temperature monitored in the tumour during a treatment session, (iii) the equivalent time at 43°C derived according to the empirical formula of Field and Morris (1983) and (iv) the equivalent time at 43°C calculated from the area beneath the time-temperature curve which lay above 40°C. The heterogeneous temperature distributions usually observed during treatments resulted in a wide range of values for dose within the tumour. To account for this, van der Zee et al introduced into their analysis 3 more parameters which could be considered to represent the hyperthermia 'dose' delivered to the tumour. These were 'the average dose level in the tumour' and the maximum and minimum doss achieved in the tumour. The analysis showed that only the parameters representing the minimum hyperthermia dose showed a significant positive association with response. None of the four parameters chosen to express heat dose showed a significantly greater association with response compared with the remaining three paremeters. There was a strong positive association between response and the total dose of radiotherapy and the minimum hyperthermia dose. The positive association between the minimum hyperthermia dose and response remained after taking the influences of radiotherapy dose and tumour type into account. There was also a strong negative association between tumour volume and response.

TREATMENTS INVOLVING INTERSTITIAL HYPERTHERMIA

There have been several studies involving the use of interstitial hyperthermia and radiation for treatment of advanced or recurrent tumours in a wide variety of sites including the pelvis and head and neck. Aristizabal and Oleson (1984) discussed the results of treatments given to 64 patients with bulky tumours which had recurred after previous therapy or were unlikely to respond to conventional therapy. Intraoperative hyperthermia induced by 500kHz LCF was given for 30 minutes followed 2 to 3 hours later by interstitial ^{192}I irradiation. Temperature mapping was carried out during the treatment by periodically moving thermometers through their catheters from deep to superficial peripheries of the tumours. Time averaged maximum and minimum intratumoural temperatures for the complete study were 44.6^{o}C and 41°C respectively. A complete response rate of 38% was observed and significant correlations were found between complete response and radiation dose, the average minimum temperature and tumour volume with response decreasing as the volume increased. Cosset et al (1985) reported their experiences in using LCF induced hyperthermia and brachytherapy for 29 implantations in 23 patients. Custom built plastic tubes with a central metallic region were used to localise the heated volume and improve normal tissue tolerance. The hyperthermic treatment prescribed was a minimum of 44°C for 45 minutes. This was achieved in 16 of the 25 treatments for which detailed thermal mapping was available. In the remainder of these 25 cases, a minimum of 43°C was achieved five times and of 42°C on four occasions. The variation in temperature measured within the treated volume was less than 1°C in 16 cases and up to 4°C in only one case. The ^{192}I implants were inserted 10 to 20 minutes after termination of hyperthermia and a dose of 30Gy (in the first 17 cases) or 40Gy (in the later cases) was delivered in 2 to 3 days. Of 23 evaluable cases, 19 showed complete response at 2 months after treatment. The four remaining cases showed a partial response and of these, three were cases in which a satisfactory minimum temperature and/or distribution had not been achieved. There were six relapses (all cases which had received 30Gy) from 5 to 13 months after treatment, four at the edges of the treated volume and two 'in situ'.

Emami et al (1987) reported their experiences of interstitial thermoradiotherapy in which 48 recurrent and/or persistent tumours in 46 patients were treated. Common sites were head and neck, pelvis and breast with 29, 7 and 6 cases, respectively. In nine tumours, hyperthermia was induced by an LCF technique in which 2 or more (up to 6) arrays of stainless steel stylets were implanted. The number of needles in an array varied from 3 to 8, depending upon tumour size. Heating was carried out in the operating theatre with the patient under general anaesthesia. After the hyperthermic treatment, the steel needles were replaced by plastic tubes for routine afterloading of the ^{192}I implant. In 30 tumours, hyperthermia was induced by 915MHz antennas inserted through plastic catheters. In these cases hyperthermia was given before and after ^{192}I radiation treatment. In all cases, the goal was to achieve a minimum temperature of 42.5-43°C for 60 minutes within the tumour. Of the 48 tumours treated, 26 showed complete response. In general, tumours less than 4cm in diameter had a better complete response rate than those with diameters in the range 4-10cm although this difference was not apparent if only those tumours which had received a satisfactory hyperthermic treatment were considered. Clearly, some excellent preliminary results in terms of local tumour control have been observed following interstitial hyperthermia and radiation, especially when the minimum temperature achieved within tumour is relatively high (as in the study of Cosset et al (1985)). Emami et al (1987) comment that the results obtained with interstitial thermoradiotherapy are suggestive of some improvement over interstitial radiotherapy alone. However, Kapp (1986) has drawn attention to the fact that excellent local control rates have also been reported for the use of external beam plus interstitial radiation therapy without hyperthermia and to the need for a carefully controlled trial to establish the role of hyperthermia.

There is interest in investigating the potential use of hyperthermia in the treatment of brain tumours. Salcman and Samaras (1983) treated six patients with glioblastoma multiforme and anaplastic astrocytoma with a single 2450MHz microwave antenna. After craniotomy and resection of accessible tumour, the antenna was implanted and hyperthermia delivered with the patient under general anaesthesia. Although only a small volume of tissue was heated to 42°C or above, there were no major complications and four of the six patients were alive 18 months after treatment. Roberts et al (1984) have treated six patients with malignant glioma by brachytherapy and hyperthermia. Arrays of 4 to 6 catheters were implanted to receive radioactive seeds and microwave antennas and an additional catheter was implanted for further thermometry. The goal of the treatment was to achieve 42-43°C at the tumour boundary for a period of 60 minutes followed by ^{192}I therapy to give 20Gy at the tumour boundary over a period of 5 to 11 days. A second hyperthermia session was given upon removal of the radioactive seeds, followed two weeks later by external beam therapy to the whole brain. With the exception of one patient, the hyperthermia treatments were administered without direct complication.

TREATMENTS INVOLVING REGIONAL HYPERTHERMIA

There have been several reports of the clinical use of RF capacitive heating using large electrodes (a pair of 25cm diameter or a pair with 25 cm and 21cm diameters) driven at a frequency of 8 or 13.56MHz for heating deep seated tumours. For example, Hiraoka et al (1984b) treated a total of 24 tumours in 23 patients. The tumours were located in liver, colorectum, pancreas, thorax, stomach, urinary bladder and ischium and 21 of them had a mean diameter greater than 5cm. Hyperthermia treatments were for 30-60 minutes once or twice per week. A total of 133 temperature measurements were made in the 24 tumours and in 50% of these measurements, a maximum tumour temperature of at least 43°C was achieved. The authors did not indicate the lowest temperatures and duration of temperature elevation they observed in this study. Rectal temperatures of 39-40°C and tachycardia were usually observed at the end of each treatment. In another study, Song et al (1986) used hyperthermia alone or in combination with radiation in the treatment of 9 superficial tumours with varying histologies and with sizes ranging from 5 x 4 x 3 cm^3 (in the dorsum of the hand) to 10 x 8 x 8 cm^3 (in the thorax). Temperatures were measured within one or two catheters which were inserted through tumour. Thermal mapping was carried out by withdrawing the thermocouple probes in 5mm steps. Intratumoural temperatures above 42°C were achieved in seven of the nine tumours. In some tumours, temperatures declined as treatment progressed and in all tumours, considerable temperature gradients were observed so that parts of the tumour, particularly those adjacent to normal tissue, remained lower than 42°C.

Considerable clinical experience of deep heating has been gained with the Annular Array device. Sapozink et al (1984) and Gibbs et al (1984) reported their initial experiences with this device in the treatment of 46 patients with advanced tumours in the abdomen or pelvis. On average, a total of five hyperthermia treatments were administered twice weekly. Positioning of the bolused patient within the array was facilitated by measuring relative electric fields at a few points at the skin/bolus interface. Such measurements provide a qualitative assessment of the SAR distribution produced within the patient (Turner, 1984a) and often a 'balanced' condition is achieved by suitable choice of frequency and position. Temperatures were measured by high resistance thermistors (Bowman, 1976) inserted through catheters which were implanted whilst the patient was under general anaesthesia. From time to time during a treatment, the probes were moved through known distances along the catheter to obtain a 'thermal map' (Gibbs, 1983). Extensive temperature measurements were analysed for 12 pelvic and 11 abdominal cases. All of these patients achieved a temperature of 42°C or higher at some point within the tumour during the course of their treatments. In the 8 pelvic and 5 abdominal cases allowed to exceed 43°C, 54% and 61%, respectively, of intratumoural temperature measurements were above 43°C during the best heating achieved during a course of treatments. However, in terms of the whole course of treatments, only about 25% of the points monitored within the tumour reached 43°C and these temperatures were maintained on average for 14 minutes (pelvic cases) and 12 minutes (abdominal cases). The mean maximum normal tissue temperatures were 42.6 $\pm$ 0.8°C in the pelvis and 41.0 $\pm$ 1.1°C in the abdomen. At the operating frequencies chosen (60-70MHz), whole-body standing wave resonances can be induced which may lead to localised maxima in SAR in tissues outside the aperture of the array. This can be particularly troublesome in the neck region and in the lower extremities (Turner, 1984a). Such hot spots, which give rise to localised pain, can be reduced by placing saline filled bags around these regions. Local toxicity including bladder spasm, pelvic and abdominal pain was experienced in 30% of 28 patients undergoing pelvic treatment and in the cases of 7 patients, these acute local effects limited the amount of power which could be applied. Although acute local effects were less of a problem during abdominal treatments, systemic side effects, particularly systemic heating, were common and often limited the power which could be applied. Sapozink et al (1984) concluded that encouraging objective responses were derived from external beam radiotherapy and regional hyperthermia.

Some magnetic induction applicators have undergone clinical evaluation. For example, Storm et al (1981), in an extensive clinical study, have used a single turn coil, carrying a 27MHz current, which is placed concentrically around the patient. No details of the minimum temperatures obtained during treatments were given. In contrast, Oleson et al (1983) did report detailed temperature measurements obtained during treatments in which a similar device was used and suggested that its clinical usefulness is probably limited to treating tumours in relatively superficial locations in the trunk or in the shoulder girdle. Comparisons of the performance of the annular array and the concentric coil magnetic induction applicator in the same patients (Sapozink et al, 1985; Oleson et al, 1986) have shown that the array is superior to the concentric coil for treating pelvic tumours and is probably better in the case of abdominal tumours. These clinical experiences are in close agreement with the predictions of 2-dimensional theoretical models (Paulsen et al, 1985).

At operating frequencies of 60 to 70MHz, equal phase and amplitude excitation of all four quadrants together with the approximately central positioning of the patient within the array could be expected to lead to a fairly uniform distribution of electric field throughout the tissue within the aperture of the device (Strohbehn et al, 1986). Such field distributions are likely to result in unwanted heating of large volumes of normal tissues and may lead to treatment limiting systemic effects. Since deep tumours are often located eccentrically within the body, treatments might be improved if the field distribution could be matched better to the requirements of individual patients.

Sathiaseelan et al (1986) demonstrated the feasibility of improving the temperature distribution produced within a patient by adjusting the relative phases and amplitudes of the fields from the eight pairs of aperture sources forming the array. Samulski et al (1987) have compared the results of treatments to twelve patients, most of whom had tumours located within the lower abdomen or pelvis, in which power was applied to the complete array or only to the two adjacent quadrants close to the tumour. The patients were positioned centrally within the array and full water bolusing was used in each case. The aim in these treatments was to elevate temperature within the tumour to a minimum of 42.5°C for 45 minutes whilst maintaining the temperature of normal tissues below 43°C. The fraction of temperature measurements made within tumour which were above 42.5°C was 26% for treatments in which all four quadrants were energised and 31% for those in which only two quadrants were used. This slight improvement in the temperatures monitored within tumour was achieved without increasing systemic stress or pain although higher normal tissue temperatures were observed during the two quadrant treatments. The authors concluded that tissue heterogeneity and variations in blood flow seemed to be dominant in determining temperature distributions and that any improvement resulting from modifications of the SAR distribution is unlikely to be large.

SUMMARY

This paper has discussed some recent technical developments and trends in clinical hyperthermia. A variety of RF and microwave techniques is available for non-invasive treatment of tumours within 2 to 3cm of the body surfaces. Clinical experience with these devices is that considerable temperature gradients can result within tissues. The introduction of array applicators in which some control over the pattern of energy deposition is possible should improve superficial treatments in this respect. The ability to achieve relatively high minimum temperatures with interstitial hyperthermia makes this heating technique attractive when brachytherapy is considered to be applicable. Clinical experience to date is that a number of intraoperative uses of the technique are encouraging and complications are minimal.

A number of electromagnetic devices for regional hyperthermia are being developed and evaluated. The limited clinical experience with these is that it is often difficult to achieve therapeutic temperatures for sufficient periods of time throughout the tumour volume, a finding which is in general agreement with theoretical predictions of the performance of such devices. There is a need to tailor the distribution of energy deposition to particular cases rather than producing a relatively uniform distribution throughout the tissues. Such studies are now underway. However, the resulting temperature distributions will remain highly dependent upon blood flow within the tissues and future work may need to consider the possibility of manipulating blood flow in attempts to improve treatments.

The efficacy of hyperthermic treatments appears to be strongly correlated with the minimum temperature achieved within the tumour, so care must be taken when designing or selecting an applicator to ensure that all regions of a tumour can be heated adequately. This is particularly important in the peripheral regions of tumours.

Despite the variety of techniques used and their obvious limitations, clinical results indicate the combination of hyperthermia with radiotherapy may be useful. Some clinical trials have been suggested or are underway (eg Kapp, 1986; Overgaard, 1987). In the near future these will be restricted to relative superficial tumours but in time the goal must be to include deep seated tumours. The ability to heat deep seated tumours adequately and to acquire detailed knowledge of their temperatures remain challenging problems to all involved in this field.

REFERENCES

Abe, M., Hiraoka, M., Takahashi, M., Ono, K. and Nohara, H., 1982, Clinical experience with microwave and radiofrequency thermotherapy in the treatment of advanced cancer., in "National Cancer Institute Monograph 61", L.A.Dethlefsen and W.C.Dewey, eds., National Cancer Institute, Bethesda, Md., pp411-414.

Arcangeli, G., Arcangeli, G., Guerra, A., Lovisolo, G., Cividalli, A., Marino, C. and Mauro F., 1985, Tumour response to heat and radiation: prognostic variables in the treatment of neck mode metastases from head and neck cancer., Int. J. Hyperthermia, 1:207-217

Arcangeli, G., Cividalli, A., Lovisolo, G. and Mauro, F., 1987, The combination of heat and radiation in cancer treatment., in "Physics and Technology of Hyperthermia", S.B.Field and C.Franconi, eds., Martinus Nijhoff, Dordrecht, pp574-585.

Aristizabal, S.A. and Oleson, J.R., 1984, Combined interstitial irradiation and localized current field hyperthermia: results and conclusions from clinical studies., Cancer Research(Suppl), 44:4757s-4760s.

Armitage, D.W., LeVeen, H.H., and Pethig, R., 1983, Radiofrequency-induced hyperthermia: computer simulation of specific absorption rate distributions using realistic anatomical models., Phys. Med. Biol., 28:31-42.

Atkinson, W.J., Brezovich, I.A., and Chakraborty, D.P., 1984, Usable frequencies in hyperthermia with thermal seeds., IEEE Trans. Biomed. Engng., BME-31:70-75.

Bach Andersen, J., 1987, Electromagnetic power deposition: inhomogenous media, applicators and phased arrays., in "Physics and Technology of Hyperthermia", S.B.Field and C.Franconi, eds., Martinus Nijhoff, Dordrecht, pp159-188.

Bach Andersen, J., Baun, A., Harmark, K., Heinzl, L., and Raskmark, P., 1984, A hyperthermia system using a new inductive applicator., IEEE Trans. Biomed. Engng., BME-31:21-27.

Bach Andersen, J. and Raskmark, P., 1985, A regional hyperthermia phased array system., in "Proceedings 7th Annual Conference of the Engineering in Medicine and Biology Society (IEEE 85-CH2198-0)", IEEE, New York, pp331-333.

Bahl, I.J., and Stuchly, S.S, 1980, Analysis of a microstrip covered with a lossy dielectric., IEEE Trans. Microwave Theory & Tech., MTT-28:104-109.

Bowman, R.R., 1976, A probe for measuring temperatures in radiofrequency heated material., IEEE Trans. Microwave Theory & Tech., MTT-24:43-45.

Brezovich, I.A., Lilly, M.B., Durant, J.R., Richards, D.B., 1981, A practical system for clinical radiofrequency hyperthermia., Int. J. Radiat. Oncol. Biol. Phys, 7:423-430.

Brezovich, I.A., Young, J.H., Atkinson, J.W., and Wang, M.T., 1982, Hyperthermia consideration for a conducting cylinder heated by an oscillating electric field parallel to the cylinder axis., Medical Physics 9:746-748.

Brezovich, I.A., Atkinson, W.J., and Chakraborty, D.P., 1984, Temperature distributions in tumor models heated by self-regulating nickel-copper alloy thermoseeds., Medical Physics 11:145-152.

Cosset, J-M., Dutreix J., Haie, C., Gerbaulet, A., Janoray, P. and Dewar, J.A., 1985, Interstitial thermoradiotherapy: a technical and clinical study of 29 implanations performed at the Institute Gustave Roussy., Int. J. Hyperthermia, 1:3-13.

Dahl, O., 1986, Hyperthermia and drugs. in "Hyperthermia" D.J.Watmough and W.M.Ross, eds., Blackie Ltd, Glasgow pp121-153.

Doss, J.D., and McCabe, C.W., 1976, A technique for localized heating in tissue: an adjunct to tumour therapy., Medical Instrum., 10:16-21.

Dubois, J.B., Bordure, G., Delauzin, J.P. and Hay, M., 1983, Treatment of superficial tumours by electrontherapy and hyperthermia with 2450MHz microwaves: thermal dosimetry and preliminary clinical results. (Abstract), Strahlentherapie 159:371.

Dunlop, P.R.C., Hand, J.W., Dickinson, R.J. and Field, S.B., 1986, An assessment of local hyperthermia in clinical practice. Int. J. Hyperthermia, 2:39-50.

Emani, B., Perez, C.A., Leybovich, L., Straube, W. and Vongerichten, D., 1987, Interstitial thermoradiotherapy in treatment of malignant tumours., Int. J. Hyperthermia, 3:107-118.

Engelhardt, R., 1987, Hyperthermia and Drugs. Recent Results in Cancer Research, 104:136-203.

Field, S.B., 1987, Biological aspects of hyperthermia., in "Physics and Technology of Hyperthermia", S.B.Field and C.Franconi, eds., Martinus Nijhoff, Dordrecht, pp19-53.

Field, S.B. and Morris C.C., 1983, The relationship between heating time and temperature: its relevance to clinical hyperthermia., Radiotherapy and Oncology, 1:179-186.

Franconi, C., Tiberio, C.A., Raganella, L., and Begnozzi, L., 1986, Low frequency RF twin-dipole applicator for intermediate depth hyperthermia., IEEE Trans. Microwave Theory & Tech., MTT-34:612-619.

Gerner, E.W., 1985, Definition of thermal dose. Biological isoeffect relationships and dose for temperature induced cytotoxicity. in "Hyperthermic Oncology (Vol 2)", J. Overgaard, ed., Taylor and Francis, London, pp245-251.

Gibbs, F.A., 1983, 'Thermal mapping' in experimental cancer treatment with hyperthermia: description and use of a semiautomatic system., Int. J. Radiat.Oncol.Biol.Phys., 9:1057-1063.

Gibbs, F.A., Sapozink, M.D., Gates, K.S. and Stewart, J.R., 1984 Regional hyperthermia with an annular phased array in the experimental treatment of cancer: report of work in progress with a technical emphasis., IEEE Trans. Biomed. Engng., BME-31:115-119.

Gonzalez Gonzalez, D., van Dijk, J.D.P., Blank, L.E.C.M. and Rümke, Ph., 1986, Combined treatment with radiation and hyperthermia in metastatic malignant melanoma., Radiotherapy and Oncology, 6:105-113.

Hagmann, M.J. and Levin, R.L., 1984, Analysis of the helix as an RF applicator for hyperthermia., Electronics Letters, 20:337-338.

Hahn, G.M., 1982, "Hyperthermia and Cancer", Plenum Press, New York.

Halac, S., Roemer, R.B., Oleson, J.R. and Cetas, T.C., 1983, Magnetic induction heating of tissue: numerical evaluation of tumour temperature distributions., Int. J. Radiat. Oncol. Biol., 9:881-891.

Hand, J.W., 1987,Electromagnetic applicators for non-invasive local hyperthermia., in "Physics and Technology of Hyperthermia", S.B.Field and C.Franconi, eds. Martinus Nijhoff, Dordrecht., pp189-210.

Hand, J.W., and Hind, A.J., 1986, A review of RF and microwave applicators for localised hyperthermia. in "Physical Techniques in Clinical Hyperthermia", J.W.Hand and J.R.James, eds., Research Studies Press, Letchworth, UK, pp98-148.

Hand, J.W. and Johnson, R.H., 1986, Field penetration from electromagnetic applicators for localized hyperthermia., Recent Results in Cancer Research, 101:7-17.

Hand, J.W., Johnson, R.H., and James, J.R., 1987, A microwave hyperthermia system with multi-element applicator for treatment of superficial tumours. (Abstract), Int. J. Hyperthermia, 3(6):in press.

Hiraoka, M., Jo, S., Dodo, Y., Ono, K., Takahashi, M., Nishida, H. and Abe, M., 1984a, Clinical results of radiofrequency hyperthermia combined with radiation in the treatment of radioresistant cancers., Cancer, 54:2898-2904.

Hiraoka, M., Jo, S., Takahashi, M. and Abe, M., 1984b, Thermometry results of RF capacitive heating for human deep-seated tumours., in "Hyperthermic Oncology (Vol 1)", J. Overgaard, ed., Taylor and Francis, London, pp609-612.

Johnson, C.C. and Guy, A.W., 1972, Non-ionizing electromagnetic wave effects in biological materials and systems.,Proc. IEEE, 60:692-718.

Johnson, R.H., James, J.R., Hand,J.W., Hopewell, J.W., Dunlop, P.R.C., and Dickinson, R.J., 1984, New low profile applicators for local heating of tissue., IEEE Trans. Biomed. Engng., BME-31:28-37.

Johnson, R.H., Preece, A.W., Hand, J.W. and James, J.R., 1987, A new type of lightweight low frequency electromagnetic hyperthermia applicator, in "IEEE MTT-S International Microwave Symposium Digest (IEEE 87-CH2395-2)", IEEE, New York, pp239-242.

Jouvie, F., Bolomey, J.C., and Gaboriaud, G., 1986, Discussion of capabilities of microwave phased arrays for hyperthermia treatment of neck tumours., IEEE Trans Microwave Theory & Tech., MTT-34:495-501.

Kapp, D.S., 1986, Site and disease selection for hyperthermia clinical trials., Int. J. Hyperthermia, 2:139-156.

Kapp, D.S., Bagshaw, M.A., Meyer, J.L., Hahn, G.M., Samulski, T.V., Fessenden, P., Lee, E.R. and Lohrbach, A.W., 1986, Hyperthermia as an adjuvant to radiation in the treatment of superficial metastases: A randomized trial of 2 vs. 6 heat treatments.(Abstract), in "Abstract Book of 34th Annual Meeting of Radiation Research Society, Las Vegas, 1986", Radiation Research Society, Philadelphia, p24.

Kato, H., and Ishida, T., 1983, A new inductive applicator for hyperthermia. J. Microwave Power 18:331-336.

Kim, J.H. and Hahn, E.W., 1979,Clinical and biological studies of localized hyperthermia., Cancer Research, 39:2258-2261.

Kim, J.H., Hahn, E.W. and Ahmed, S.A., 1982a, Combination of hyperthermia and radiation therapy for malignant melanoma, Cancer, 50:478-482.

Kim, J.H., Hahn, E.W. and Antich, P.E., 1982b, Radiofrequency hyperthermia for clinical cancer therapy, in, "National Cancer Institute Monograph 61", L.A.Dethlefsen and W.C.Dewey, eds., National Cancer Institute, Bethesda, Md., pp339-342.

Kim, J.H., Hahn, E.W., Ahmed, S.A. and Kim, Y.S., 1984, Clinical study of the sequence of combined hyperthermia and radiation therapy of malignant melanoma, in, "Hyperthermic Oncology (Vol 1)", J.Overgaard, ed., Taylor and Francis, London, pp387-390.

Lagendijk, J.J.W., 1987, Heat transfer in tissues, in, "Physics and Technology of Hyperthermia", S.B.Field and C.Franconi, eds., Martinus Nijhoff, Dordrecht, pp517-552.

Lagendijk, J.J.W., and de Leeuw, A.A.C., 1986, The development of applicators for deep body hyperthermia. Recent Results in Cancer Res., 101:18-35.

Lee, E.R., Samulski, T.V., Fessenden, P. and Kapp, D.S., 1986, Control scan surface heating. (Abstract), in, "Abstract Book of 34th Meeting of Radiation Research Society, Las Vegas, 1986", Radiation Research Society, Philadelphia, p31.

Li, R-Y., Zhang,T-Z., Lin, S-Y. and Wang, H-P., 1984, Effect of hyperthermia combined with radiation in the treatment of superficial malignant lesion in 90 patients, in, "Hyperthermic Oncology (Vol 1)", J.Overgaard, ed., Taylor and Francis, London, pp395-397.

Lilly, M.B., Brezovich, I.A. and Atkinson, W.J., 1985, Hyperthermia induction with thermally self-regulating ferromagnetic implants, Radiology, 154:243-244.

Lin, J.C. and Wang, Y-J., 1987, Interstitial microwave antennas for thermal therapy, Int. J. Hyperthermia, 3:37-47.

Lindholm, C-E., Kjellen, E., Nilsson, P. and Herzman, S., 1987, Microwave-induced hyperthermia and radiotherapy in human superficial tumours. Clinical results with a comparative study of combined treatment versus radiotherapy alone, Int. J. Hyperthermia, 3(5):in press.

Lovisolo, G.A., Adami, M., Cividalli, A., Mauro, F., Borrani, A., Calami, G. and Arcangeli, G., 1984, Performance of a new waveguide applicator: optimization by suitable variation of frequency, power and water cooling, in, "Hyperthermic Oncology (Vol 1)", J.Overgaard, ed., Taylor and Francis, London, pp679-682.

Luk, K.H., Purser, P.R., Castro, J.R., Meyler, T.S. and Phillips, T.L., 1981, Clinical experiences with local microwave hyperthermia, Int. J. Radiat. Oncol. Biol. Phys., 7:615-619.

Manning, M.R., Cetas, T.C., Miller, R.C., Oleson, J.R., Connor, W.G. and Gerner, E.W., 1982, Clinical hyperthermia: results of a phase 1 trial employing hyperthermia alone or in combination with external beam or interstitial radiotherapy, Cancer, 49:205-216.

Meyer, J.L., 1984, The clinical efficacy of localized hyperthermia, Cancer Res., (Supp)44:4745s-4751s.

Milligan, A.J., 1984, Whole-body hyperthermia induction techniques, Cancer Res., (Supp)44:4869s-4872s.

Oleson, J.R., 1982, Hyperthermia by maganetic induction: I.Physical characteristics of the technique, Int. J. Radiat. Oncol. Biol. Phys., 8:1747-1756.

Oleson, J.R., Heusinkveld, R.S. and Manning, M.R., 1983, Hyperthermia by magnetic induction: II.Clinical experience with concentric electrodes, Int. J. Radiat. Oncol. Biol. Phys., 9:549-556.

Oleson, J.R., Sim, D.A., Conrad, J., Fletcher, A.M. and Gross, E.J., 1986, Results of a Phase 1 regional hyperthermia device evaluation: microwave annular array versus radiofrequency induction coil, Int. J. Hyperthermia, 2:327-336.

Overgaard, J., 1981, Fractionated radiation and hyperthermia. Experimental and clinical studies, Cancer, 48:1116-1123.

Overgaard, J., 1987, The design of clinical trials in hyperthermic oncology, in, "Physics and Technology of Hyperthermia", S.B.Field and C.Franconi, eds., Martinus Nijhoff, Dordrecht, pp598-620.

Overgaard, J. and Overgaard, M., 1984, "A clinical trial evaluating the effect of simultaneous or sequential radiation and hyperthermia in the treatment of malignant melanoma, in, "Hyperthermic Oncology (Vol 1)", J.Overgaard, ed., Taylor and Francis, London, pp383-386.

Paglione, R., Sterzer, F., Mendecki, J., Friedenthal, E. and Botstein, C., 1981, 27MHz ridged waveguide applicators for localised hyperthermia treatment of deep seated malignant tumors, Microwave Journal, 24:71-80.

Paulsen, K.D., Strohbehn, J.W. and Lynch, D.R., 1984, Theoretical temperature distributions produced by an annular phased array type system in CT-based patient models, Radiat. Res., 100:536-552.

Paulsen, K.D., Strohbehn, J.W. and Lynch, D.R., 1985, Comparative theoretical performance for two types of regional hyperthermia systems, Int. J. Radiat. Oncol. Biol. Phys., 11:1659-1671.

Perez, C.A., Kopecky, W., Baglan, R., Rao, D.V. and Johnson, R., 1981, Local microwave hyperthermia in cancer therapy. Preliminary report, Henry Ford Hospital Medical Journal, 29:23-27.

Perez, C.A., Kuske, R.R., Emani, B. and Fineberg, B., 1986, Irradiation alone or combined with hyperthermia in the treatment of recurrent carcinoma of the breast in the chest wall: a nonrandomized comparison, Int. J. Hyperthermia, 2:179-188.

Perez, C.A. and Meyer, J.L., 1985, Clinical experience with localized hyperthermia and irradiation, in, "Hyperthermic Oncology (Vol 2)", J.Overgaard, ed., Taylor and Francis, London, pp181-198.

Roberts, D.W., Coughlin, C.T., Wong, T.Z., Douple, E.B. and Strohbehn, J.W., 1984, A pilot study on the feasibility of using interstitial antenna array hyperthermia and irridium brachytherapy for treatment of glioblastomas (Abstract), in, "Abstract Book of 32nd Annual Meeting of the Radiation Research Society, Orlando Fl.", Radiation Research Society, Philadelphia, p18.

Roos, D. and Hamnerius, Y., 1985, An interstitial microwave applicator with improved heating volume (Abstract), Strahlentherapie, 161:548.

Ruggera, P.S. and Kantor, G., 1984, Development of a family of RF helical coil applicators which produce transversely uniform axially distributed heating in cylindrical fat-muscle phantoms, IEEE Trans. Biomed. Engng., BME-31:98-106.

Salcman, M. and Samaras, G.M., 1983, Interstitial microwave hyperthermia for brain tumours, Journal of Neuro-Oncology, 1:225-236.

Samaras, G.M., 1984, Intracranial microwave hyperthermia: heat induction and temperature control, IEEE Trans. Biomed. Engng., BME-31:63-69.

Samulski, T.V., Kapp, D.S., Fessenden, P. and Lohrback, A., 1987, Heating deep seated eccentrically located tumors with an annular phased array system: a comparative clinical study using two annular array operating configurations, Int. J. Radiat. Oncol. Biol. Phys., 13:83-94.

Sandhu, T.S. and Kolozsvary, A., 1984, Conformal hyperthermia applicators, in, "Hyperthermic Oncology (Vol 1)", J.Overgaard, ed., Taylor and Francis, London, pp675-678.

Sapareto, S.A. and Dewey, W.C., 1984, Thermal dose determination in cancer therapy, Int. J. Radiat. Oncol. Biol. Phys., 10:787-806.

Sapozink, M.D., Gibbs, F.A., Gates, K.S. and Stewart, J.R., 1984, Regional hyperthermia in the treatment of clinically advanced, deep seated malignancy: results of a pilot study employing an annular array applicator, Int. J. Radiat. Oncol. Biol. Phys., 10:775-786.

Sapozink, M.D., Gibbs, F.A., Thomson, J.W. and Stewart, J.R., 1985, A comparison of deep regional hyperthermia from an annular phased array and a concentric coil in the same patients, Int. J. Radiat. Oncol. Biol. Phys., 11:179-190.

Sathiaseelan, V., Iskander, M.F., Howard, G.C.W. and Bleehen, N.M., 1986, Theoretical analysis and clinical demonstration of the effect of power control using the annular phased-array hyperthermia system, IEEE Trans. Microwave Theory & Tech., MTT-34:514-519.

Scott, R.S., Johnson, R.J.R., Story, K.V. and Clay, L., 1984, Local hyperthermia in combination with definitive radiotherapy: increased tumor clearance, reduced recurrence rate in extended follow-up, Int. J. Radiat. Oncol. Biol. Phys., 10:2119-2123.

Song, C.W., Lokshina, A., Rhee, J.G., Patten, M. and Levitt, S.H., 1984, Implication of blood flow in hyperthermic treatment of tumours, IEEE Trans. Biomed. Engng., BME-31:9-16.

Song C.W., Rhee, J.G., Lee, C.K.K. and Levitt, S.H., 1986, Capacitive heating of phantom and human tumors with an 8MHz radiofrequency applicator, Int. J. Radiat. Oncol. Biol. Phys., 12:365-372.

Stauffer, P.R., Cetas, T.C. and Jones, R.C., 1984, Magnetic induction heating of ferromagnetic implants for inducing localized heating in deep seated tumors, IEEE Trans. Biomed. Engng., BME-31:235-251.

Stein, A.D. and Trembly, B.S., 1985, Improved temperature distributions for hyperthermic treatment of cancer through the use of variably phased antenna arrays, in, "Proceedings of the 11th Annual Northeast

Bioemgineering Conference (IEEE 85CH2203-8)", W.S.Kuklinski and W.J.Ohley, eds., IEEE, New York, pp62-65.

Sterzer, F., Paglione, R.W., Wozniak, F.J., Friedenthal, E. and Mendecki, J., 1986, A robot-operated microwave hyperthermia system for treating large malignant surface lesions, Microwave Journal, 29(7):147-152.

Storm, F.K., Harrison, W.H., Elliott, R.S., Kaiser, L.R., Silberman, A.W. and Morton, D.L., 1981, Clinical radiofrequency hyperthermia by magnetic-loop induction, J. Microwave Power, 16:179-184.

Strohbehn, J.W. and Mechling, J.A., 1986, Interstitial techniques for clinical hyperthermia, in, "Physical Techniques in Clinical Hyperthermia", J.W.Hand and J.R.James, eds., Research Studies Press, Letchworth, UK, pp210-287.

Strohbehn, J.W., Paulsen, K.D. and Lynch, D.R., 1986, Use of the finite element method in computerized thermal dosimetry, in, "Physical Techniques in Clinical Hyperthermia", J.W.Hand and J.R.James, eds., Research Studies Press, Letchworth, UK, pp383-451.

Tanabe, E., McEuen, A.H., Norris, C.S., Fessenden, P. and Samulski, T.V., 1983, A multielement microstrip antenna for local hyperthermia, in, "IEEE MTT-S International Microwave Symposium Digest (IEEE 83 CH 1871-3)" IEEE, New York, pp183-185.

Taylor, L.S., 1978, Electromagnetic syringe, IEEE Trans. Biomed. Engng., BME-25:303-305.

Trembly, B.S., 1985, The effects of driving frequency and antenna length on power deposition within a microwave antenna array used for hyperthermia, IEEE Trans. Biomed. Engng., BME-32:152-157.

Turner, P.F., 1984a, Hyperthermia and inhomogeneous tissue effects using an annular phased array, IEEE Trans. Microwave Theory & Tech., MTT-32:874-882.

Turner, P.F., 1984b, Regional hyperthermia with an annular phased array IEEE Trans. Biomed. Engng., BME-31:106-114.

Turner, P.F., 1986a, Mini-annular phased array for limb hyperthermia, IEEE Trans. Microwave Theory & Tech., MTT-34:508-513.

Turner, P.F., 1986b, Interstitial equal-phased arrays for EM hyperthermia, IEEE Trans. Microwave Theory & Tech., MTT-34:572-578.

U, R., Noell, T., Woodward, K.T., Worde, B.T., Fishburn, R.I. and Miller, L.S., 1980, Microwave-induced local hyperthermia in combination with radiotherapy of human malignant tumors, Cancer, 45:638-646.

Valdagni, R., Kapp, D.S. and Valdagni, C., 1986, N_3(TNM-UICC) metastatic neck nodes managed by combined radiation therapy and hyperthermia: clinical results and analysis of treatment parameters, Int. J. Hyperthermia, 2:189-200.

van der Zee, J., van Rhoon, G.C., Wike-Hooley, J.L., van den Berg, A.P. and Reinhold, H.S., 1984, Thermal enhancement of radiotherapy in breast carcinoma, in, "Hyperthermic Oncology (Vol 1)", J.Overgaard, ed., Taylor and Francis, London, pp345-348.

van der Zee, J., van Putten, W.L.J., van den Berg, A.P., van Rhoon, G.C., Wike-Hooley, J.L., Broekmeyer-Reurink, M.P. and Reinhold, H.S., 1986, Retrospective analysis of the response of tumours in patients treated with a combination of radiotherapy and hyperthermia, Int. J. Hyperthermia, 2:337-349.

van der Zee, J., Faithfull, N.S., van Rhoon, G.C. and Reinhold, H.S., 1987, Whole body hyperthermia as a treatment modality, in, "Physics and Technology of Hyperthermia", S.B.Field and C.Franconi, eds., Martinus Nijhoff, Dordrecht, pp420-440.

van Rhoon, G.C., Visser, A.G., van den Berg, P.M. and Reinhold, H.S., 1984, Temperature depth profiles obtained in muscle equivalent phantoms using the RCA 27MHz ridged waveguide, in, "Hyperthermic Oncology (Vol 1)", J.Overgaard, ed., Taylor and Francis, London, pp499-502.

Wait, J.R., 1985, Focused heating in cylindrical targets: part 1. IEEE Trans. Microwave Theory & Tech., MTT-33:647-649.

Wait, J.R. and Lumori, M., 1986, Focused heating in cylindrical targets: part 2. IEEE Trans. Microwave Theory & Tech., MTT-34:357-359.

Wong, T.Z., Strohbehn, J.W., Jones, K.M., Mechling, J.A. and Trembly, B.S., 1986, SAR patterns from an interstitial microwave antenna-array hyperthermia system, IEEE Trans. Microwave Theory & Tech., MTT-34:560-567.

PART II

BIOLOGICAL EFFECTS AND MECHANISMS

BIOLOGICAL RESPONSES TO STATIC AND TIME-VARYING MAGNETIC FIELDS

T.S. Tenforde*

Research Medicine and Radiation Biophysics Division
Lawrence Berkeley Laboratory, University of California
Berkeley, California 94720

INTRODUCTION

The numerous sources of high-intensity magnetic fields used in industry, research and medicine have led to an increased interest in determining the effects of these fields on biological systems. In this chapter a
detailed description is given of the physical mechanisms through which
magnetic fields interact with living matter at the tissue, cellular and
molecular levels. The biological interactions of static magnetic fields,
extremely-low-frequency (ELF) magnetic fields with frequencies below
300 Hz, and combined static and ELF fields are considered. A general
summary is given of laboratory studies and possible human health effects
associated with exposure to static and ELF magnetic fields.

STATIC MAGNETIC FIELDS

Interaction Mechanisms

Three classes of physical interactions of static magnetic fields
with biological systems are now well established on the basis of experimental data: (1) electrodynamic interactions with ionic conduction currents; (2) magnetomechanical effects, including the orientation of magnetically anisotropic structures in uniform fields and the translation of
paramagnetic and ferromagnetic materials in magnetic field gradients;
(3) effects on electronic spin states of the reaction intermediates in
certain types of charge transfer processes. Each of these physical
interaction mechanisms, along with relevant experimental data, will be
described in the following paragraphs.

Electrodynamic interactions. Ionic currents interact with static
magnetic fields as a result of the Lorentz forces exerted on moving charge
carriers. This electrodynamic interaction gives rise to an induced elec-
tric field $\vec{E}_i = - \vec{v} \times \vec{B}$, where $\vec{v}$ is the velocity of current flow and $\vec{B}$
is the magnetic flux density. This phenomenon is the physical basis of
the Hall effect in solid state materials, and it also occurs in several

---

*Present address: Life Sciences Center (K4-14), Battelle Pacific
Northwest Laboratories, P.O. Box 999, Richland, Washington 99352.

biological processes that involve the flow of electrolytes in an aqueous medium. Examples of such processes are the ionic current flows that occur in the circulatory system, in nerve impulse propagation and in visual phototransduction processes.

A well-studied example of electrodynamic interactions that leads to measurable biological effects is the induction of electrical potentials as a result of blood flow in the presence of a static magnetic field. It is a direct consequence of the Lorentz force exerted on moving ionic currents that blood flowing through a cylindrical vessel of diameter, d, will develop an electrical potential, ψ, given by the equation (Tenforde, 1985a):

$$\psi = |\vec{E_i}|d = |\vec{v}||\vec{B}|d \sin \Theta \qquad (1)$$

where Θ is the angle between $\vec{B}$ and $\vec{v}$.

The induced blood flow potentials within the central circulatory systems of several species of mammals exposed to large static magnetic fields have been characterized from electrocardiogram (ECG) records obtained with surface electrodes (Beischer and Knepton, 1964; Togawa et al., 1967; Beischer, 1969; Gaffey and Tenforde, 1979, 1981; Tenforde et al., 1983, 1985). As demonstrated by the data shown in Fig. 1 for a <u>Macaca</u> monkey exposed to a 1.5-Tesla static field [1 Tesla (T) = 10,000 Gauss], the primary change in the ECG is an augmentation of the signal amplitude at the locus of the T-wave. Based on its temporal sequence in the ECG record, this change in T-wave amplitude has been attributed to the electrical potential that is induced within the aortic vessel during pulsatile blood flow in the presence of a magnetic field. This induced electrical signal is completely reversible upon termination of the magnetic field exposure. In small animal species such as rats, the aortic blood flow potential can be detected in the ECG when the magnetic flux density exceeds 0.3 T (Gaffey and Tenforde, 1981). For larger animal species such as dogs, monkeys and baboons, the threshold field level that induces a measurable potential is approximately 0.1 T (Tenforde et al., 1983, 1985). The linear dependence of the aortic blood flow potential on magnetic field strength and its variation as a function of animal orientation within the field [see Eq. (1)] have been confirmed experimentally (Gaffey and Tenforde, 1981; Tenforde et al., 1983, 1985).

The linear dependence of a magnetically-induced blood flow potential on vessel diameter [see Eq. (1)] leads to the prediction that the induced aortic flow potential should have a greater magnitude in humans than in the smaller animal species that have been studied in the laboratory. This prediction is supported by a calculation based on Eq. (1) that the maximum magnitudes of the induced aortic blood flow potentials in a rat and a man placed within a 2 T field are 0.5 and 13.4 mV, respectively. An increase in the magnitude of magnetically-induced blood flow potentials as a function of animal size has also been demonstrated directly by experimental data obtained for rats, baboons, monkeys and dogs (Tenforde, 1985b; Tenforde et al., 1985).

Using a combination of phonocardiography and echocardiography to study the temporal sequence of cardiac valve displacements in relation to the timing of signals recorded in the ECG (see Fig. 1), it has been possible to identify magnetically-induced potentials that are associated with pulsatile blood flows through the mitral, tricuspid, aortic, and pulmonary valves. The correlation between magnetically-induced potentials detectable in the ECG and specific pulsatile blood flows within the heart and the major vessels of the central circulatory system provides a useful

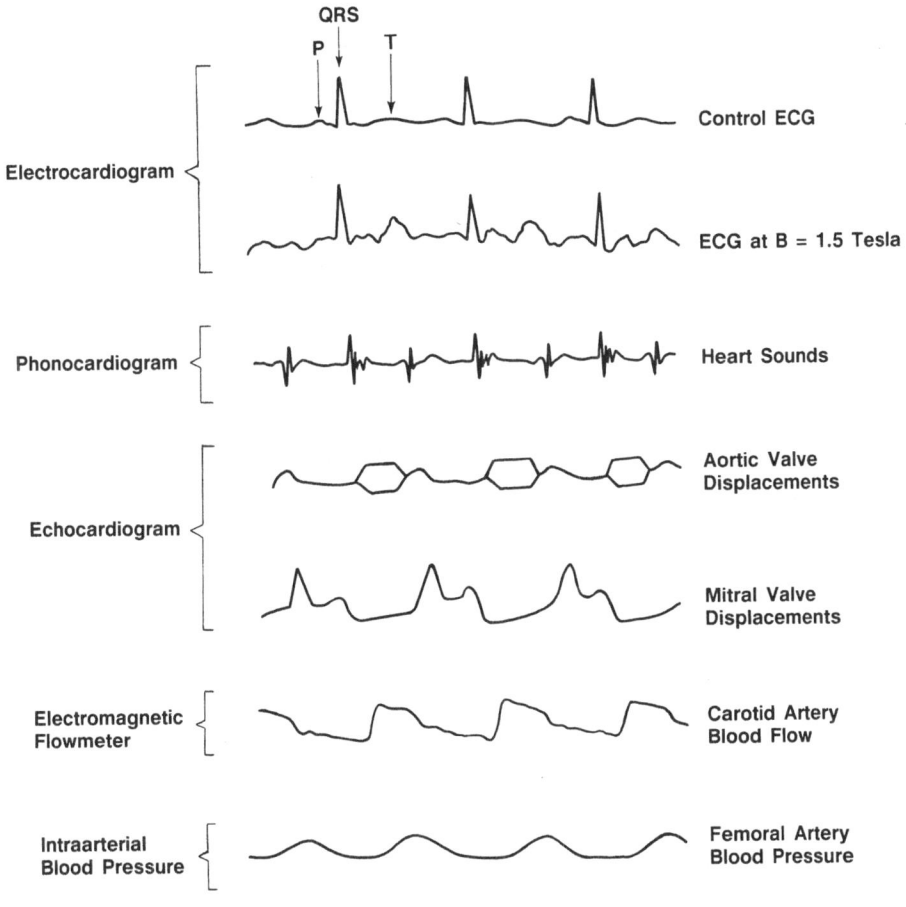

Fig. 1.   Illustration of the cardiovascular parameters measured in _Macaca_
          monkeys exposed to a 1.5 T static magnetic field. The second
          trace from the top demonstrates the superimposed blood flow po-
          tentials that are recorded in the ECG during exposure of animals
          to high-intensity magnetic fields. The most prominent magneti-
          cally-induced potential appears at the locus of the T-wave in
          the ECG, and is associated with pulsatile aortic blood flow in
          the presence of a magnetic field. The magnetohydrodynamic in-
          teraction is sufficiently weak that no measurable hemodynamic
          alterations occur at a field level of 1.5 T. [Source: C.T.
          Gaffey and T.S. Tenforde, unpublished data].

noninvasive method for the study of cardiovascular dynamics (Tenforde et al., 1985).

Magnetohydrodynamic effects on the rate of arterial blood flow and intra-arterial blood pressure have also been studied in laboratory animals exposed to high-intensity magnetic fields. The electrodynamic interaction between an applied magnetic field and a flowing electrolyte solution such as blood creates a net volume force within the fluid. The magnetohydro-dynamic consequence of this electrical force is a reduction in the axial flow velocity of the fluid (Tenforde, 1985a). A combination of arterial blood flow velocity measurements and intra-arterial blood pressure mea-surements have been carried out in beagle dogs and Macaca monkeys exposed to static magnetic fields with flux densities up to 1.5 T (Fig. 1). In accord with theoretical predictions (Tenforde, 1985a), these experimental results have demonstrated that magnetohydrodynamic interactions in a 1.5-T field do not produce a measurable alteration in blood flow dynamics.

Another important physiological process that is potentially sensitive to electrodynamic interactions with static magnetic fields is the conduc-tion of nerve impulses. Simple theoretical calculations, however, demon-strate that the interaction of a magnetic field with the ionic currents in an axonal membrane is extremely weak (Liboff, 1980; Wikswo and Barach, 1980). For example, it has been estimated that a static magnetic field in excess of 24 T would be required to produce a Lorentz force on nerve ionic currents equal to one-tenth the force they experience from the electric field of the nerve membrane (Wikswo and Barach, 1980). The absence of a measurable interaction of a 2 T static field with the ionic currents of an isolated sciatic nerve has been demonstrated experimen-tally. In various studies with isolated neurons, fields of 1.2-2.0 T applied in either a parallel or perpendicular configuration relative to the nerve axis have been found to have no influence on the amplitude or conduction velocity of evoked action potentials (Schwartz, 1978, 1979; Gaffey and Tenforde, 1983). Static magnetic fields were also found to have no effect on other bioelectric properties of sciatic nerves, includ-ing the threshold for nerve excitation and the duration of the absolute and relative refractory periods that follow the passage of an action po-tential (Gaffey and Tenforde, 1983).

Magnetomechanical interactions. There are two basic mechanisms through which static magnetic fields exert mechanical forces and torques on objects. In the first type of magnetomechanical interaction, rota-tional motion of a substance occurs in a uniform field until it achieves a minimum energy state. The second mechanism involves the translational force exerted on a paramagnetic or ferromagnetic substance placed in a magnetic field gradient. These two phenomena will be discussed sepa-rately.

Macromolecules and structurally ordered molecular assemblies with a high degree of magnetic anisotropy will experience a torque in a uniform field and rotate until they reach an equilibrium orientation that rep-resents a minimum energy state. Macromolecules such as DNA that exhibit this property generally have a cylindrical symmetry, and magneto-orienta-tion occurs as a result of anisotropy of the diamagnetic susceptibility tensor along the axial and radial coordinates. The extent to which these molecules orient is a function of their magnetic interaction energy, U, relative to the Boltzmann thermal energy, kT, where k is the Boltzmann constant (1.382×10^{-23} J/K) and T is the Kelvin temperature. The ratio U/kT is referred to as the "order parameter," and for a cylindrical mole-cule is given by the equation:

$$U/kT = - VB^2(\chi_{\|} - \chi_{\perp})/2\mu_okT \qquad (2)$$

In Eq. (2) V is the molecular volume, $(\chi_\| - \chi_\perp)$ is the difference in magnetic susceptibility along directions parallel and perpendicular to the axis of the cylinder, and μ_0 is the magnetic permeability $(4\pi \times 10^{-7}$ H/m).

For individual macromolecules the order parameter given by Eq. (2) is much less than unity and the extent of orientation of individual molecules in strong magnetic fields is very small. For example, optical birefringence measurements on calf thymus DNA in solution have demonstrated that a field of 13 T is required to produce orientation of 1% of the molecules (Maret et al., 1975). In contrast, there are several examples of macromolecular assemblies that can be completely oriented by fields on the order of 1 T (Maret and Dransfeld, 1985). These assemblies behave as structurally coupled units in which the summed magnetic anisotropy is large, thus giving rise to a large magnetic interaction energy. Examples of molecular assemblies that exhibit magneto-orientation include retinal rod outer segments, muscle fibers, photosynthetic systems (chloroplast grana, photosynthetic bacteria and Chlorella cells), purple membranes of Halobacteria, and filamentous virus particles (Tenforde, 1985a).

Although the magneto-orientation of biologically important structures such as retinal rod outer segments can be demonstrated by optical techniques when these units are suspended in a aqueous medium (Becker et al., 1978), extensive studies have failed to reveal any influence of this effect on visual functions in vivo. This fact is illustrated by the experimental studies presented in Fig. 2, in which a series of electroretinogram (ERG) recordings were made from a Macaca monkey before, during and after exposure to static magnetic fields with flux densities up to 1.5 T. Neither the A-wave (receptor field potential) or the B-wave (postsynaptic potential) elicited by flashes of white light were altered during magnetic field exposure. Similar results were obtained in studies on three monkeys and six cats (Gaffey and Tenforde, 1984; Tenforde et al., 1985). These data indicate that magneto-orientational forces exerted by fields up to 1.5 T have no significant influence on visual phototransduction processes in vivo. The most likely reason for this lack of effect is the motional restriction imposed on retinal photoreceptors by virtue of being embedded in a rigid structural matrix within the intact retina.

Although the biological implications of magneto-orientational effects demonstrated in the laboratory are presently unclear, a recent finding indicates that this type of magnetic field interaction can have profound effects on membrane properties when the exposure is conducted at temperatures approaching a thermal phase transition. A significant increase in the transport of a chemotherapeutic drug, cytosine arabinofuranoside, through the phospholipid bilayer membranes of unilamellar liposomes was observed in uniform, static magnetic fields greater than 10 mT when the exposure was made at temperatures in the prephase transition range of 40.2 - 40.7°C (Liburdy et al., 1986). This permeability effect was found to have a sigmoidal dependence on magnetic flux density, with a 50% level at 15 mT. A theoretical model of this magnetic field interaction has been formulated (Tenforde and Liburdy, 1988). The model predicts that a rapid change in bilayer permeability can be produced by the magnetic orientation of phospholipid clusters (domains) that form within bilayer membranes at temperatures in the prephase transition region. When the average radius of curvature of the ensemble of phospholipid domains approaches a critical value equal to one-sixth of the spherical liposome's radius, the membrane passes through an unstable equilibrium state in which weak magnetic forces produce major deformations of the liposome's shape. This interaction, in turn, facilitates the formation of boundary layer separations between phospholipid domains, and thereby leads to a rapid release of encapsulated solutes. The results of these recent experimental and theoretical studies

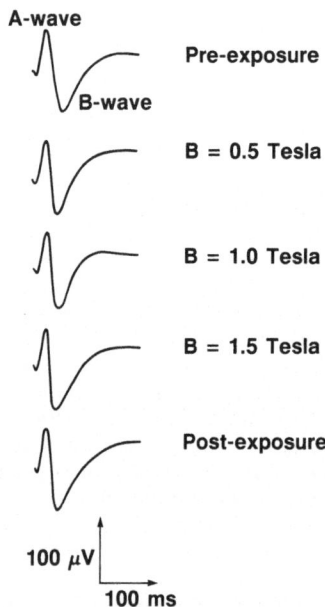

Fig. 2. Electroretinogram traces recorded in vivo from a <u>Macaca</u> monkey before, during and after exposure to static magnetic fields with flux densities up to 1.5 T. The magnitudes and durations of the A-wave and B-wave signals were unaffected by exposure to the field. [Source: C.T. Gaffey and T.S. Tenforde, unpublished data].

on phospholipid bilayer membranes suggest that it would be worthwhile to investigate the magnetic field sensitivity of eukaryotic cell membranes at prephase transition temperatures.

A second type of magnetomechanical interaction is the translation of paramagnetic and ferromagnetic substances in static magnetic field spatial gradients. Denoting the magnetic susceptibility as χ and the volume as V, the force, $F(z)$, experienced in a linear magnetic field gradient, dB/dz, is equal to the product of the net magnetic moment and the field gradient:

$$F(z) = \frac{\chi VB}{\mu_0} (dB/dz) \tag{3}$$

The forces exerted on paramagnetic and ferromagnetic substances by strong static magnetic field gradients provide the physical basis for a number of useful biological and biochemical processes. Examples of the application of magnetic forces include the targeting of drugs encapsulated in magnetic microcarriers (Widder et al., 1982), the separation of deoxygenated erythrocytes from whole blood (Melville et al., 1975; Paul et al., 1978), and the separation of antibody-secreting cells from a suspension of bone marrow cells (Poynton et al., 1983).

In contrast to their useful applications in biology, the forces exerted by strong magnetic field gradients can pose a significant physical hazard in the workplace and in magnetic resonance imaging facilities. There is a risk of large tools and other metallic objects becoming projectiles in the proximity of a high-field magnet. In addition, significant magnetic forces are exerted on many types of implanted medical devices, including aneurysm clips, dental amalgam, prostheses, and pacemaker cases (Tenforde and Budinger, 1986).

Magnetic field effects on electronic spin states. Several classes of organic chemical reactions can be influenced by static magnetic fields in the range of 10-100 mT as a result of effects on the electronic spin states of the reaction intermediates (Frankel, 1986; Schulten, 1982). One example of such reactions that has been studied extensively is the photo-induced charge transfer reaction in bacterial photosynthesis (Hoff, 1981). This reaction involves a radical pair intermediate state through which electron transfer occurs to the ultimate acceptor molecule, a ubiquinone-iron complex. Under natural conditions the electron transfer occurs within 200 picosec following flash excitation of the bacteriochlorophyll. However, chemical reduction of the acceptor molecules extends the lifetime of the intermediate state to about 10 nanosec. With an extended lifetime, the singlet state of the radical pair intermediate evolves into a triplet state via the hyperfine interaction mechanism. In the presence of an external magnetic field greater than approximately 10 mT, however, the triplet channels are blocked and the resulting yield of triplet product is expected to decrease by two thirds. This predicted blocking of triplet channels by a weak magnetic field has been confirmed experimentally using laser pulse excitation and optical absorption measurements (Michel-Beyerle et al., 1979).

It should be emphasized that the magnetic field effect on the photo-induced electron transfer in photosynthesis occurs only when the photosynthetic system is placed in an abnormal state by chemical reduction of the electron acceptor molecules. The possibility cannot be excluded, however, that similar magnetic field effects may occur in other radical-radiated biological processes under naturally occurring conditions. For example, it has been proposed that an anisotropic Zeeman interaction with a radical-mediated reaction system could provide a basis for geomagnetic

direction finding (Schulten et al., 1978). Several types of enzymatic
reactions also involve radical intermediate states that may exhibit sensitivity to static magnetic fields (Tenforde, 1985b).

Resonance interactions. Several recent studies have provided experimental evidence that the combination of a weak static magnetic field,
comparable in strength to the geomagnetic field, and a time-varying magnetic field in the ELF frequency range can produce resonance interactions
that influence ion movements through membrane channels and other biological phenomena. The physical mechanism underlying this effect has been
suggested to be ion cyclotron resonance (Liboff, 1985; McLeod and Liboff,
1986; Liboff and McLeod, 1988). In this process a resonant transfer of
energy from a time-varying magnetic field occurs when its frequency
matches the cyclotron resonance frequency of an ion moving in a circular
or helical path within the earth's magnetic field. The resonance condition is formally expressed by the equation:

$$f_c = qB/2\pi m \tag{4}$$

where f_c is the ion cyclotron resonance frequency, q = ion charge,
m = ion mass, and B = flux density of the local geomagnetic field. For
the typical range of the geomagnetic field over the surface of the earth
(30 - 70 µT), the resonant frequencies of many biologically important
ions such as Na+, K+ and Ca++ fall within the ELF range.

Several lines of experimental evidence suggest that ion cyclotron
resonance interactions can influence biological processes. Four recent
publications on this subject have reported that certain combinations of
static magnetic field flux density and time-varying magnetic field frequency can alter (1) the rate of calcium ion release from brain tissue
(Blackman et al., 1985), (2) the operant behavior of rats in a timing
discrimination task (Thomas et al., 1986), (3) calcium-dependent diatom
mobility (Smith et al., 1987), and (4) calcium ion uptake by human lymphocytes (Liboff et al., 1987). The frequencies of the time-varying
fields used in these various experiments matched the ion cyclotron resonance frequencies for potassium, lithium and calcium ions.

Although these experimental results are suggestive of a resonance
mechanism through which weak static and ELF fields could produce measurable biological effects, there are a number of theoretical difficulties
with the interpretation of this work. Four major problems with the ion
cyclotron resonance theory are the following: (1) the collision frequency
of ions undergoing cyclotron resonance motion in membrane channels is required to be orders of magnitude less than the typical collision frequency
in an aqueous solution at physiological temperatures; (2) the interaction
energy of the weak static magnetic field with biological ions is several
orders of magnitude less than the Boltzmann thermal energy, kT (= 4.28 x
10^{-21} J at 310 K); (3) the thermally-generated electrical noise (Nyquist
noise) present in ion transport channels that traverse biological membranes is approximately two orders of magnitude greater than the electric
field established in these channels by the resonant time-varying magnetic
field (Tenforde and Kaune, 1987); and (4) for ion motion that is con-
strained to lie along a prescribed path, such as the helical path envi-
sioned by Liboff (1985) for ion transport through membrane channels, it
follows directly from the equation of motion for the particle that a
static magnetic field cannot influence the ion movement and establish
a resonance condition. The ion cyclotron resonance interaction is thus
limited to unconstrained ion movements through membrane channels. All
of these factors would interfere with the establishment of ion cyclotron
resonance conditions in combined static and time-varying magnetic fields.

Obviously there is a need to refine the theoretical description of this phenomenon before it can form a plausible basis for weak field interactions with biological systems.

Organisms With Unique Sensitivity to Static Magnetic Fields

Several types of organisms have been demonstrated to possess sensitivity to extremely weak magnetic fields, comparable in intensity to the geomagnetic field. In several instances, there is direct experimental evidence indicating that this magnetic sensitivity is linked to direction-finding ability. The two basic mechanisms of magnetoreception are (1) magnetic induction of weak electrical signals in specialized sensory receptors, and (2) magnetomechanical interactions with localized deposits of single-domain magnetite crystals (Tenforde, 1988). These two mechanisms of geomagnetic field detection are described in the following paragraphs.

Elasmobranch fish. A well-established example of electrodynamic interactions with weak magnetic fields is the electromagnetic guidance system of elasmobranch fish, a class of marine animals that includes sharks, skates, and rays. The heads of these fish contain long jelly-filled canals with a high electrical conductivity, known as the ampullae of Lorenzini. As an elasmobranch swims through the lines of flux of the geomagnetic field, small voltage gradients are induced in its ampullary canals. These induced electric fields can be detected at levels as low as 0.5 μV/m by the sensory epithelia that line the terminal ampullary region (Kalmijn, 1982). The polarity of the induced field in an ampullary canal depends upon the relative orientation of the geomagnetic field and the compass direction along which the fish is swimming. As a consequence, the weak electric fields induced in the ampullae of Lorenzini provide a sensitive directional cue for the elasmobranch fish.

Magnetotactic bacteria. An example of a cellular structure in which significant magnetic orientational effects occur in response to the geomagnetic field is the magnetotactic bacterium (Blakemore, 1975). Approximately two percent of the dry mass of these aquatic organisms is iron, which has been shown by Mössbauer spectroscopy to be predominantly in the form of magnetite (Frankel et al., 1979). Magnetite crystals are synthesized within the magnetotactic bacterium ("biogenic magnetite"), and they are arranged as a chain of approximately 20-30 single domain crystals. The orientation of the net magnetic moment is such that magnetotactic bacteria in the northern hemisphere migrate towards the north pole of the geomagnetic field, whereas strains of these bacteria that grow in the southern hemisphere move towards the south magnetic pole (Blakemore et al., 1980). Magnetotactic bacteria that have been found at the geomagnetic equator are nearly equal mixtures of south-seeking and north-seeking organisms (Frankel et al., 1981). Because of the polarities of their magnetic moments, the magnetotactic bacteria in both the northern and southern hemispheres migrate downwards in response to the vertical component of the geomagnetic field. It has been proposed that this downward directed motion, which carries the bacteria into the bottom sediments of their aquatic environment, may be essential for the survival of these microaerophilic organisms (Blakemore, 1975; Blakemore et al., 1980). Recent studies indicate that fossil bacterial magnetite may be responsible for the natural remanent magnetization in deep-sea sediments (Peterson et al., 1985; Stolz et al., 1986).

Birds, bees, mollusks, and other animal species. Evidence has been presented that weak magnetic fields exert an influence on the migratory patterns of birds (Keeton, 1971), the waggle dance of bees (Martin and Lindauer, 1977), and the kinetic movements of mollusks (Ratner, 1976). A possible mechanism underlying the magnetic sensitivity of these organisms is related to the discovery of small deposits of magnetite in the cranium and neck muscles of pigeons (Walcott et al., 1979; Presti and Pettigrew, 1980), the abdominal region of bees (Gould et al., 1978), and the tooth denticles of mollusks (Lowenstam, 1962; Kirschvink and Lowenstam, 1979). It has been argued on theoretical grounds that weak magnetic interactions with the small magnetite deposits identified in avians and in bees could provide somatosensory inputs that convey directional information (Yorke, 1981).

Magnetite has recently been reported to be localized in various anatomical sites in several other species of animals, including dolphins, tuna, salmon, butterflies, turtles, mice and humans (Tenforde, 1988). The possible role of magnetite in the geomagnetic direction-finding mechanism apparently possessed by some of these species has not as yet been established. In addition, it is not clear that a sensitivity to the geomagnetic field direction exists for all of the mammalian species (e.g., humans) in which magnetite deposits have been reported to occur (Gould and Able, 1981; Fildes et al., 1984).

General Summary of Laboratory Studies on Static Magnetic Field Effects

As discussed in the preceding section of this chapter, several species of marine animals and various lower life forms possess an inherent sensitivity to static magnetic fields with intensities as low as that of the geomagnetic field. In higher organisms, however, laboratory studies of static magnetic field interactions have produced numerous contradictory findings, and the only in vivo effect that is well established at the present time is the induction of electrical potentials in the central circulatory system. There are also numerous instances in which contradictory results have been reported in the literature. For example, the report by Malinin et al. (1976) of cell transformation resulting from exposure to a 0.5 T field at 4 K was shown to be an artifact resulting from unconventional culture techniques (Frazier et al., 1979). A second example is the finding that thermoregulation in rodents is influenced by strong magnetic field gradients (Sperber et al., 1984), which could not be replicated in a second laboratory (Tenforde, 1986a).

During the past decade a large number of studies have been conducted in which the biological effects of static magnetic fields were examined under well-controlled laboratory conditions, including the use of precise dosimetry, large numbers of experimental subjects, quantitative biochemical and physiological end points, and careful control of ambient environmental conditions that could influence the experimental results. Based on laboratory studies involving field levels of 1 T or higher, the following important biological processes appear not to be altered by exposure to static magnetic fields at high intensities (Tenforde, 1985b): (1) cell growth and morphology, (2) DNA structure and gene expression, (3) reproduction and development (pre- and post-natal), (4) visual functions, (5) nerve bioelectric activity, (6) cardiovascular dynamics, (7) hematological indices, (8) immune responsiveness, (9) physiological regulation and circadian rhythms, and (10) animal behavior.

It should also be noted that in a large number of studies, no biological effects were observed as a result of exposing cellular, tissue, and animal specimens to static magnetic fields with flux densities less than 1 T. In general, there is no convincing evidence to support the existence of narrow ranges of static magnetic field strengths ("windows") below 1 T in which biological responses might occur that would not be detected at higher field levels.

Human Health Studies of Static Magnetic Field Effects

One of the earliest studies of the possible effects of exposure to static magnetic fields on human health was conducted in the Soviet Union by Vyalov (1974). The exposure group consisted of 645 workers whose hands were routinely exposed to static fields of 2 to 5 mT, and whose chest and head were in fields of 0.3 to 0.5 mT under normal working conditions. It was estimated that the magnetic field exposure levels were 10 to 50 times larger than the typical values during 10 to 15 percent of the workday. The control group in this study consisted of 138 supervisors in a machine-building plant who were not in contact with magnets. A number of subjective symptoms were reported among the exposed group, including headache, fatigue, dizziness, unclear vision, noise in the ears, and itching and sweating on the palms of the hands. Edema and desquamation on the palms of the hands were also reported. In addition, minor physiological effects including decreased blood pressure and changes in hematological parameters were noted in the exposed group. These studies were qualitative in nature and statistical analysis was not performed on the clinical data. There was also no attempt to assess the possible effects of stressful environmental factors such as high ambient temperature, airborne metallic particles, or the chemical agents used for degreasing and other procedures.

In contrast to the Soviet study, three recent epidemiological surveys in the United States and Europe failed to reveal any significant health effects associated with chronic exposure to static magnetic fields. Marsh et al. (1982) conducted a study on the health data of 320 workers in plants using large electrolytic cells for chemical separation processes. The average static field level in the work environment was 7.6 mT and the maximum field was 14.6 mT. The study included a control group of 186 unexposed workers. Among the exposed group, slight decreases were found in the blood leukocyte count and the percent of monocytes, while a small increase occurred in the lymphocyte percentage. However, the mean value of the white cell count for the exposed group remained within the normal range. There was also a slight tendency for elevated systolic and diastolic blood pressure levels among the black workers in the study. None of the observed changes in blood pressure or hematologic parameters was considered indicative of a significant adverse effect associated with magnetic field exposure.

A similar finding of no adverse health effects was reported by Barregård et al. (1985) for employees during the period 1951-1983 in a chloralkali plant in Sweden, where a direct current of 100 kA is used in the production of chlorine by electrolysis. The exposed group consisted of 157 men who worked in static magnetic fields with flux densities ranging from 4 to 29 mT. As compared with the Swedish male population, these workers had no excess cancer incidence and the mortality rate from all causes was similar to that of the general population.

A third study characterized the prevalence of disease among 792 workers at U.S. National Laboratories who were exposed occupationally to static magnetic fields (Budinger et al., 1984). The control group consisted of 792 unexposed workers matched for age, race and socioeconomic status. The range of magnetic field exposures was from 0.5 mT for long durations to 2 T for periods of several hours. No significant increase or decrease in the prevalence of 19 categories of disease was observed in the exposed group relative to the controls. Of the 792 exposed subjects, 198 had experienced exposures of 0.3 T or higher for periods of 1 hr or longer. No difference in the prevalence of disease was found between this subgroup and the remainder of the exposed population or the matched controls. No trends were observed in the health data suggestive of a dose-response relationship.

In contrast to the results of the studies described above, Milham (1982) reported that workers exposed to large static magnetic fields in the aluminum industry have an elevated leukemia mortality rate. The proportionate mortality ratios (PMR) for all forms of leukemia and for acute leukemia among these workers were compared with general population values determined from 438,000 death records of adult males in the state of Washington during the period 1950–1979. The PMR values for all types of leukemia and for acute leukemia were reported to be 189 and 258, respectively, both of which differed significantly from the no-effect level of 100. Because of the large magnetic fields associated with the aluminum reduction process, which have been measured using a magnetic field personal dosimeter and found to be as high as 57 mT during anode changes on prebake cells (Tenforde, 1986d), Milham suggested that a correlation may exist between exposure to these fields and leukemogenesis. The excess of leukemias observed in Milham's study was confirmed in a subsequent study involving 21,829 workers in 14 aluminum reduction plants (Rockette and Arena, 1983). In this second study, an excess incidence of pancreatic, genitourinary and benign tumors was also found among the aluminum workers.

In contrast to the studies by Milham (1982) and Rockette and Arena (1983), a recent survey of 6455 French aluminum plant workers showed their cancer mortality and mortality from all causes not to differ significantly from that observed for the general male population of France (Mur et al., 1987). The mortality statistics on these workers were analyzed for the period 1950 through 1976. The only finding of significance was an elevated risk of lung cancer among workers who had been employed for 10 years of less (standardized mortality ratio = 1.94). However, this elevated risk of lung cancer was not associated with any particular electrolysis process used in aluminum production.

Although two of the three epidemiological studies on persons directly involved in aluminum production have found an increased cancer risk, there is at present no clear evidence to indicate the responsible carcinogenic factors within the work environment. The process used for aluminum reduction creates coal tar pitch volatiles, fluoride fumes, sulfur oxides and carbon dioxide. The presence of hydrocarbon particulates, and perhaps other environmental contaminants, must be taken into account in any attempt to relate magnetic field exposure and increased cancer risk among persons working in the aluminum industry.

EXTREMELY-LOW-FREQUENCY MAGNETIC FIELDS

Interaction Mechanisms

The primary physical interaction of time-varying magnetic fields with living systems is the induction of electric fields and currents in tissue. For the specific case of a circular loop of tissue with radius R intersected by a spatially uniform, time-varying magnetic field orthogonal to the loop, Faraday's law gives for the magnitude of the average electric field tangent to the loop surface:

$$E = (R/2) \frac{dB}{dt} \tag{5}$$

If the magnetic field is sinusoidal with an amplitude, B_0, and a frequency, f, then $B = B_0 \sin(2\pi ft)$ and from Eq. (5):

$$E = \pi fRB_0 \sin(2\pi ft) \tag{6}$$

From Ohm's law, the current density, $\vec{J}$, induced in tissue with an average conductivity, σ, is given by:

$$\vec{J} = \sigma\vec{E} \qquad\qquad (7)$$

Various physical characteristics of time-varying magnetic fields are of importance in assessing their biological effects, including the fundamental field frequency, the maximum and average flux densities, the presence of harmonic frequencies, and the waveform and polarity of the signal. Several types of waveforms have been used in biological research with ELF magnetic fields, including sinusoidal, square-wave, and pulsed waveforms. Two characteristics that are of key importance in analyzing the effects of square-wave and pulsed fields are the rise and decay times of the magnetic field waveform. These parameters determine the maximum time rate of change of the magnetic field, and hence the maximum instantaneous electric field and resulting current density induced in tissue [see Eqs. (5)-(7)].

Although the initial physical interaction of ELF magnetic fields with living systems is the induction of electric currents in tissue, a number of secondary events may occur that involve biochemical and structural alterations at the cellular and subcellular levels. A substantial amount of evidence indicates that the pericellular currents established by ELF fields produce alterations in components of the cell membrane surface (Adey, 1981). These effects, in turn, set up transmembrane signalling events that alter cellular biochemistry and functions. The various mechanisms through which ELF fields could influence membrane properties have recently been reviewed (Tenforde and Kaune, 1987). Two general classes of phenomena have been proposed: (1) long-range cooperative interactions at the cell surface, e.g., coherent electric dipole oscillations, and (2) effects on specific membrane structures, e.g., alterations in ligand-receptor binding or ion transport channels. At the present time, there is no clear experimental evidence to support any single interaction mechanism through which ELF fields could influence cell membrane properties. It is important that this aspect of ELF field interactions at the cellular level should receive increased emphasis in future research.

Magnetophosphenes

An effect of time-varying magnetic fields on humans that was first described by d'Arsonval (1896) is the induction of a flickering illumination within the visual field known as magnetophosphenes. This phenomenon occurs as an immediate response to stimulation by either pulsed or sinusoidal magnetic fields with frequencies less than 100 Hz, and the effect is completely reversible with no apparent influence on visual acuity. The maximum visual sensitivity to sinusoidal magnetic fields has been found at a frequency of 20 Hz in human subjects with normal vision. At this frequency the threshold magnetic field flux density found by Lövsund et al. (1980a) to elicit phosphenes is approximately 10 mT. The corresponding time rate of change of the sinusoidal field is 1.26 T/sec. In recent studies Silny (1986) has observed thresholds for magnetophosphene perception in human volunteers as low as 5 mT with 18-Hz sinusoidal fields. In studies with pulsed fields having a rise time of 2 msec and a repetition rate of 15 Hz, the threshold values of dB/dt for eliciting phosphenes ranged from 1.3 to 1.9 T/sec in five adult subjects (Budinger et al., 1984). There was a trend in the data which suggested that the threshold was lower among younger subjects. In related studies it was also observed that the stimulus duration is an important parameter, since pulses of 0.9 msec duration with dB/dt = 12 T/sec did not evoke phosphenes.

Several types of experimental evidence indicate that the magnetic field interaction leading to magnetophosphenes occurs in the retina: (1) magnetophosphenes are produced by time-varying magnetic fields applied in the region of the eye, and not by fields directed toward the visual cortex in the occipital region of the brain (Barlow et al., 1947); (2) pressure on the eyeball abolishes sensitivity to magnetophosphenes (Barlow et al., 1947); (3) the threshold magnetic field flux density required to elicit magnetophosphenes in human subjects with defects in color vision was found to have a different dependence on the field frequency than that observed for subjects with normal color vision (Lövsund et al., 1980a); (4) in a patient in whom both eyes had been removed as the result of severe glaucoma, phosphenes could not be induced by time-varying magnetic fields, thereby precluding the possibility that magnetophosphenes can be initiated directly in the visual pathways of the brain (Lövsund et al., 1980a).

Although experimental evidence has clearly implicated the retina as the site of magnetic field action leading to phosphenes, it is not as yet resolved whether the photoreceptors or the neuronal elements of the retina are the sensitive substrates that respond to the field. In a series of experiments on in vitro frog retinal preparations, extracellular electrical recordings were made from the ganglion cell layer of the retina immediately following termination of exposure to a 20-Hz, 60-mT field (Lövsund et al., 1980b). It was found that the average latency time for response of the ganglion cells to a photic stimulus increased by 5 msec ($p < 0.05$) in the presence of the magnetic field. In addition, the ganglion cells that exhibited electrical activity during photic stimulation ("on" cells) ceased their activity during magnetic field stimulation (i.e., they became "off" cells). The converse behavior of ganglion cells was also observed. These observations indicate that stimulation of the retina by light and by a time-varying magnetic field elicits responses in similar post-synaptic neural pathways.

The direct involvement of electric currents induced in the retina by ELF magnetic fields in the stimulation of postsynaptic neurons is also suggested by the close similarity of electrophosphenes and magnetophosphenes. As demonstrated by Lövsund et al. (1980c), the frequency response of the human eye to electrically-generated phosphenes is similar to that observed for magnetically-induced phosphenes. In a recent study with human subjects, Carstensen et al. (1985) found that electrophosphenes were generated by a 25-Hz electric current of 4 mA passing along a path from the left eye to the right hand. At 60 Hz the current required to generate electrophosphenes increased by a factor of 5. These investigators estimated that the threshold 60-Hz electric field that must be induced in the retina to produce phosphenes is approximately 1 V/m. Based on a finite-element analysis of the field distribution in the head, the threshold 60-Hz current density required to induce electrophosphenes was estimated to be on the order of 10 mA/m^2.

General Summary of Laboratory Studies on ELF Magnetic Field Effects

Time-varying magnetic fields that induce current densities above 1 A/m^2 in tissue lead to neural excitation and are capable of producing irreversible biological effects such as cardiac fibrillation. Several investigators have achieved direct neural stimulation using pulsed or sinusoidal magnetic fields that induced tissue current densities in the range of 1-10 A/m^2. In one study involving electromyographic recordings from the human arm (Polson et al., 1982), it was found that a pulsed field with dB/dt greater than 10^4 T/sec was required to stimulate the median nerve trunk. The duration of the magnetic stimulus has also been found to be an important parameter in the excitation of nerve and nerve-muscle specimens.

Using a 20-kHz sinusoidal field applied in bursts of 0.5 and 50 msec duration, Öberg (1973) found that a progressive increase in the magnetic flux density was required to stimulate the frog gastrocnemius neuromuscular preparation when the burst duration was reduced to less than 2-5 msec. A similar rise in threshold stimulus strength has been observed for frog neuromuscular stimulation using pulsed magnetic fields with pulse durations less than approximately 1 msec (Ueno et al., 1978, 1984).

ELF magnetic fields that induce peak current densities greater than approximately 1-10 mA/m^2 have been reported to produce various alterations in the biochemistry and physiology of cells and organized tissues. One example is the effect of the bidirectional pulsed fields used to facilitate bone fracture reunion in humans (Bassett et al., 1982). A large number of laboratory investigations have also led to reports of a broad spectrum of alterations in cellular, tissue and animal systems in which current densities exceeding 1-10 mA/m^2 were induced by ELF magnetic fields (Tenforde, 1986b, 1986c; Tenforde and Budinger, 1986). These effects include: (1) altered cell growth rate, (2) decreased rate of cellular respiration, (3) altered metabolism of carbohydrates, proteins and nucleic acids, (4) effects on gene expression and genetic regulation of cell functions, (5) teratological and developmental effects, (6) morphological and other nonspecific tissue changes in animals, frequently reversible with time following exposure, (7) endocrine alterations, (8) altered hormonal responses of cells and tissues, including effects on cell surface receptors, and (9) altered immune response to antigenic stimulation. In assessing these reported effects of ELF magnetic fields, it is important to recognize that very few of the observations have been independently replicated in a second laboratory. In some instances where attempts at replication were carried out, the results were contradictory. For example, Maffeo et al. (1984) were unable to replicate the finding by Delgado et al. (1982) that pulsed magnetic fields with low intensities produce teratological effects in developing chick embryos.

Time-varying magnetic fields that induce tissue current densities less than approximately 1-10 mA/m^2 have been found to produce few, if any, biological effects. This general observation is not surprising, since the endogenous current densities present in many organs and tissues lie in the range of 0.1-10 mA/m^2 (Bernhardt, 1979).

Human Health Studies of ELF Field Effects

One of the most controversial issues related to the interaction of ELF fields with humans is the reported link between residential and occupational exposure to power-frequency fields and cancer risk. The first report on this subject was published by Wertheimer and Leeper (1979), who found that cancer deaths (primarily leukemia and nervous system tumors) in children less than 19 years of age in the Denver, Colorado area was correlated with the presence of high-current primary and secondary wiring configurations in the vicinity of their residences. This retrospective epidemiological study was based on 344 fatal childhood cancer cases during the period 1950 to 1973 and an equal number of age-matched controls chosen from birth records. The electrical power lines near the birth and death residences of the cancer cases and the residences of the controls were inspected and classified as being either high-current configurations (HCC) or low-current configurations (LCC), which were assumed to reflect the local intensity of the 60-Hz magnetic fields within the homes of the subjects. The percentage of the cancer cases whose birth and death residences were near HCC was found to be significantly greater than for the control subjects, from which Wertheimer and Leeper concluded that an association may exist between the strength of magnetic fields from the residential power distribution lines and the frequency of childhood cancer. In a subsequent publication, these authors reported that a similar association exists for the incidence of adult cancer (Wertheimer and Leeper, 1982). This later study was based on 1179 cancer cases (78% fatal cancers) in Denver, Boulder and Longmont, Colorado, during the period 1967 to 1977.

Following the initial report of Wertheimer and Leeper on childhood cancer, four other epidemiological studies have been made to determine whether a relationship exists between residential magnetic fields from power line sources and the incidence of leukemia in children. In the first of these studies, Fulton et al. (1980) used methodology that was matched as closely as possible to that of Wertheimer and Leeper, including the designation of HCC and LCC power lines. This study involved 119 leukemia patients with ages of onset from 0 to 20 years, whose address histories were obtained from medical records at Rhode Island Hospital, and 240 control subjects chosen from Rhode Island birth certificates. In their study, Fulton et al. (1980) concluded that no statistically significant correlation existed between the incidence of leukemia and the residential power line configurations. Wertheimer and Leeper (1980) were critical of the study by Fulton et al. (1980) on the basis that the case and control groups had not been matched for interstate migration, for years of occupancy at residences, or for the ages of the children at the time their residential addresses were determined from birth records and hospital medical records. In a reevaluation of the data obtained by Fulton et al. (1980), Wertheimer and Leeper (1980) excluded cases and controls aged 8 and above in order to define a complete residential history for the remaining subjects (53 cases and 71 controls). In this subset of the total population studied by Fulton and associates, Wertheimer and Leeper found a weakly significant correlation (p = 0.05) between the incidence of leukemia and residential HCC wiring configurations.

Another study of childhood leukemia incidence was conducted in the county of Stockholm by Tomenius (1986), who analyzed the residential 50-Hz magnetic fields for 716 cases that had a stable address from the time of birth to the time of leukemia diagnosis, and for 716 controls that were matched for age, sex, and birth location. An evaluation was made of the electrical wiring configurations near the residences of the study population, and measurements were also made at the entrance door to each residence of the magnetic field flux density in the frequency range above 30 Hz. Among the residences where a magnetic field level exceeding 0.3 μT was recorded, the incidence of leukemia was greater by a statistically significant amount than the expected level. The most frequently observed types of cancer were nervous system tumors and leukemia.

In contrast to the findings of Wertheimer and Leeper (1979) and Tomenius (1986), Myers et al. (1985) found no relationship between the risk of childhood cancer and residential proximity to overhead power lines. This study was conducted in the Yorkshire Health Region in England, and included 376 cancer cases diagnosed in children less than 15 years of age during the period 1970 to 1979. A total of 590 age-matched controls were included in the study. Magnetic fields at the birth addresses were calculated on the basis of data from the electrical load records for the overhead lines. The results of this epidemiological study showed no significant elevation in the cancer risk ratio with increasing field strength, and no dependence of the risk ratio on distance from the overhead lines was observed.

Another recent case-control epidemiological study by Savitz et al. (1987) attempted to verify the initial findings of Wertheimer and Leeper (1979) on childhood cancer in the Denver, Colorado area. This study involved a total of 357 cancer cases diagnosed between 1976 and 1983, and the cancer incidence data were analyzed on the basis of both the Wertheimer/Leeper wiring code and spot measurements of power-frequency magnetic fields in the homes. A correlation between cancer risk in children less than 14 years of age and the proximity of their residences to high-current wiring configurations was found. However, no association was observed between the measured household fields and childhood cancer incidence. The results of this study are therefore ambiguous, and indicate that a more precise estimate of exposure than the power-line configuration or a point-in-time measurement of household fields is needed.

Two other recent epidemiological surveys have failed to find an association between residential exposure to power-frequency fields and cancer risk. An epidemiological study conducted by McDowall (1986) in England found no correlation between cancer mortality and residential exposure to the fields from electrical utility installations (substations and overhead power lines). This study involved a retrospective analysis of the mortality from 1971 to 1983 among a population of 7631 persons in East Anglia who were identified as living in the proximity of electrical installations. The standardized mortality ratios for this large study population were computed for three major causes of death: cancer, cardiovascular disease, and respiratory disease. The overall mortality rates were lower than expected for all disease classes, and the results of this study did not support the earlier claims of an elevated cancer risk associated with residential exposure to power-frequency fields.

A recent case-control study of the incidence of acute nonlymphocytic leukemia (ANL) in three counties in Washington State also failed to find a correlation between residential exposure to power-frequency fields and cancer risk (Stevens et al., 1986). This epidemiological survey involved 164 cases of ANL and 204 controls from the same geographic area. Residential wiring codes were analyzed by the Wertheimer/Leeper technique, and this information was supplemented by direct measurements of the residential electric and magnetic fields (Kaune et al., 1987). In addition, confounding variables such as smoking habits and socio-economic status of the case and control subjects were analyzed. The overall results of this study provided no evidence for a possible association between residential exposure to power-frequency fields and the risk of contracting ANL.

The controversy surrounding the issue of ELF field exposure and cancer risk has been increased by a large number of epidemiological reports published since 1982 in which an apparent association was found between employment in various electrical occupations and cancer risk (primarily leukemia). Many of these studies have been reviewed previously (Savitz, 1986; Tenforde, 1986b). Savitz and Calle (1987) recently attempted to collate the data from 11 of these published studies in order to estimate the average relative risk of all leukemias, acute leukemias, and acute myelogenous leukemias among workers in 12 different classes of electrical occupations. The overall relative risk and 95 percent confidence intervals for leukemia mortality were found to be the following: (1) total leukemias: 1.2 (1.1-1.3), (2) acute leukemias: 1.4 (1.2-1.6), and (3) acute myelogenous leukemias: 1.5 (1.2-1.8). Savitz and Calle (1987) conclude from their analysis that a correlation exists between employment in electrical occupations and leukemia risk. However, they point out that none of the epidemiological surveys conducted thus far have established that exposure to power-frequency electromagnetic fields is the causal factor leading to an elevated cancer risk among electrical workers.

Overall, the epidemiological studies on the possible correlation between cancer risk and residential exposure to electromagnetic fields do not support the conclusion of a strong association. In the earlier studies on this subject, especially those conducted by Wertheimer and Leeper (1979, 1982) in the Denver, Colorado area, the control groups were chosen in a nonblind manner. In addition, quantitative measurements of the power-frequency fields within the residences of the case and control subjects have been made only in the recent studies by Savitz et al. (1987) and Stevens et al. (1986). Finally, with the exception of these two studies, no attempt was made to analyze the role of confounding variables (e.g., smoking habits) in the overall cancer risk of the case and control populations. These considerations, as well as the primarily negative outcomes of the most recent epidemiological studies, suggest that ELF fields in the residential environment pose little, if any, cancer risk. For workers in electrical occupations, the available information indicates that these individuals have a small elevation in the risk of cancer, especially leukemia. However, a causal link between occupational exposure to power-frequency fields and cancer has not been established. There exists a clear need for additional epidemiological surveys on large populations of electrical workers, in which efforts are made to analyze the possible role of confounding variables and to conduct proper dosimetry measurements for exposure assessment.

Several studies have also been made of the general health profiles of individuals who work in electrical occupations or who were exposed to ELF magnetic fields under controlled laboratory conditions. Medical examinations of 379 workers in electrical substations in Italy revealed no adverse clinical symptoms relative to a control group of 133 workers (Baroncelli et al., 1986). Laboratory studies on humans exposed to ELF magnetic fields have also failed to reveal any adverse physiological or psychological symptoms in the exposed subjects. The strongest field used in these experiments was a 5-mT, 50-Hz field to which subjects were exposed for 4 hr by Sander et al. (1982). No field-associated changes were observed in serum chemistry, blood cell counts, blood gases and lactate concentration, electrocardiogram, pulse rate, skin temperature, circulating hormones (cortisol, insulin, gastrin, thyroxin), and various neuronal measurements including visually evoked potentials recorded in the electroencephalogram.

A recent study by Wertheimer and Leeper (1986) led to evidence of seasonal changes in fetal growth and in abortion rate among women that used electrically heated beds during the winter months. The authors contend that these adverse effects on fetal development could result from exposure to the power-frequency electromagnetic fields present at the surfaces of electrically heated beds. They point out, however, that the potentially harmful effect of excessive heat on fetal growth cannot be excluded on the basis of their data.

In general, there is no convincing evidence that exposure to ELF fields influences fetal development or leads to adverse human health effects when end points other than cancer are considered.

SUMMARY AND CONCLUSIONS

The interactions of static magnetic fields with living systems are well understood on the basis of both theoretical models and extensive laboratory data. The principal interaction mechanisms that produce measurable biological effects include electrodynamic interactions with ionic currents such as those occurring in the circulatory system, magneto-orientational effects on magnetically anisotropic assemblies of macromolecules, translational forces on paramagnetic and ferromagnetic materials in strong magnetic field gradients, and Zeeman interactions with electronic spin states in radical-mediated charge transfer reactions. Electrodynamic interactions have also been demonstrated to form the basis for the electromagnetic guidance system in elasmobranch fish. Magnetomechanical forces exerted by the geomagnetic field influence the movements of magnetotactic bacteria, mollusks, and possibly other animal species such as bees and avians that contain deposits of magnetite crystals. In higher animal species, however, there is no evidence for the existence of a sensitivity to the geomagnetic field. Similarly, there is no evidence based on either laboratory studies or human health surveys that static magnetic fields at levels up to 2 T produce any adverse physiological or behavioral effects.

A number of different effects of ELF magnetic fields have been reported to occur at the cellular, tissue, and animal levels. Certain effects, such as the induction of magnetophosphenes in the visual system, have been established through replication in several laboratories. A number of other effects, however, have not been independently verified or, in some cases, replication efforts have led to conflicting results. A substantial amount of experimental evidence indicates that the effects of ELF magnetic fields on cellular biochemistry, function, and structure can be related to the induced current density, with a majority of the reported effects occurring at current density levels in excess of 10 mA/m^2. These effects therefore occur at induced current density levels that exceed the endogenous currents normally present in living tissues. From this perspective, it is extremely difficult to interpret the results of recent epidemiological studies that have found an apparent correlation between exposure to ELF fields and cancer incidence. The levels of current density induced in tissue by occupational or residential exposure to ELF fields are, in nearly all circumstances, significantly lower than the levels found in laboratory studies to produce measurable perturbations in biological functions. There is a clear need for additional epidemiological research to clarify whether exposure to ELF fields is, in fact, causally linked to cancer risk. In addition, more studies of both a theoretical and experimental nature are needed to elucidate the molecular and cellular mechanisms through which low-intensity ELF fields can influence living systems. Finally, for many of the biological effects observed in the laboratory, there exists a need to determine the threshold field levels that lead to reproducible perturbations in biological functions.

ACKNOWLEDGMENTS

Research in the author's laboratory is supported by the U.S. Department of Energy under contract DE-AC03-76SF00098 with the Lawrence Berkeley Laboratory.

REFERENCES

Adey, W. R., 1981, Tissue interactions with nonionizing electromagnetic fields, Physiol. Rev., 61:435.

Barlow, H. B., Kohn, H. I., and Walsh, E. G., 1947, Visual sensations aroused by magnetic fields, Am. J. Physiol., 29:124.

Baroncelli, P., Battisti, S., Checcucci, A., Comba, R., Grandolfo, M., Serio, A., and Vecchia, P., 1986, A health examination of railway high-voltage substation workers exposed to ELF electromagnetic fields, Am. J. Indust. Med., 10:45.

Barregård, L., Järvholm, B., and Ungethum, E., 1985, Cancer among workers exposed to strong static magnetic fields, Lancet, 2(8460):892.

Bassett, C. A. L., Mitchell, S. N., and Gaston, S. F., 1982, Pulsing electromagnetic field treatment in ununited fractures and failed arthrodeses, J. Am. Med. Assoc., 247:623.

Becker, J. F., Trentacosti, F., and Geacintov, N. E., 1978, A linear dichroism study of the orientation of aromatic protein residues in magnetically oriented bovine rod outer segments, Photochem. Photobiol., 27:51.

Beischer, D. E., 1969, Vectorcardiogram and aortic blood flow of squirrel monkeys (Saimiri sciureus) in a strong superconductive electromagnet, pp. 241-259 in: "Biological Effects of Magnetic Fields," Vol. 2, M. F. Barnothy, ed., Plenum, New York.

Beischer, D. E. and Knepton, J. C., 1964, Influence of strong magnetic fields on the electrocardiogram of squirrel monkeys (Saimiri sciureus), Aerospace Med., 35:939.

Bernhardt, J., 1979, The direct influence of electromagnetic fields on nerve and muscle cells of man within the frequency range of 1 Hz to 30 MHz, Radiat. Envir. Biophys., 16:309.

Blackman, C. F., Benane, S. G., Rabinowitz, J. R., House, D. E., and Joines, W. T., 1985, A role for the magnetic field in the radiation-induced efflux of calcium ions from brain tissue in vitro, Bioelectromagnetics, 6:327.

Blakemore, R., 1975, Magnetotactic bacteria, Science, 190:377.

Blakemore, R. P., Frankel, R. B., and Kalmijn, A. J., 1980, South-seeking magnetotactic bacteria in the Southern Hemisphere, Science, 212:1269.

Budinger, T. F, Bristol, K. S., Yen, C. K., and Wong, P., 1984, Biological effects of static magnetic fields, pp. 113-114 in: Proc. 3rd Ann. Meeting Soc. Mag. Res. Med., New York, Aug. 4-6, 1984.

Budinger, T. F., Cullander, C., and Bordow, R., 1984, Switched magnetic field thresholds for the induction of magnetophosphenes, pp. 118-119 in: Proc. 3rd Ann. Meeting Soc. Mag. Res. Med., New York, Aug. 4-6, 1984.

Carstensen, E. L., Buettner, A., Genberg, V. L., and Miller, M. W., 1985, Sensitivity of the human eye to power frequency electric fields, IEEE Trans. Biomed. Engin., BME-32:561.

D'Arsonval, M. A., 1896, Dispositifs pour la measure des courants alternatifs á toutes frequencies, C.R. Soc. Biol. (Paris), 3(100 Ser.):450.

Delgado, J. M. R., Leal, J., Monteagudo, J. L., and Garcia, M. G., 1982, Embryological changes induced by weak, extremely low frequency electromagnetic fields, J. Anat., 134:533.

Fildes, B. N., O'Loughlin, B. J., and Bradshaw, J. L., 1984, Human orientation with restricted sensory information: no evidence for magnetic sensitivity, Perception, 13:229.

Frankel, R. B., 1986, Biological effects of static magnetic fields, pp. 169-196 in: "Handbook of Biological Effects of Electromagnetic Fields," C. Polk and E. Postow, eds., CRC Press, Boca Raton, Florida.

Frankel, R. B., Blakemore, R. P., and Wolfe, R. S., 1979, Magnetite in freshwater magnetotactic bacteria, Science, 203:1355.

Frankel, R. B., Blakemore, R. P., Torres de Araujo, F. F., and Esquival, D. M. S., 1981, Magnetotactic bacteria at the geomagnetic equator, Science, 212:1269.

Frazier, M. E., Andrews, T. K., and Thompson, B. B., 1979, In vitro evaluation of static magnetic fields, pp. 417-435 in: "Biological Effects of Extremely Low Frequency Electromagnetic Fields," R. D. Phillips, M. F. Gillis, W. T. Kaune, and D. D. Mahlum, eds., NTIS Rep. No. CONF-781016, Springfield, Virginia.

Fulton, J. P., Cobb, S., Preble, L., Leone, L., and Forman, E., 1980, Electrical wiring configurations and childhood leukemia in Rhode Island, Am. J. Epidemiol., 111:292.

Gaffey, C. T. and Tenforde, T. S., 1979, Changes in the electrocardio-
grams of rats and dogs exposed to DC magnetic fields, Lawrence
Berkeley Laboratory Rep. No. 9085, University of California,
Berkeley, California.

Gaffey, C. T. and Tenforde, T. S., 1981, Alterations in the rat electrocardiogram induced by stationary magnetic fields, Bioelectromagnetics, 2:357.

Gaffey, C. T. and Tenforde, T. S., 1983, Bioelectric properties of frog
sciatic nerves during exposure to stationary magnetic fields, Radiat.
Envir. Biophys., 22:61.

Gaffey, C. T. and Tenforde, T. S., 1984, Electroretinograms of cats and
monkeys exposed to large stationary magnetic fields, p. 7 in: Abst.
6th Ann. Meeting Bioelectromagnetics Soc., Atlanta, Georgia, July
15-19, 1984.

Gould, J. L., Kirschvink, J. L., and Deffeyes, K. S., 1978, Bees have
magnetic remanence, Science, 201:1026.

Gould, J. L. and Able, K. P., 1981, Human homing: an elusive phenomenon,
Science, 214:2061.

Hoff, A. F., 1981, Magnetic field effects on photosynthetic reactions,
Quart. Rev. Biophys., 14:599.

Kalmijn, A. J., 1982, Electric and magnetic field detection in elasmobranch fishes, Science, 218:916.

Kaune, W. T., Stevens, R. G., Callahan, N. J., Severson, R. K., and
Thomas, D. B., 1987, Residential magnetic and electric fields,
Bioelectromagnetics, 8:315.

Keeton, W. T., 1971, Magnets interfere with pigeon homing, Proc. Natl.
Acad. Sci. USA, 68:102.

Kirschvink, J. L. and Lowenstam, H. A., 1979, Mineralization and magnetization of chiton teeth: paleomagnetic, sedimentologic and biologic
implications of organic magnetite, Earth Planet. Sci. Lett., 44:193.

Liboff, A. R., 1985, Geomagnetic cyclotron resonance in living cells,
J. Biol. Phys., 13:99.

Liboff, A. R. and McLeod, B. R., 1988, Kinetics of channelized membrane
ions in magnetic fields, Bioelectromagnetics, 9:39.

Liboff, A. R., Rozek, R. J., Sherman, M. L., McLeod, B. R., and Smith,
S. D., 1987, Ca^{2+}-45 cyclotron resonance in human lymphocytes,
J. Bioelect., 6:13.

Liboff, R. L., 1980, Neuromagnetic thresholds, J. Theor. Biol., 83:427.

Liburdy, R. L., Tenforde, T. S., and Magin, R. L., 1986, Magnetic field-
induced drug permeability in liposome vesicles, Radiat. Res.,
108:102.

Lövsund, P., Öberg, P. Å., Nilsson, S. E. G., and Reuter, T., 1980a,
Magnetophosphenes: a quantitative analysis of thresholds, Med. Biol.
Engin. Comput., 18:326.

Lövsund, P., Nilsson, S.E.G., and Öberg, P. Å., 1980b, Influence on frog
retina of alternating magnetic fields with special reference to
ganglion cell activity, Med. Biol. Engin. Comput., 19:679.

Lövsund, P., Öberg, P. Å., and Nilsson, S. E. G., 1980c, Magneto- and
electrophosphenes: a comparative study, Med. Biol. Engin. Comput.,
18:758.

Lowenstam, H. A., 1962, Magnetite in denticle capping in recent chitons
(Polyplacophora), Geol. Soc. Am. Bull., 73:435.

Maffeo, S., Miller, M., and Carstensen, E. L., 1984, Lack of effect of
weak low frequency electromagnetic fields on chick embryogenesis,
J. Anat., 139:613.

Malinin, G. I., Gregory, W. D., Morelli, L., Sharma, V. K., and Houck,
J. C., 1976, Evidence of morphological and physiological transformation of mammalian cells by strong magnetic fields, Science, 194:844.

Maret, G. and Dransfeld, K., 1985, Biomolecules and polymers in high
magnetic fields, Top. Appl. Phys., 57:143.

Maret, G., vonSchickfus, M., Mayer, A., and Dransfeld, K., 1975, Orientation of nucleic acids in high magnetic fields, Phys. Rev. Lett., 35:397.

Marsh, J. L., Armstrong, T. J., Jacobson, A. P., and Smith, R. G., 1982, Health effect of occupational exposure to steady magnetic fields, Am. Indust. Hyg. Assoc. J., 43:387.

Martin, H. and Lindauer, M., 1977, The effects of the earth's magnetic field on gravity orientation in the honeybee (Apis mellifica), J. Comp. Physiol., 122:145.

McDowall, M. E., 1986, Mortality of persons resident in the vicinity of electricity transmission facilities, Br. J. Cancer, 53:271.

McLeod, B. R. and Liboff, A. R., 1986, Dynamic characteristics of membrane ions in multifield configurations of low-frequency electromagnetic radiation, Bioelectromagnetics, 7:177.

Melville, D., Paul, F., and Roath, S., 1975, Direct magnetic separation of red cells from whole blood, Nature, 255:706.

Michel-Beyerle, M. E., Scheer, H., Seidlitz, H., Tempus, D., and Haberkorn, R., 1979, Time-resolved magnetic field effect on triplet formation in photosynthetic reaction center of Rhodopseudomonas sphaeroides R-26, FEBS Lett., 100:9.

Milham, S., Jr., 1982, Mortality from leukemia in workers exposed to electrical and magnetic fields, New Engl. J. Med., 307:249.

Mur, J. M., Moulin, J. J., Meyer-Bisch, C., Massin, N., Coulon, J. P., and Loulergue, J., 1987, Mortality of aluminum reduction plant workers in France, Int. J. Epidemiol., 16:257.

Myers, A., Cartwright, R. A., Bonnell, J. A., Male, J. C., and Cartwright, S. C., 1985, Overhead power lines and childhood cancer, in: Abst. Int. Conf. Elec. Mag. Fields in Med. Biol., London, England, Dec. 4-5, 1985.

Öberg, P. Å., 1973, Magnetic stimulation of nerve tissue, Med. Biol. Engin., 11:55.

Paul, F., Roath, S., and Melville, D., 1978, Differential blood cell separation using a high gradient magnetic field, Br. J. Hematol., 38:273.

Petersen, N., vonDobeneck, T., and Vali, H., 1986, Fossil bacterial magnetite in deep-sea sediments from the South Atlantic Ocean, Nature, 320:611.

Polson, M. J. R., Barker, A. T., and Freeston, I. L., 1982, Stimulation of nerve trunks with time-varying magnetic fields, Med. Biol. Engin. Comput., 20:243.

Poynton, C. H., Dicke, K. A., Culbert, S., Frankel, L. S., Jagannath, S., and Reading, C. L., 1983, Immunomagnetic removal of CALLA positive cells from human bone marrow, Lancet, 1(8323):524.

Presti, D. and Pettigrew, J. D., 1980, Ferromagnetic coupling to muscle receptors as a basis for geomagnetic field sensitivity in animals, Nature, 285:99.

Ratner, S. C., 1976, Kinetic movements in magnetic fields of chitons with ferromagnetic structures, Behav. Biol., 17:573.

Rockette, H. E. and Arena, V. C., 1983, Mortality studies of aluminum reduction plant workers: potroom and carbon department, J. Occup. Med., 25:549.

Sander, R., Brinkmann, J., and Kuhne, B., 1982, Laboratory studies on animals and human beings exposed to 50 Hz electric and magnetic fields, in: Proc. Int. Cong. Large High Voltage Elect. Syst., Paper 36-01, Paris, Sept. 1-9, 1982.

Savitz, D. A., 1986, Human health effects of extremely low frequency electromagnetic fields: critical review of clinical and epidemiological studies, pp. 49-64 in: "Proc. IEEE Power Engin. Soc. Panel Session on Biological Effects of Power Frequency Electric and Magnetic Fields," Publ. No. 86TH0139-6-PWR, IEEE Service Center, Piscataway, New Jersey.

Savitz, D. A. and Calle, E. E., 1987, Leukemia and occupational exposure to electromagnetic fields: review of epidemiological surveys, J. Occup. Med., 29:47.

Savitz, D. A., Wachtel, H., Barnes, F. A., John, E. M., and Tyrdik, J. G., 1987, Final results of case-control study of childhood cancer and electromagnetic field exposure, in: "Biological Effects from Electric and Magnetic Fields, Air Ions and Ion Currents Associated with High Voltage Transmission Lines," Abst. Ann. DOE/EPRI Contractors Review, Kansas City, Missouri, Nov. 2-5, 1987. Also available as Final Report: "Case-Control Study of Childhood Cancer and Exposure to Electromagnetic Fields," New York State Power Lines Project, Health Research Inc., Albany, New York.

Schulten, K., 1982, Magnetic field effects in chemistry and biology, pp. 61-83 in: "Festkorperprobleme XXII: Advances in Solid State-Physics," P. G. Aachen, ed., Proc. 46th Ann. Meeting German Phys. Soc., Mar. 29-Apr. 2, 1982.

Schulten, K., Swenberg, C. E., and Waller, A., 1978, A biomagnetic sensory mechanism based on magnetic field modulated coherent electron spin motion, Zeit. Phys. Chem., 111:1.

Schwartz, J.-L., 1978, Influence of a constant magnetic field on nervous tissues: I. Nerve conduction velocity studies, IEEE Trans. Biomed. Engin., BME-25: 467.

Schwartz, J.-L., 1979, Influence of a constant magnetic field on nervous tissues: II. Voltage-clamp studies, IEEE Trans. Biomed. Engin., BME-26: 238.

Silny, J., 1986, The influence threshold of the time-varying magnetic field in the human organism, pp. 105-115 in: "Biological Effects of Static and Extremely Low Frequency Magnetic Fields," J. H. Bernhardt, ed., MMV Medizin Verlag, Munich, West Germany.

Smith, S. D., McLeod, B. R., Liboff, A. R., and Cooksey, K., 1987, Calcium cyclotron resonance and diatom mobility, Bioelectromagnetics, 8:215.

Sperber, D., Oldenbourg, E., and Dransfeld, K., 1984, Magnetic field induced temperature change in mice, Naturwissenschaften, 71:100.

Stevens, R. G., Severson, R. K., Kaune, W. T., and Thomas D. B., Epidemiological study of residential exposure to ELF electric and magnetic fields and risk of acute non-lymphocytic leukemia, in: "Biological Effects from Electric and Magnetic Fields, Air Ions and Ion Currents Associated with High Voltage Transmission Lines," Abst. Ann. DOE/EPRI Contractors Review, Denver, Colorado, Nov. 18-20, 1986. Also available as Final Report: "Epidemiological Studies of Cancer and Residential Exposure to Electromagnetic Fields," New York State Power Lines Project, Health Research Inc., Albany, New York.

Stolz, J. F., Chang, S.-B. R., and Kirschvink, J. L., 1986, Magnetotactic bacteria and single-domain magnetite in hemipelagic sediments, Nature, 321:849.

Tenforde, T. S., 1985a, Mechanisms for biological effects of magnetic fields, pp. 71-92 in: "Biological Effects and Dosimetry of Static and ELF Electromagnetic Fields," M. Grandolfo, S. M. Michaelson, and A. Rindi, eds., Plenum, New York.

Tenforde, T. S., 1985b, Biological effects of stationary magnetic fields, pp. 93-127 in: "Biological Effects and Dosimetry of Static and ELF Electromagnetic Fields," M. Grandolfo, S. M. Michaelson, and A. Rindi, eds., Plenum, New York.

Tenforde, T. S., 1986a, Thermoregulation in rodents exposed to high-intensity stationary magnetic fields, Bioelectromagnetics, 7:341.

Tenforde, T. S., 1986b, Biological effects of extremely-low-frequency magnetic fields, pp. 21-40 in: "Proc. IEEE Power Engin. Soc. Panel Session on Biological Effects of Power Frequency Electric and Magnetic Fields," Publ. No. 86TH0139-6-PWR, IEEE Service Center, Piscataway, New Jersey.

Tenforde, T. S., 1986c, Interaction of ELF magnetic fields with living matter, pp. 197-225 in: "CRC Handbook of Biological Effects of Electromagnetic Fields," C. Polk and E. Postow, eds., CRC Press: Boca Raton, Florida.

Tenforde, T. S., 1986d, Magnetic field applications in modern technology and medicine, pp. 23-35 in: "Biological Effects of Static and Extremely Low Frequency Magnetic Fields," J. H. Bernhardt, ed., MMV Medizin Verlag, Munich, West Germany.

Tenforde, T. S., 1988, Electroreception and magnetoreception in simple and complex organisms, Bioelectromagnetics, in press.

Tenforde, T. S., and Budinger, T.F., 1986, Biological effects and physical safety aspects of NMR imaging and in vivo spectroscopy, pp. 493-548 in: "NMR in Medicine: Instrumentation and Clinical Applications," S. R. Thomas and R. L. Dixon, eds., Medical Monograph No. 14, Amer. Assoc. Physicists in Med., New York.

Tenforde, T. S., and Kaune, W. T., 1987, Interaction of extremely low frequency electric and magnetic fields with humans, Health Phys., 53:585.

Tenforde, T. S., and Liburdy, R. P., 1988, Magnetic deformation of phospholipid bilayers: effects of liposome shape and solute permeability at prephase transition temperatures, J. Theor. Biol., in press.

Tenforde, T. S., Gaffey, C. T., and Raybourn, M. S., 1985, Influence of stationary magnetic fields on ionic conduction processes, pp. 205-210 in: "Proc. 6th Symposium and Technical Exhibition on Electromagnetic Compatibility," T. Dvořák, ed., Zurich, Switzerland, Mar. 5-7, 1985.

Tenforde, T. S., Gaffey, C. T., Moyer, B. R., and Budinger, T. F., 1983, Cardiovascular alterations in Macaca monkeys exposed to stationary magnetic fields: experimental observations and theoretical analysis, Bioelectromagnetics, 4:1.

Thomas, J. R., Schrot, J., and Liboff, A. R., 1986, Low-intensity magnetic fields alter operant behavior in rats, Bioelectromagnetics, 7:349.

Togawa, T., Okai, O., and Oshima, M., 1967, Observation of blood flow E.M.F. in externally applied strong magnetic fields by surface electrodes, Med. Biol. Engin., 5:169.

Tomenius, L., 1986, 50-Hz electromagnetic environment and the incidence of childhood tumors in Stockholm County, Bioelectromagnetics, 7:191.

Ueno, S., Lövsund, P., and Öberg, P. Å., 1978, Capacitive stimulatory effect in magnetic stimulation of nerve tissue, IEEE Trans. Mag., MAG-14:958.

Ueno, S., Harada, K., Ji, C., and Oomura, Y., 1984, Magnetic nerve stimulation without interlinkage between nerve and magnetic flux, IEEE Trans. Mag., MAG-20:1660.

Vyalov, A. M., 1974, Clinico-hygienic and experimental data on the effects of magnetic fields under industrial conditions, pp. 163-174 in: "Influence of Magnetic Fields on Biological Objects," Y. A. Kholodov, ed., Rep. No. JPRS 63038, National Technical Information Service, Springfield, Virginia.

Walcott, C., Gould, J. L., and Kirschvink, J. L., 1979, Pigeons have magnets, Science, 205:1027.

Wertheimer, N. and Leeper, E., 1979, Electrical wiring configurations and childhood cancer, Am. J. Epidemiol., 109:273.

Wertheimer, N. and Leeper, E., 1980, Re: Electrical wiring configurations and childhood leukemia in Rhode Island, Am. J. Epidemiol., 111:461.

Wertheimer, N. and Leeper, E., 1982, Adult cancer related to electrical wires near the home, Int. J. Epidemiol., 11:345.

Wertheimer, N. and Leeper, E., 1986, Possible effects of electric blankets and heated waterbeds on fetal development, Bioelectromagnetics, 7:13.

Widder, K. J., Senyei, A. E., and Sears, B., 1982, Experimental methods
    in cancer therapeutics, J. Pharm. Sci., 71:379.
Wikswo, J. P. and Barach, J. P., 1980, An estimate of the steady magnetic
    field strength required to influence nerve conduction, IEEE Trans.
    Biomed. Engin., BME-27:722.
Yorke, E. D., 1981, Sensitivity of pigeons to small magnetic field varia-
    tions, J. Theor. Biol., 89:533.

BIOLOGICAL EFFECTS OF

RADIO FREQUENCY ELECTROMAGNETIC RADIATION

W. Ross Adey

Veterans Administration Medical Center and
Loma Linda University School of Medicine
Loma Linda, C.A. 92357 USA

INTRODUCTION

In the last 50 years, there has been an exponential growth of man-made electromagnetic (EM) fields, associated with comunication systems that now blanket the earth and with a vast and ever-increasing network of electric power distribution systems (Nat. Acad. Sci. USA, 1977). It is therefore curious that although many of the most important fundamental observations on the physical effects of light and other nonionizing EM radiations were made more than 100 years ago, knowledge of their biological effects has grown more slowly. Tissue interactions with EM fields have been extensively studied in terms of two quite different endpoints; in their thermal effects and in ionization of atoms in biomolecular systems.

Ionizing radiation poses hazards to living organisms through its destructive effects on key macromolecular systems, where damage to DNA in the cell nucleus, for example, leads to permanent genetic changes transmitted as mutations to succeeding generations of cells. It is implicit for ionizing radiation that it must have sufficiently high photon energies to disrupt the atomic organization of the exposed macromolecular systems; and since EM radiation with wavelengths longer than the ultraviolet region of the spectrum (photon energies less than about 12 eV) does not possess sufficient energy to cause ionization, there has been a persisting view in the physical and engineering sciences that nonionizing EM fields are incapable of inducing bioeffects, other than by heating; or more explicitly, that evidence for important "athermal" effects remains in the realm of disjointed observations and inconsistent experiments that fail to provide a connected picture of either major sequences in transductive coupling or of physiological mechanisms influenced by exposure to EM fields at levels below those producing biologically significant heating.

This inadequate view (Foster and Guy, 1986; Foster and Pickard; 1987) overlooks the existence of cooperative organization in biomolecular systems, and the profoundly important role that cooperativity appears to play in the detection of tissue components of nonionizing EM fields (Adey, 1981a and b; Adey and Lawrence, 1984; Lawrence and Adey, 1982; Schmitt, Schneider and Crothers, 1975). Also, the nature of these interactions isso far removed from concepts and models that have guided research in ionizing radiation that expertize in the latter area can offer little in the search for underlying mechanisms. Equilibrium thermodynamics and the classical models of the statistical mechanics of matter appear equally inappropriate in their applications to most key questions on the biological effects of nonionizing EM fields. It is in the essence of cooperativity that nonequilibrium processes are involved, with energy exchanges likely to be confined within relatively narrow intensity and frequency bands. Familiar biological doctrines of excitatory processes based on ionic shifts toward or away from equilibrium conditions do not pertain in these models (Frohlich, 1968, 1975, 1980, 1986b; Illinger, 1981; Kaiser, 1984), even though the sequence of excitatory events beyond the first transductive coupling can be so described.

In most cells, there is a major electrical gradient (the membrane potential) of about 0.1 V across the 40 A^o width of the double layer of fat molecules (the plasma membrane) that forms the essential structure of the cell membrane. This gradient of 10^5 V/cm has been considered an effective electrical barrier against cell stimulation by weak EM fields in surrounding fluid, based on equilibrium models of cellular excitation that focus on depolarization of this membrane potential and associated massive changes in ionic equilibria across the cell membrane. It has also been assumed that these equilibrium models of cellular excitation offer an adequate explanation of the first events in cell membrane transductive coupling of molecular and electrochemical stimuli at the cell surface. The Hodgkin-Huxley (1952) equilibrium model has been generally accepted as an appropriate description of both the sequence and the energetics of excitatory shifts in ionic equilibria. However, this elegant thesis was originally offered only in the context of a mathematical description of major perturbations in Na^+ and K^+ ionic equilibria that occur at a certain epoch in excitation of squid giant nerve fibers. There has been a philosophic incubus for more than 30 years from the early and seemingly inappropriate extrapolation of this model for nerve fibers to threshold phenomena in excitation of nerve cells (Eccles, 1953).

At that time, excitatory mechanisms in nerve fibers and nerve cells were grouped under a common rubric of ionic equilibrium mechanisms. There was little interest in the possibility that functional organization of membranes of cell bodies might involve threshold sensitivities to both oscillating EM fields and to molecular stimuli at energy levels substantially lower than predicted by Hodgkin-Huxley models, and substantially below typical thresholds in nerve fibers. Much recent research cited below has shown that imposed weak low frequency fields (and radiofrequency fields amplitude-modulated at ELF frequencies) that are many orders of magnitude weaker in the pericellular fluid than the membrane potential gradient can modulate actions of hormone, antibody, neurotransmitter and cancer-promoter molecules at their cell surface receptor sites. From their electrical characteristics, these sensitivities appear to involve nonequilibrium and highly cooperative processes that mediate a major amplification of initial weak triggers associated with binding of these molecules at their specific cell surface receptor sites (Adey, 1983, 1986, 1987; Adey and Lawrence, 1984; Lawrence and Adey, 1982).

In recent years, there has been an increasing focus on the role of Ca^{2+} ions in excitatory processes. There is clear evidence that Ca^{2+} ions (Adey et al., 1982; Bawin and Adey, 1976; Bawin et al., 1978a; Bawin et al., 1978b; Blackman et al., 1979, 1985a and b; Lin-Liu and Adey, 1982) and calcium-dependent processes at cell membranes (Dixey and Rein, 1981; Kaczmarek and Adey, 1974; Lyle et al., 1983) initiate intracellular enzyme responses (Adey, 1986; Luben et al., 1982; Luben and Cain, 1984; Byus et al., 1987a and b) that are modulated by extracellular EM fields that induce transmembrane currents millions of times weaker than required for threshold excitation with a Hodgkin-Huxley model. We shall note that cell membranes function as powerful amplifiers of their first weak interactions with both

EM fields and humoral stimuli. As revealed by field effects, these interactions are nonequilibrium in character and consistent with quantum processes involving long range interactions between electric charges on cell surface macromolecules. Observed sensitivities are as low as 10^{-7}V/cm in the ELF spectrum.

Many of these interactions are "windowed" with respect to field frequency and amplitude, and to duration of field exposure. Imposition of EM fields in these studies has led to identification of some of the key sequences and the energetics of steps that couple signals from the outside to the inside of the cell, and points the way to further studies of their cooperative nature. Windowed aspects of many of these responses support their nonlinear and nonequilibrium character, and focuses current research on physical substrates for these interactions (Adey, 1975, 1977, 1981a and b, 1983, 1984, 1986; Lawrence and Adey, 1982; Maddox, 1986).

THE RADIOFREQUENCY ENVIRONMENT AND ITS BIOPHYSICAL IMPLICATIONS

There has been an exponential increase over the past half century in the use of devices employing EM energy in the workplace, in the home, and in external environments. They cover a very broad spectrum from below 100 Hz to the millimeter microwave bands and the far infrared region (Adey, 1981a). These artificial fields typically exceed intensity of the natural background by many orders of magnitude, and their proliferation has dramatically increased the potential for public exposure at many parts of this broad spectrum. For example, in the USA, permitted leakage at the door of a microwave oven is 5 mW/cm^2 (130 V/m); a hand-held radio transmitter operating close to the head may produce a similar field; and in the suburbs of American cities, FM radio broadcast stations operating in the vicinity of 100 MHz produce fields up to 1-4 microwatts/cm^2 (2-4 V/m).

Resulting tissue field levels depend on body geometry with respect to field orientation and field frequency (Johnson et al., 1977; Gandhi et al., 1975, 1976, 1977). For ELF electric fields, a simple capacitance model of coupling suffices. In consequence, coupling is poor at these low frequencies. Thus, a 10 kV/m field from a 60 Hz power line would only produce tissue gradients in man of about 0.1 mV/cm. On the other hand, radiofrequency fields couple much more efficiently, particularly where body dimensions approximate the wavelength of the incident field. At frequencies of 100-400 MHz, field intensities of 1.0 mW/cm^2 (61 V/m in air) are associated with tissue gradients of 10-100 mV/cm in mammalian organisms.

Whole-Body Resonance at Radio- and Microwave Frequencies

Absorption of EM energy increases the kinetic energy of molecular constituents of the absorbing medium. Increased rates and energies of collisional events are associated with raised temperature (Straub, 1977). However, the mammalian cardiovascular and respiratory systems are capable of efficient thermal exchange with the environment, so that the observed rise in temperature may be substantially below the expected increment in an equivalent volume of nonliving tissue. On the other hand, enhanced field levels may occur in resonant interactions with body segments or with the whole body as a function of the relationship between body geometry and the wavelength of the imposed field. These relationships are important in induction of hyperthermic states for therapeutic purposes.

Energy absorption is proportional to f^2 for body lengths less than 0.05 wavelength, increasing at a rate greater than f^2 near, but below, resonance; and decreasing at a rate less than f^2 at frequencies beyond resonance (Johnson et al., 1977).

The head behaves resonantly at around 400 MHz in the adult and in the region of 700 MHz in the infant (Gandhi and Hagmann, 1977). At 2450 MHz, there is a band of enhanced absorption in the outer 1.0 cm, whereas at 918 Mhz there is a converse pattern of higher absorption in the center of the head (Kritikos and Schwan, 1972). In the long axis of the body, maximum absorption occurs at frequencies around 35 MHz for the grounded subject and closer to the free space resonant frequency of 70 MHz for man standing on or above a perfectly conducting ground plane. In the transverse and anteroposterior axes, maximum absorption occurs at frequencies of 135-165 MHz (Gandhi, 1975). Rates of energy deposition in the presence of a ground plane or a reflector, rates of energy deposition at resonant frequencies are sharply increased, and with grounding and reflection combined, "they are truly staggering" (Gandhi et al., 1976).

Just as distortion of radiofrequency fields within tissues occurs in the presence of a metallic sensor or in the vicinity of any dielectric material substantially different from tissue, so too is a free field modified by the presence of humans or other biota (Beischer and Reno, 1975). This is important in assessing field effects on a group of subjects in changing proximity to each other. Measurements on humans at frequencies from 1 to 12 GHz indicate substantial perturbations in front of the subjects, with body reflections producing a series of standing waves. Sharp field enhancements occur at antinodes in this wave pattern.

Tissue Determinants of Radiofrequency Energy Absorption

EM radiation has a kinetic energy that is transferred to any particle encountered by the propagating field, inducing movement of charged particles. Currents so generated lead to two phenomena (Little, 1977). They may be dissipated as eddy currents, producing heat. Also, their movement as a current generates a magnetic field that interacts with the intrinsic field of the EM wave to resist change in the applied magnetic field (Lenz's law). Penetration of an EM wave into a conducting material will therefore be limited to a depth (d) that is a function of the frequency of the wave (e) and the conductivity of the material (s)

$$d = 1//\overline{e.s}$$

EM waves therefore exhibit <u>skin effects</u> in remaining near the surface of most conducting materials. This frequency-dependent penetration and preponderant energy deposition at superficial levels is important in evaluating prospects for successful therapeutic utilization of millimeter microwaves in oncology, even where tumor-specific absorption bands may exist at these frequencies (Frohlich, 1978).

Energy absorption in the mammalian body has a spatial distribution that is sensitive to layering of skin, fat, muscle and bone. This layering is associated with a resonance for three-dimensional bodies that is quite distinct from the geometrical resonance discussed above. Calculations for a multilayered prolate spheroidal model of man predict a whole body layering resonance at 1.8 GHz with an energy absorption 34 percent greater than predicted by a homogeneous model (Barber et al., 1979). A related model for a specific skin-fat-muscle cylindrical model of man predicts a layyering resonance frequence of 1.2 GHz, with an averaged energy absorption twice that calculated for a corresponding homogeneous model. The layering frequency is the same for incident waves polarized either perpendicularly or parallel to the cylinder axis. Energy deposited in different body layers will clearly be different, and further modeling of this differential distribution has been attempted by considering the human body as a series of small blocks (Hagmann et al., 1979).

Energy deposition at a particular place in body tissue is based on calculation of the mass-normalized rate of energy absorption in W/kg (SAR). Caution is desirable in evaluating use of the SAR in terms of averaged values. A major issue is the vitally important question of levels of energy deposition at interfaces represented by tissue discontinuities. these may be structurally gross and macroscopic, as at the interface between muscle and bone. They may also occur in a microcosm of molecular dimensions, as at the margins of macromolecular domains on membrane surfaces, or on the surface. In either case, they might be expected to be the site of enhanced energy deposition, not effectively modeled in the general SAR concept and even less in the "averaged SAR." Local electrical fields up to 100 times larger than average fields can be induced around microscopic wedge-shaped boundaries between regions with different dielectric constants likely to be present in the human body (Nilsson and Petterson, 1979).

BIOMOLECULAR MECHANISMS IN INTERACTIONS WITH RADIOFREQUENCY FIELDS

In this brief survey, it will be useful to examine the biomolecular basis of interactions with radiofrequency fields from two distinct but related viewpoints: in comparison of thermal and athermal modes of interaction; and in comparison of effects attributable to direct molecular interactions at the field carrier frequency with those dependent on low-frequency amplitude-modulation of the carrier wave.

Dielectric Behavior of Tissue Elements and Thermal Responses to Radio Frequency Fields

The cellular nature of tissue establishes a vast number of dielectric partitions that separate strongly conducting intracellular and extracellular compartments. Thus cells are enclosed by poorly conducting membranes with a typical resistance of 3-100 kohm.cm^{-2} (Coombs et al., 1959) and lie within an extracellular space with a specific resistance as low as 4 ohm.cm^{-1} (Nicholls and Kuffler, 1964). In this respect, tissues may be modeled in accordance with Maxwell's experiment with nonconducting sea urchin eggs as the dispersed phase in a bucket of seawater (Cole, 1940). Membrane structures contribute complexly to the observed dielectric behavior of tissues, and water itself is an important additional factor.

The dielectric properties of tissues, cell suspensions and macromolecules have been extensively studied (Schwan 1957-77). The dielectric properties of tissues indicate that polarization from induced charges exceeds contributions from permanent dipoles. Dielectric dispersion studies indicate that most water is in the free state. Schwan has identified three major dispersions: alpha, beta and gamma. Each dispersion is fairly well defined by either a single relaxation time or by a small spectrum of relaxation times:

Alpha-dispersion: dielectric relaxation of free water (frequency near 20 GHz, dielectric constant 10-50).
Beta-dispersion: Maxwell-Wagner type of relaxation resulting from charging of cell membranes (frequency near 1.0 MHz, dielectric constant of 10^2-10^4).
Gamma-dispersion: variability with frequency of the apparent outer cell membrane capacitance (frequency near 100 Hz, dielectric constant around 10^5).

Schwan suggests that this frequency dependence of membrane capacitance may result from several causes: a frequency-dependent access to inner membrane structures that connect with the outer membrane; a frequency-dependent surface admittance, tangential to the cell membrane caused by counterion displacement from fixed-charge sites on membrane surface macromolecules; and a frequency-dependent intrinsic membrane capacitance arising in ionic gating currents.

Dispersions of dielectric constants of ice, protein-bound water and free water show that ice relaxes at audio frequencies, bound water between 100 and 1000 MHz, and free water near 20 GHz (Schwan et al., 1976). These data were interpreted as indicating that a major fraction of all biopolymers is surrounded by free water. Although the measurements were made with field levels that offer an excellent model for tissue heating induced by these fields, the techniques may not reveal small amounts of structured bound water, as for example, in intimate contact with cell surface macromolecules as a highly viscous layer (Grant, 1977; Ling et al., 1972). Though difficult to detect, existence of water in this special relationship to macromolecular substrates (Drost-Hansen, 1978) may be important in the transductive coupling of weak EM fields. For example, binding of calcium to macromolecular charge sites may occur directly to carboxyl groups, whereas O-sulphate groups offer a looser electrostatic attraction through an atmosphere of water molecules (Perlin, 1977).

<u>Biomolecular Organization in Responses to Continuous Wave and Amplitude-Modulated Radiofrequency Fields at "Athermal" Levels</u>

<u>Resonant molecular interactions with continuous wave fields</u>. Illinger (1981) has evaluated the collisional basis of molecular interactions with microwave and far-infrared fields. The duration of the collisional perturbation is crucial to the form of the dielectric response function (complex permittivity) of a molecule at the field frequency in the presence of collisional perturbations, and to the attenuation function that describes interaction with the field (Frohlich, 1946; Van Vleck and Weisskopf, 1945). Where the collisional perturbations are very brief for one period of the impressed field, every collision is effective in interrupting the absorption-emission process. In a fluid with many collisions per unit time, a collision-broadened relaxation-type spectrum results. Conversely, where collisional perturbations are very long compared to the period of the EM field, there is a resonant-type spectrum, even in a fluid where there are numerous collisions per unit time. Since the duration of a typical collision in a molecular fluid is fixed at a given temperature and pressure, the field frequency determines whether there is a relaxation or resonance type spectrum.

Illinger concludes that no compelling evidence exists for resonant absorption in ordinary molecular fluids below 3000 GHz. Nevertheless, experimental evidence suggests resonant interactions in growing cells exposed to 41 GHz fields (Grundler et al., 1977) and in DNA at 11 GHz Edwards ot al., 1984), as discussed below. Attenuation of microwave fields at these frequencies is dominated by the ubiquitous presence of water, which also shields other possible biomolecular absorption processes, including quasi-lattice vibrations in biopolymers and the vibrations of hydrogen-bonded bridgeheads.

<u>Amplitude-modulation dependence of tissue interactions with radio-frequency fields; quantum mechanical models of long-range interactions</u>. As discussed below, there is now a broad body of evidence showing an absence of bioeffects from exposure to unmodulated radiofrequency fields at athermal levels, and a sharply contrasting sensitivity to fields at the same carrier frequency when these fields are sinusoidally amplitude- or pulse-modulated at ELF frequencies below 1.0 KHz, and particularly at frequencies below 100 Hz. No known mechanisms explain these ELF-dependent effects on the basis of direct interactions with component dipoles of molecular systems oscillating

at these low frequencies. Therefore a structural and functional basis must reside in properties of molecular systems.

These unexpected biological sensitivities have led to predictive models that address two of the more baffling aspects of these interactions: sensitivity to low incident field energy and possible bases for molecular interactions in the ELF spectrum below 100 Hz. There are at least four major groups of models (see Adey, 1981a for review; also discussion on modeling below). They have emphasized phase transitions at ELF in systems that may exhibit entrainment and chaotic behavior (Grodsky, 1976; Kaiser, 1984), charge "pumping" and Lotka-Volterra models of charge population transitions (Frohlich, 1977; Noyes and Field, 1974), limit cycle phenomena (Kaczmarek, 1976), and possible tunneling effects at the boundary separating membrane surface fixed-charge zones that exhibit coherence and incoherence (Adey and Bawin, 1980; Bawin et al., 1978a).

Modulation-dependent responses imply mechanisms for demodulation of the low-frequency envelope of the carrier waveform. We have proposed that this may occur in the strands of cell surface glycoproteins, where their dense polyanionic terminal amino sugars with the associated counterion sheet of cations may form a demodulation system through charge separation (Bawin et al., 1978a). This view is supported by intracellular recording in Aplysia neurons, where preliminary studies have indicated no modulation components of a 450 MHz field (10.0 mW/cm^2) sinusoidally amplitude- modulated at frequencies of 2.1, 10.1 and 20.1 Hz (Sheppard et al., 1984).

MILLIMETER MICROWAVE INTERACTIONS WITH CELLS AND MACROMOLECULES

In the difficult search for resonant interactions at millimeter wavelengths, there have been studies in macromolecules, bacteria and simple cellular systems. Studies at these frequencies present major technical difficulties in the engineering of suitable exposure systems, in ensuring biocompatible exposure devices, and in evaluation of data for physical and biological artifacts. Studies by Webb (Webb and Dodds, 1968; Webb, 1980) were the first to report inhibition of bacterial growth by 136 GHz microwaves. Grundler, Keilmann and their coworkers (Grundler, Keilmann and Frohlich, 1977) described a resonant growth rate response in yeast cells irradiated with 41.5 GHz microwaves. These studies led to an examination of millimeter wave absorption in molecular systems that would elucidate underlying physical mechanisms. Microwave absorption in DNA has been the focus of several studies (Edwards et al., 1984; Swicord et al., 1984; Gabriel et al., 1987), based on initial observations by Edwards and colleagues that a resonant absorption occurs in polyclonal DNA at 11 GHz. Other studies have examined the possibility of soliton vibrations of the Davydov type (Davydov, 1979; Scott, 1981) in the polyamide spines of DNA and helical proteins initiated by millimeter wave fields (Layne et al., 1985; Lomdahl et al., 1982).

Absorption of Millimeter Microwaves in Bacterial Cultures

Webb (1968, 1980) exposed E. coli cultures to a spectrum of millimeter wave fields (59-143 GHz) at field intensities of 5 uW/cm^2 to 5 mW/cm^2, using thin film (200 um) techniques to minimize field attenuation by excess water. He observed effects on growth rates, DNA and RNA synthesis and on protein synthesis that depended on the state of the organisms (dormant, rapidly growing or senescent), and exhibited sharp frequency dependencies. This frequency dependency showed two sets of frequencies that affected these metabolic processes, one separated by integral multiples of 7 GHz and the other by 5 GHz. Raman spectroscopy revealed no activity in resting, living cells, but Raman spectra with sharp peaks appeared in cells activated by added nutrients containing an oxidizable carbon source. The spectrum changed continuously as the cells progressed through their life cycle. Sets of lines between 200 and 3400 cm^{-1} moved gradually to higher frequencies while those between 5 and 200 cm^{-1} moved to lower frequencies.

Scott (1981) suggested that Webb's Raman spectral data might arise in soliton vibrations. Subsequent detailed numerical analysis by Lomdahl et al. (1982) support this view. However, six studies in other laboratories have yielded conflicting results (see Layne et al. (1985) for review). All these laser-Raman studies of metabolically active bacteria were performed with scanning monochromators, a technique in which variations in bacterial flow through the laser beam and in bacterial fluorescence with time may cause variations in spectral baselines. This baseline variation can result in false Raman peaks, as the wavelength setting of a monochromator is scanned in time. Using a spectrometer equipped with an optical multichannel analyzer that would avoid these artifacts, Layne et al. (1985) found no Raman lines attributable to the metabolic process nor to the cells themselves. However, they noted that synchronous E. coli cultures become more fluorescent during a limited phase of the division cycle. The effect is reproducible in synchronous cultures but is not seen in asynchronous ones. Layne et al. ascribe this transient increase in fluorescence to a variation in the redox state of a chemical species in the bacteria or to variations in the intracellular optical field.

Millimeter Wave Absorption in Aqueous Solutions of DNA

A series of studies by Swicord, Edwards and Davis has examined microwave absorption in aqueous solutions of DNA exposed to 8-12 GHz fields, using two quite different methods; an optical heterodyne detection technique and dielectrometry with an automated network analyzer. These studies were conducted with low-level fields producing "nonsignificant thermal induction" (Swicord et al., 1983, 1984).

These experimental approaches were based on theoretical studies of van Zandt and Prohofsky (Prohofsky, 1979; Kohli et al., 1981; Van Zandt et al., 1982) which considered longitudinal and transverse vibration modes (acoustic or optical) in dehydrated double-stranded DNA of varying base pair numbers. The resonant modes were base-pair-number or chain-length-dependent, with the longitudinal acoustic vibrational modes being the most likely in the microwave frequency range. Van Zandt et al. (1982) have extended these models to consider critically damped absorption in DNA in aqueous solutions, since damping of vibration by surrounding water molecules may be important in determining vibrational modes in biological systems. More recently, Scott (1985) has developed a nonlinear soliton model which assumes that the excited acoustic wave is anharmonic.

Initial studies by Swicord et al. used phase fluctuation optical heterodyne spectroscopy (PFLOH spectroscopy) with a He-Ne single frequency laser source. Absorption of microwaves in aliquid medium results in thermal expansion of the liquid, and thus a fluctuation in the index of refraction. This fluctuation causes an effective path length change for the signal beam. Using DNA extracted from E. coli, this method revealed a 10-20 percent increase in absorption at 11 GHz due to a DNA weight increment of only 1 percent in the solution (Swicord et al., 1984).

Their later studies with dielectrometry techniques addressed physical problems of microwave absorption in water and saline; the associated biophysical problems of measuring differential effects attributable to the presence of DNA in solution; techniques of DNA sample preparation; and the significance of DNA chain length.

Emphasizing the importance of DNA sample preparation, they noted a strong relationship between average DNA chain length and absorption in the 8-12 GHz frequency range (Swicord et al., 1983) at 45 to 85 min of incubation of a single DNA sample in the presence of a DNA endonuclease enzyme which progressively segments the chain into shorter lengths. After 85 min, the absorption coefficient had increased by 70 percent at 11 GHz as a result of adding only 0.175 percent DNA by weight; suggesting that the absorption coefficient for the DNA molecule in this sample was about 400 times that of the solvent at this frequency. As enzyme activity continued, absorption decreased back to the level of the solvent. In subsequent studies with monoclonal DNA in ring-shaped and rod-shaped forms in aqueous solution, they observed sharp resonances corresponding to relaxation times of hundreds of milliseconds (Edwards et al., 1984).

Other studies have failed to detect evidence of resonant absorption of millimeter waves. Foster et al. (1983) reported negative results which they explained as arising in a "mixing" effect based on a Maxwell-Wagner model, where a slight increase in absorption occurs due to interaction of two materials. Swicord et al. (1984) consider that their results cannot be interpreted on the basis of classical absorption theory and that difficulties in DNA preparation, including such factors as strand length and the development of ring- and rod-shaped forms, are important determinants of absorption characteristics. A joint British and Swedish study (Gabriel et al., 1987) in independent laboratories also failed to detect resonant or enhanced absorption in the frequency range 1-10 GHz. Clearly, further research must take account of critically important steps in handling DNA for spectroscopic studies; and must consider aspects of spectroscopic system sensitivities that may differ between laboratories.

CELL MEMBRANE SUBSTRATES FOR TRANSMEMBRANE SIGNALING AND ENERGY TRANSFER

Initial evidence that interactions with radiofrequency fields at athermal tissue levels might be sensitive to low-frequency amplitude modulation came from studies of effects on calcium binding in isolated cerebral tissue (Bawin et al., 1975, 1978a and b). A striking aspect of these initial studies was the _windowing_ of binding sensitivities with respect to both modulation frequency and field intensity.

These observations have been substantially extended by Blackman and his colleagues (Blackman et al., 1979, 1985a and b), as discussed below. In turn, these findings led to a search for calcium-dependent cellular mechanisms that might be initiated or modified by these modulation-dependent interactions, or by ELF fields directly. Three enzyme systems have been identified as molecular markers inside cells in their responses to field interactions initiated on cell surfaces: 1) the metabolic enzyme adenylate cyclase; 2) messenger protein kinase enzymes; and 3) ornithine decarboxylase essential for cell growth and synthesis of DNA. Also, immune mechanisms involving lymphocytes in destruction of tumor cells by surface contact (allogeneic cytotoxicity) are modified by amplitude-modulated microwave fields. Details of these interactions are discussed below.

From these studies, it has become clear that the cell membrane is a prime site in detecting and transducing amplitude-modulation components of RF and microwave fields, as well as in transductive coupling of ELF fields.

Structural Substrates of RF Field Interactions; the Fluid Mosaic Model of Cell Membranes in Transductive Coupling from Surface Receptors to Cell Interior

Much has been learned in the past decade about mechanisms of athermal tissue interactions with low-frequency EM fields and with RF/microwave fields amplitude-modulated at low frequencies. Although it remains for future research to reveal detailed aspects of physical mechanisms underlying these interactions at athermal tissue levels, it appears from these studies that cell membranes are the prime site of transductive coupling of ELF and ELF-modulated environmental fields. What is the structural basis for these interactions at cell membranes? Is it possible to account for membrane amplification inherent in detection of these weak fields at cell surfaces? For example, what is the basis for their ensuing influence on intracellular mechanisms, despite the huge electric barrier of the membrane potential (10^5V/cm, discussed above), since these intracellular mechanisms function at energy levels millions of times higher than in the imposed fields?

With the electron microscope, the cell membrane is defined as a lipid bilayer 40A^o in width, with polar head groups of these fat molecules on external and internal surfaces of the double layer. This is the plasma membrane. More importantly, these studies of membrane ultrastructure have revealed numerous strands of protein (intramembranous particles, IMPs) inserted into the plasma membrane (Fig. 1). These IMPs span the membrane from the inside to the outside. They have external protrusions into the fluid surrounding the cell, with terminal glycoprotein strands that sense electric fields and form receptor sites for chemical stimuli, including hormones, antibodies, neurotransmitters and certain chemical cancer promoters. Inside the cell, they make contact with key elements in the cell machinery, including enzymes and the numerous fine tubes and filaments of the cytoskeleton. They "float" in the sea of lipid molecules of the plasma membrane, leading to the generally accepted fluid mosaic model of the cell membrane (Singer and Nicolson, 1972). Thus, these intramembranous protein strands form signaling pathways by which external stimuli are sensed and conveyed to the cell interior.

The fluid mosaic model evolved from earlier concepts of a protein-lipid-protein sandwich and dispersion of proteins within the lipid bilayer (Davson and Danielli, 1952; Benson, 1965). The IMPs "float" in the lipid bilayer. Their external protrusions terminate in strands of sialic acids (amino sugars) that are polyanionic, forming a huge negatively chargwed sheet on the membrane surface. In addition to participating in cell surface receptor mechanisms for hormones, etc., they attract cations, principally hydrogen and calcium, in a counterion layer.

The mosaic of IMPs is extensively rearranged during their participation in interactions with hormones, antibodies and other stimulating molecules, such as chemical cancer promoters (see below), when these molecules bind at their cell surface receptor sites. The membrane is thus coded longitudinally in the plane of the membrane surface. This is important in interactions with extracellular EM fields in which the predominant current flow is along the membrane surface. Strands of protein cross the bilayer and may have protrusions inside and outside the bilayer (Fig.1). External protrusions with sialic acid (amino sugar) terminals form a huge negatively charged sheet on the membrane surface.

The Cellular Microenvironment; Extracellular Spaces and Pathways for Current Flow in Tissue

Concepts of a cell emphasize the role of a bounding membrane surrounding an organized interior that participates in the chemistry

118

essential for all terrestrial life. This enclosing membrane is the
organism's window on the world around it. As a sensor, it detects altered
chemistry in the surrounding fluid. It offers a pathway for inward signals
generated on its surface by a wide variety of stimulating ions and
molecules. As effectors, cell membranes may induce cell movement or secrete
substances synthesized internally. Both inward and outward transactions at
cell membranes are susceptible to manipulation by intrinsic and imposed EM
fields.

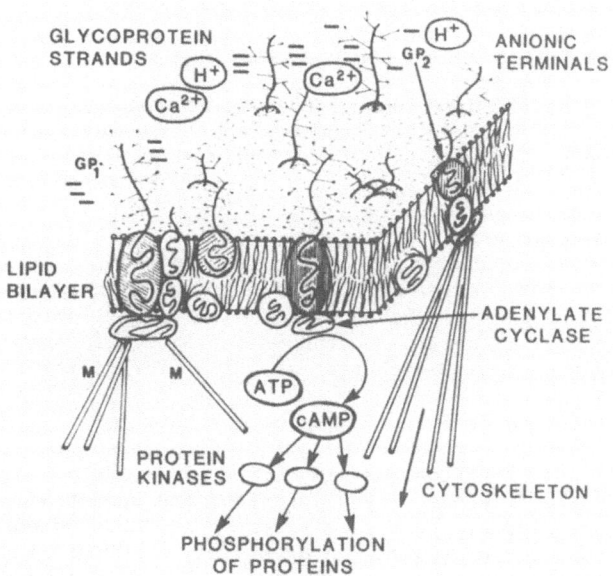

Fig. 1. Fluid mosaic cell membrane model describes structural substrates
         of major tissue interactions with imposesd EM fields. IMPs in
         plasma membrane (lipid bilayer) have protruding external
         glycoprotein strands that are highly negatively charged on their
         amino sugar terminals. They form receptor sites for antibodies,
         hormones and neurotransmitters and attract Ca ions. Inside the
         cell, IMPs make contact with enzymes and with tubes and filaments
         of the cytoskeleton that convey signals to the nucleus and other
         elements.

Cells are separated by narrow fluid channels that take on special
importance in signaling from cell to cell. These channels act as windows on
the electrochemical world surrounding the cell. Hormones, antibodies and
neurotransmitter molecules move along them to reach binding sites on cell
membrane receptors, formed on the protruding strands of IMPs as described
above. These tiny fluid gutters (extracellular spaces, ECS), typically no
more than 150 A^{o} wide, are also preferred pathways for tissue components
of EM fields, since they offer a much lower electrical impedance than cell
membranes. They are also the channels through which all cellular nutrients
must pass and products of cellular metabolism are removed. Glycoprotein
strands of IMPs protruding into these gutters offer an anatomical substrate
for the first detection of weak electrochemical oscillations in pericellular

fluid, including field potentials arising in adjoining cells or as tissue components of environmental EM fields.

Distribution of current flow in tissue is determined by the high membrane resistance of cells and the strongly conducting fluid in which they are bathed (Cole, 1940). Typical cell membrane resistances are in the range 3,000-100,000 $ohm.cm^2$. Extracellular fluid has a specific resistance of only 50 $ohm.cm^{-1}$. From a consideration of dielectric dispersions, cell membranes exhibit reactive behavior at frequencies as high as the low gigahertz range (Schwan, 1974). Thus, although the ECS forms only about 10 percent of the conducting cross-section of typical tissue, it is clearly a preferred pathway, carrying at least 90 percent of any imposed or intrinsic current and <u>directing it along cell membrane surfaces</u>.

Intercellular Communication through Gap Junctions

There are specialized regions of contact between membranes of adjacent cells (Robertson, 1963) that serve to couple cells electrically (Furshpan and Furakawa, 1962). When stained with lanthanum, high resolution electron microscopy reveals a thin 2-3 nm cleft containing protein to which Revel and Karnovsky (1967) assigned the term <u>gap junction</u>. These junctions are perforated by numerous tiny tubes (<u>connexons</u>) 1.5 nm in diameter that span the entire membrane; so that when connexons of adjacent cell membranes come into register, the interior of adjacent cells are effectively in continuity. These connexons thus provide the physicl substrate for ionic coupling and transfer of essential metabolic substances between cells in the process of <u>metabolic cooperation</u> (see Fletcher et al., 1987 for review).

Disruption of intercellular communication through gap junctions leads to serious disorders in growth control, including tissue repair and neoplastic transformation (Loewenstein, 1966, 1979, 1981). Controlled cell growth occurs in the presence of gap junctions and uncontrolled growth in their absence. Our studies indicate a synergic action of chemical cancer promoters and modulated microwave fields at cell membranes in modification of gap junction functions (Fletcher et al., 1986). As discussed below, <u>disruption of gap junction communication is now viewed as a prime factor in cancer promotion and tumor formation</u> (Adey, 1987a and b, 1988a; Trosko, 1987; Yamasaki, 1987).

Cell Membrane Receptor Proteins as Substrates for Transmembrane Signaling and Energy Transfer

Much attention now focuses on the long strands of membrane receptor proteins for the human epidermal growth factor (EGF) and the nerve growth factor (NGF) as models of coupling proteins in studies of the nature of transmembrane signals.

The entire 1210 amino acid sequence of the EGF receptor protein has been deduced by Ullrich et al. (1985), with striking findings on the sequences that make up the extracellular, intramembranous and cytoplasmic portions of the chain. Extracellular and intracellular segments are each composed of approximately 600 <u>hydrophilic</u> amino acids. The molecule appears to cross the membrane only once. The salient and surprising finding is the extremely short length of the intramembranous segment of 23 amino acids, predominantly <u>hydrophobic</u>, and with only a single amino acid with a side chain capable of hydrogen bonding.

Subsequent studies have shown that the NGF receptor protein also has a strikingly similar segment of 23 hydrophobic amino acids wwithin the membrane (Radeke et al., 1987), suggesting that this configuration plays a fundamental role in processes of transmembrane signaling. This view is strengthened by studies with a chimaeric protein constructed of the extracellular portion of the insulin receptor protein joined to the transmembrane and intracellular domains of the EGF receptor protein (Riedel et al., 1986). In this molecule, the EGF receptor kinase domain of the chimaeric protein is activated by insulin binding. The authors conclude that insulin receptors and EGF receptors employ closely related or identical mechanisms for signal transduction across the plasma membrane.

What can be inferred from these studies about the essential nature of the transmembrane signal? Ullrich et al. (1985) point out that this 23 amino acid sequence is probably too short to be involved in conformation changes; and that its hydrophobic character makes unlikely its participation in either ion or proton translocation. As an alternative, they have suggested that an EGF-induced conformation change in the extracellular segment of the receptor protein may be transmitted to the cytoplasmic domain of the protein strand by movement of this short intramembranous segment in and out of the lipid bilayer, or by receptor aggregation.

As a further option, we have hypothesized that this transmemnrane signaling may involve nonlinear vibration modes in helical proteins and generation of Davydov-Scott soliton waves (Lawrence and Adey, 1982). Although evidence for soliton waves in DNA and helical proteins is not conclusive and has been criticized on theoretical grounds (Lawrence et al., 1987), we have strong supporting evidence for nonlinear, nonequilibrium processes at critical steps in transmembrane coupling, based on windows in EM field frequency, amplitude and time of exposure that determine stimulus effectiveness (Adey, 1988b).

Despite the hydrophobic character of this short transmembrane coupling segment, addition of EGF to human epidermal cell cultures causes a 2-4-fold increase in cytoplasmic free Ca^{2+} within 30-60 sec (Moolenaar et al., 1986). This EGF-induced signal appears to result from Ca^{2+} entry via a voltage-independent channel, since it is not accompanied by a change in membrane potential. It is completely dependent on extracellular Ca^{2+} moving into the cell interior. This action is inhibited by cancer- promoting phorbol esters which have a specific membrane receptor (Ca^{2+}-dependent protein kinase C). EM field interactions with these cancer promoters are discussed below.

Summary: Three Stages in Signaling of EM Fields from Cell Surface to Interior

In summary, protein particles (IMPs) placed within cell membranes provide an essentially direct __inward__ signaling path between the cell surface and intracellular enzymatic systems and organelles. There is a minimal sequence of three steps in this transductive coupling and each is calcium-dependent:

1. Cell surface glycoproteins sense the first weak electrochemical events associated with EM fields and with binding of stimulating molecules at their receptor sites.

2. Transmembrane portions of IMPs signal these surface events to the cell interior .

3. Internally, there is coupling of these signals to intracellular enzyme systems, and through the tubes and filaments of the cytoskeleton, to the nucleus and to other organelles.

There is also an outward stream of electrical and chemical signals passing through cell membranes and linking adjoining cells through gap junctions. Both inward and outward signals are sensitive to a broad spectrum of weak EM fields, including amplitude-modulated RF fields.

EXPERIMENTAL EVIDENCE ON AMPLITUDE-MODULATION DEPENDENT TISSUE INTERACTIONS WITH RF FIELDS

If this model of transmembrane signaling is correct, it should meet certain experimental criteria, particularly with respect to the strong emphasis that it places on the role of calcium in the initial events in stimulus recognition at the cell membrane surface, including detection of amplitude-modulated RF fields. These criteria are posed in the following questions. Is the binding of calcium in cerebral tissue influenced by modulated RF fields at the same tissue electric gradients as natural oscillations, such as the electroencephalogram (EEG) in brain tissue? What is the evidence that sensitivity of calcium binding to imposed fields is consistent with nonequilibrium processes, involving long range interactions between charge sites on surface macromolecules? Would this nonequilibrium behavior manifest itself in some form of frequency sensitivity to the "biological spectrum" below 100 Hz? Would there also be "windows" of sensitivity with respect to the intensity of the imposed field?

Cooperative Modification of Calcium Binding by RF Fields at Cell Surfaces with Amplification of Initial Signals

Initial stimuli associated with weak pericellular EM fields and with binding of stimulating molecules at their membrane receptor sites elicit a highly cooperative modification of Ca^{2+} binding to glycoproteins along the membrane surface. As noted above, a longitudinal spread is consistent with the direction of extracellular current flow associated with physiological activity and with imposed EM fields. This cooperative modification of surface Ca^{2+} binding is an amplifying stage, with evidence from concurrent manipulation of initial molecular binding events by imposed RF fields that there is a far greater increase or decrease in Ca^{2+} efflux than is accounted for in the events of receptor-ligand binding (Bawin and Adey, 1976; Bawin et al., 1975; Lin-Liu and Adey, 1982). There is further striking evidence for the nonequilibrium character of this modification in Ca^{2+} binding in its occurrence in quite narrow frequency and amplitude windows (Adey, 1981a and b; Bawin et al., 1978a and b; Blackman et al., 1979).

The pioneering observations of Bawin et al. (Bawin and Adey, 1976; Bawin et al., 1975) were the first to show "tuning curves" (frequency windows) of altered Ca^{2+} efflux from tissue as a function of low frequencies in imposed EM fields, either as simple low frequency fields or with low-frequency amplitude-modulation of RF fields (Fig. 2).Sensitivities are maximal at modulation frequencies around 16 Hz and less at higher and lower frequencies. Field carrier waves at 147 and 450 MHz produced similar results and unmodulated carriers had no effect.

Most studies of EM field effects on tissue Ca binding have used cerebral tissue, including cerebral cortex in awake cats (Adey et al., 1982), isolated chick cerebral hemisphere (Bawin and Adey, 1976; Bawin et al., 1975, 1978a and b; Blackman et al., 1979, 1982, 1985), cultured neurons (Dutta et al., 1984), and cerebral synaptosome fractions (Lin-Liu and Adey, 1982). In cultured nerve cells in a 915 MHz field, increased Ca efflux occurred at specific, lower or intermediate levels (Dutta et al., 1984). The response at 0.05 mW/kg was dependent on 16 Hz modulation, but at the higher level was not. Bawin's finding of a low-frequency modulation

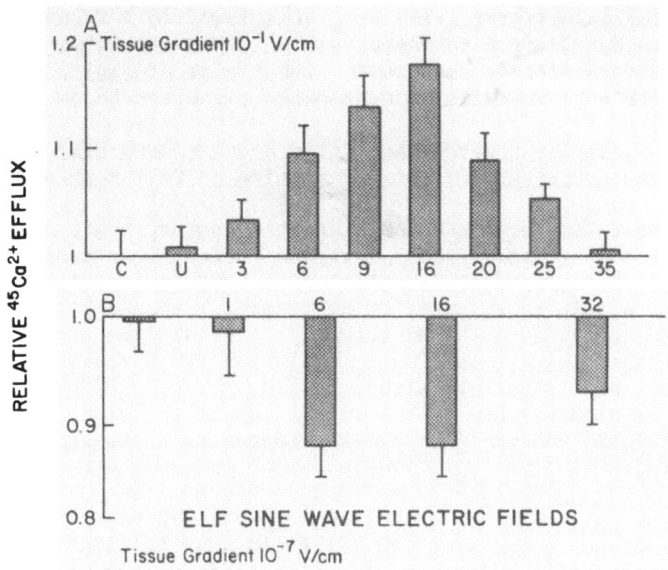

Fig. 2. <u>Upper curve</u>: $^{45}Ca^{2+}$ efflux from isolated chick cerebral hemispheres exposed to a weak RF field (147 MHz, 0.8 mW/cm^2), amplitude-modulated at low frequencies (abscissa). C. control Ca^{2+} efflux level; U, unmodulated carrier wave. (From Bawin et al., 1975). <u>Lower curve</u>: effects of far weaker electric fields in air (56 V/m, estimated tissue gradients 10^{-7} V/cm) in the same frequency spectrum from 1-32 Hz. (From Bawin and Adey, 1976). Tissue gradients differ by about 10^6 between upper and lower curves.

window was confirmed by Blackman et al. (1979), who also noted an <u>intensity window</u>, confirmed by Bawin et al. (1978b) for incident fields in the range 0.1-1.0 mW/cm^2 that produce EEG-level gradients in cerebral tissue.

These effects have been observed with RF fields amplitude-modulated at low frequencies; with low-frequency electric fields; with low-frequency EM fields; and with combined low-frequency EM fields and static magnetic fields.

It is important to note that these sensitivities were observed over an enormous range of physical dimensions, from intact cerebral cortex to cultured neurons and finally in isolated terminals of cerebral nerve fibers (synaptosomes) with mean diameters around 0.7 um. Clearly, neither size or geometry are primary determinants of these interactions.

The maximum tissue electric gradients induced by these imposed fields are at physiological levels. Typical gradients are at the levels of the electroencephalogram (EEG) in fluid around brain cells (0.1 V/cm), seen with exposures to RF fields in the studies of Bawin et al., Blackman et al., and Dutta et al. cited above. <u>At these levels, they are at least 6 orders of magnitude less than the electric barrier of the membrane potential</u>. However, with low-frequency fields, (in the spectrum below 1.0 kHz), coupling to tissues and cell cultures is far weaker than for RF fields at the same environmental intensities, inducing tissue electric gradients typically in the range 10^{-7}-10^{-3} V/cm. These weaker fields also modify tissue Ca^{2+} binding (Bawin and Adey, 1976) and modulate Ca-dependent cell mechanisms, including neurotransmitter release at gradients of 10^{-4} V/cm (Dixey and Rein, 1981) and bone matrix formation at 10^{-7} V/cm (Fitzsimmons et al., 1986). These interactions emphasize the importance of **amplification** in their ultimate effects on intracellular mechanisms discussed below.

Lymphocyte Cytotoxicity toward Tumor Cells and Windowed Effects of Modulation Frequency of RF Fields: Destruction on Cell Membrane Contact

Allogeneic T lymphocytes can be targeted against tumor cells, destroying the tumor cells by cell membrane rupture upon contact between the lymphocyte and the tumor cell (cytolysis). A 450 MHz field (1.5 mW/cm^2) reduced the cytolytic capacity of of allogeneic T lymphocytes by 20 percent when amplitude-modulated at 60 Hz (Fig. 3). Interactions were windowed, with less interaction at higher and lower modulation frequencies (Lyle et al., 1983). Unmodulated fields were without effect. Similar effects occur with simulated 60 Hz high voltage power line fields in the range of 0.1-10 mV/cm in the cell culture medium, with clear evidence of a threshold and field intensity effects (Lyle et al., 1987).

These findings, together with reduced protein kinase enzyme activity in lymphocytes exposed to the same 450 MHz field, may implicate immune surveillance mechanisms that protect against abnormal growth.

Intracellular Enzymes as Molecular Markers of Transductive Coupling Through Cell Membranes

Enzymes are protein molecules that function as catalysts, initiating and enhancing chemical reactions that would not otherwise occur at tissue temperatures. This ability resides in the pattern of electrical charges on the molecular surface. In the fashion of more familiar chemical catalysts, such as the hydrocarbon oxidation system which functions only at very high temperatures in automobile exhaust systems, a catalyst emerges unchanged from these reactions and is thus able to participate indefinitely in a specific reaction. Activation of these enzymes and the reactions in which they participate involve energies millions of times greater than in the cell surface triggering events initiated by the EM fields, emphasizing the membrane amplification inherent in this transmembrane signaling sequence.

We have found three groups of intracellular enzymes that respond to signals initiated at cell membranes as a response to athermal EM field exposure. These responses occur with or without concurrent cell membrane stimulation initiated chemically by physiological molecules (hormones, antibodies, etc.) and by cancer promoting substances. These enzymes are: 1) membrane-bound adenylate cyclase involved in activation of protein kinase enzymes through conversion of the metabolic fuel substance adenosine triphosphate (ATP) to cyclic-adenosine monophosphate (cAMP), as seen in bone cells exposed to low-frequency pulsed magnetic fields (Luben et al., 1982; Luben and Cain, 1984; Cain, Luben and Adey, 1987); 2) cAMP-independent protein kinases that perform messenger functions (Byus et al., 1984; 3) ornithine decarboxylase (ODC), essential for growth in all cells by its participation in synthesis of polyamines essential for DNA formation (Adey, 1986; Byus et al., 1987, 1988). All are Ca-dependent and their actions have been reviewed in detail elsewhere (Adey, 1986).

Lymphocyte Protein Kinase Responses to RF Fields Amplitude-Modulated at Low Frequencies: Windows in Frequency and Time

Some intracellular protein kinases are activated by signals arising in cell membranes that do not involve the cAMP pathway discussed above. This group includes membrane protein kinases related to actions of cancer promoting phorbol esters (see below). In human tonsil lymphocytes exposed to a 450 MHz field (1.0 mW/cm^2), cAMP-independent protein kinases showed activity windowed with respect to exposure duration and and modulation frequency (Byus et al., 1984). Reduced enzyme activity only occurred at modulation frequencies between 16 and 60 Hz, and only for the first 15-30 min of RF field exposure (Fig 4). Unmodulated fields elicited no responses.

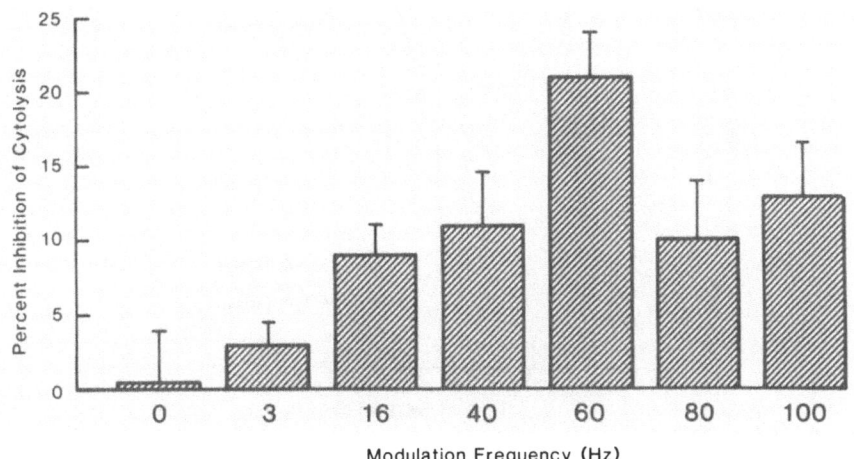

Fig. 3 Inhibition of cytoxicity of allogeneic T lymphocytes by exposure to a 450 MHz field (1.5 mW/cm^2), as sinusoidal amplitude-modulation was varied between 0 and 100 Hz, showing windowed sensitivity to modulation frequency. (From Lyle et al., 1983).

INTERCELLULAR COMMUNICATION AND CANCER PROMOTION; ENZYMATIC MARKERS OF RF FIELD INTERACTIONS WITH CHEMICAL CANCER PROMOTERS AT CELL MEMBRANES

Tumor formation as a manifestation of abnormal control of cell growth is now widely modeled as a multistep process, based on animal tumor models. These models of carcinogenesis envisage initial damage to the DNA genome within the cell nucleus. This stage of _initiation_ involves actions of mutagenic substances or agents such as ionizing radiation. Initiated or transformed cells may remain indefinitely in this condition without tumor formation. Tumor formation occurs as a second step requiring _promotion_ by agents that are not mutagenic and thus not cancer initiators by an action on DNA in the nucleus.

Tumor promoters include insecticides such as DDT, polychlorbiphenyls (PCBs) used as electrical insulators and coolants, saccharin, and plant lectins now used as cancer promoters in laboratory studies, the phorbol esters. Phorbol esters have a specific receptor at cell membranes. For that reason, we have examined cancer promotion in cultured cells in studies of joint ctions of phorbol esters and modulated RF fields, _since both act at cell membranes_ (Adey, 1986; Byus et al., 1987, 1988).

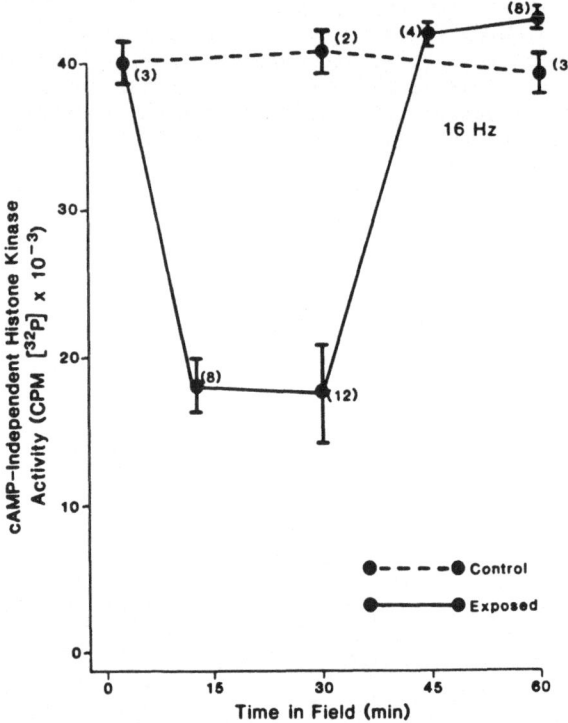

Fig. 4. cAMP-independent protein kinase activity in human lymphocytes is sharply reduced by a 450 MHz field (1.0 mW/cm^2, amplitude-modulated at 16 Hz) 15-30 min after onset of exposure, but returns to control levels at 45-60 min, despite continuing exposure - a window in time. From Byus et al., 1984).

Available evidence indicates that nonionizing EM fields do not function as classical initiators in the etiology of cancer by causing DNA damage and gene mutation. On the other hand, there is increasing evidence implicating environmental EM fields in the home and in the workplace in increased cancer risk. There is evidence of increased incidence of childhood leukemia and lymphoma from exposure to high voltage power line fields (Wertheimer and Leeper, 1979; Savitz, Wachtel and Barnes, 1986); and a greater risk of leukemia, lymphoma and pancreatic cancer amongst electrical workers (Milham, 1985); and a ten-fold increase in rates of malignant brain tumors (astrocytomas) in microwave workers with 20 years' job experience that also involved exposure to soldering fumes and/or electronic solvents (Thomas et al., 1987).

The biology of cancer therefore invites consideration of promotional processes occurring at cell membranes, rather than at the cell nucleus. We have observed strong interactions at cell membranes between cancer-promoting phorbol esters and EM fields. We hypothesize that cancer promotion may involve a distorted inward signal stream from cell membranes directed to the nucleus and other organelles (Adey, 1988a and b). At the same time, cell-cell communication through gap junctions discussed above is disrupted by chemical cancer promoters (phorbol esters, DDT, PCBs and saccharin) and is indicative of a disrupted outward signal stream from cell membranes (Yotti et al., 1979). Moreover, normal cells can exert control over those of their number that have been transformed into cancer cells. For example, differentiation is restored to carcinomatous keratinocytes by contact with normal skin fibroblasts and development of carcinomas is inhibited (Newmark, 1987). The following evidence implicates modified intercellular communication through distorted inward and outward signal streams at cell membranes in cancer promotion.

RF Modulation of Inward Signal Streams at Cell Membranes from Cancer-Promoting Phorbol Esters; Enzyme Activities of Protein Kinases and Ornithine Decarboxylase (ODC)

An important protein kinase that functions both as a receptor and as an enzyme occurs widely at cell membranes and is in highest concentration in brain tissue (Nishizuka 1983, 1984). This enzyme, phosphatidyl serine protein kinase (protein kinase C), is Ca^{2+}-dependent and normally activated by diacylglycerol formed from inositol phospholipids by the action of cell surface stimuli. Following activation of one molecule of kinase C, by a single molecule of diacylglycerol, Nishizuka has described a spreading domino effect that activates all kinase C molecules around the whole cell membrane surface. Kinase C is irreversibly activated by cancer-promoting phorbol esters. In invertebrate neurons, its activation or injection of the pure enzyme enhances inward Ca^{2+} currents (DeRiemer et al., 1984). Exhaustion of protein kinase C occurs with continued stimulation by phorbol esters and is associated with loss of cell division. This response is restored by intracellular injection of kinase C (Pasti et al, 1987), thus linking it to paths that signal from cell membranes to nuclear mechanisms mediating cell division. Protein kinase C belongs to a group of cAMP-independent protein kinases which we have identified as sensitive to modulated RF fields (Byus et al., 1984).

We have also shown that induction of ornithine decarboxylase (ODC) follows stimulation of liver and ovary cells with phorbol esters (Fig.5) and that this response is sharply enhanced by 450 MHz fields (1.0 mW/cm^2), sinusoidally amplitude-modulated at 16 Hz (Byus et al., 1988). Similar enhancement of ODC activity was noted with 60 Hz EM fields at culture fluid gradients of 0.1-10 mV/cm (Byus et al., 1987), of the same order as those induced by high voltage power line exposure.

Clinically, ODC activity in cultures of suspected cancer cells (for example human prostatic cancer) has proved a useful index of malignancy. All agents that stimulate ODC are not cancer-promoting, but all cancer promoters stimulate ODC. Although its activation pathways are not well defined, binding of phorbol esters at membrane receptors induces ODC, and ODC activity is increased by 50 percent in a 3h test period following a 1h exposure to the same 450 MHz fields tested above (Byus et al., 1988). Also, phorbol ester treatment of embryonic fibroblasts previously irradiated with X-rays and microwaves (2450 MHz, 34 mW/cm^2, 24h), increased transformation frequencies (mutation with unregulated growth) above rates in cells irradiated only with X-rays (Balcer-Kubiczek and Harrison, 1985). The findings are consistent with cell membranes as the site of persisting effects of prolonged microwave exposure. These effects may enhance promoter actions of phorbol esters.

Disruption of Intercellular Communication by Phorbol Esters and Microwaves
in Tumor Promotion: Outward Signals through Gap Junctions

Phorbol esters and other chemical cancer promoters disrupt transfer of
chemical signals between cells (Yotti et al., 1979). Trosko has further
hypothesized that two major types of intercellular communication help to
maintain "normal orchestration" of proliferation and differentiation during
development and between quiescent stem-progenitor and differentiated cells
in the adult (Trosko, 1987; Trosko and Chang, 1986): one involving transfer
of molecular signals from cells of one differentiation or tissue type to
another over an extracellular space and distance (for example, in the action
of hormones, growth factors and neurotransmitters); and the other mediated
by the transfer of relatively small molecular weight molecules and ions
through neighboring cells via gap junctions. In Trosko's model,
gap-junctional communication would involve a minimum of four steps:
recognition by neighboring cells of one another through the action of
adhesion molecules; functional gap junctions; small regulatory or signaling
molecules; and transducing protein receptors for these signaling molecules.

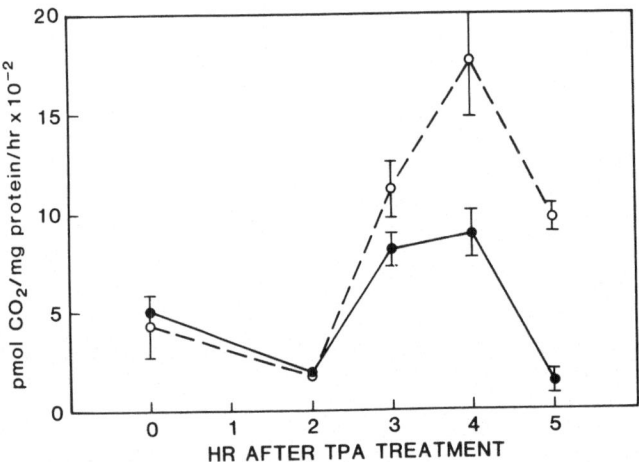

Fig. 5. Ornithine decarboxylase (ODC) activity in Chinese hamster ovary
cells stimulated by a cancer-promoting phorbol ester (tetradecanoyl
phorbol acetate, TPA), with and without concurrent exposure to a 450
MHz field (1.0 mW/cm^2 peak envelope power) amplitude-modulated at
16 Hz. (From Byus et at., 1988).

We have examined the separate and combined actions of phorbol esters
and microwave fields on gap junction communication (Fletcher et al., 1986).
In CHO hamster ovary cells, entry of alpha-lymphotoxin is dependent on their
ability to form gap junctions (Fig. 6). This capacity varies widely in
different cell strains. Entry is facilitated in strains unable to form gap
junctions; and in cells with well developed gap junctions, ability to
exclude alpha-lymphotoxin is disrupted by phorbol esters, an action enhanced
by 450 MHz fields (1.0 mW/cm^2, 16 Hz modulation). No effects were noted
with 450 MHz fields alone. These findings imply that microwave fields at
athermal levels may act synergically with chemical cancer promoters to
disrupt normal intercellular communication through gap junctions, leading to
autonomous cell growth.

THEORETICAL AND EXPERIMENTAL MODELS OF COOPERATIVE ORGANIZATION IN
PHYSIOLOGICAL SYSTEMS

It is implicit in the experiments cited above that interactions
occurring at athermal levels between biological substrates and the low
frequency components of EM fields must take place in biomolecular systems
exhibiting dynamic patterns of organization. Turing's (1952) equations carry
the prediction that a system of chemical reaction and diffusion may develop
a dynamically maintained temporal and/or spatial pattern from an initially
steady-state homogeneous distribution of matter. By choosing suitable
pattern functions that are mathematically a complete set of eigenfunctions
appropriate to the geometric configuration of the system of interest, Gmitro
and Scriven (1966) extended Turing's analysis of the origins of patterns in
a ring of cells to a whole gamut of line-like and surface-like
configurations.

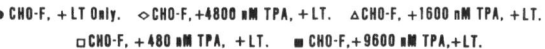

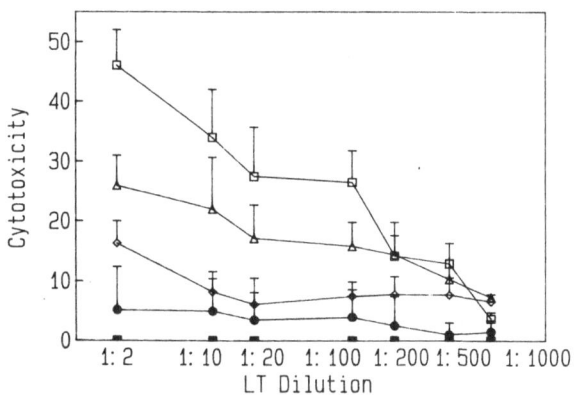

Fig. 6. Effects of cancer-promoting phorbol ester (TPA) on alpha
lymphotoxin-mediated cytolysis of CHO hamster ovary cells treated
with increasing doses of TPA in the presence of graduated
concentrations of lymphotoxin. Results are cumulated from 2 to 4
experiments (n - 8 to 16). (Redrawn from Fletcher et al., 1987b).

Othmer and Scriven (1971) applied certain of these concepts to cellular
networks, emphasizing how instability and pattern formation are influenced
by preexisting pattern. In the context of developmental biology, they chose
for a prexisting pattern the regular compartmentalizations of a system into
discrete cells, each interchanging material with its nearest neighbor and
with the surrounding bath of intercellular fluid. Here, in passing from
unicellular to multicellular systems, one encounters a new level of
organization and complexity, a level stemming from cell-to-cell
communication of various types. Through chemical contact, cells may
interchange one or more of their constituents, thereby altering metabolic

states (Fletcher et al., 1987a and b). In this model, chemical transformations taking place in individual cells may be influenced by mass transport between cells, thereby establishing differences in cell composition. There are natural instabilities inherent in this type of dynamic interaction. They may lead to spatial organization and temporal oscillations within groups of cells in the genesis of form and rhythm. Othmer and Scriven focused on these possibilities, taking chemical interactions as their model, although it may also be relevant to other modes of intercellular communication, including electromagnetic interactions discussed above.

Dissipative Processes

There is much evidence that the molecular organization in biological systems needed to perform this sensing of weak stimuli, whether thermal, chemical or electrical, may reside in joint functions of molecular assemblies or subsets of these assemblies (Katchalsky et al., 1974). Katchalsky (1974) has reviewed historic developments from the time of Heroclitus (540-480 B.C.) of concepts of dynamic patterns that develop in a population of elements as a result of complex flow patterns. These flow patterns can undergo sudden transitions to new self-maintaining arrangements that will be relatively stable over time.

Because these dynamic patterns are initiated and sustained by continuous inputs of energy, they are classed as "dissipative" processes. For this reason, they occur far from equilibrium with respect to at least on important parameter in the system (Katchalsky and Curran, 1965). As nonequilibrium processes, they may be characterized by resonant or windowed phenomena, an important aspect of their occurrence in tissue interactions with weak EM fields, as discussed above. Also, two or more quite distinct mechanisms can give rise to the same dynamic pattern. A given pattern therefore need not relate to a unique mechanism; conversely, different mechanisms may generate a common pattern (Othmer and Scriven, 1969, 1971, 1974). To biologists accustomed from their earliest training to consider cellular excitatory phenomena in terms of equilibrium processes, these concepts, though quite old in certain key areas of physics and chemistry (Rayleigh, 1916), offer new insights on possible substrates for initial events in signal transductive coupling at cell membranes.

Concepts of Cooperativity in Biomolecular Systems

A strong theme characterizing many of these functional linkages between participating elements of a dynamic pattern is cooperativity, defijned as ways in which components of a macromolecule, or a system of macromolecules, act together to switch from one stable state to another. These joint actions frequently involve phase transitions, hysteresis, and avalanche effects in input-output relationships (Schmitt et al., 1975; Wyman, 1948; Wyman and Allen, 1951). Trigger signals to coopersative processes may be weak and the amplified response may be orders of magnitude larger, as in the sharply nonlinear release of $^{45}Ca^{2+}$ from binding sites in cerebral tissue by added Ca ions (Kaczmarek and Adey, 1973) and by weak EM fields as discussed above (Bawin and Adey, 1976; Bawin et al., 1975); and in a series of Ca-dependent processes at cell membranes that include the large generation of cAMP by glucagon binding to membrane receptors (Rodbell et al., 1974); the generation of cAMP by binding of parathyroid hormone to its membrane receptors and the modulation of this process by weak EM fields (Luben et al., 1982); the amplification of immune responses in patching and capping at cell membranes (Edelman et al., 1973) and the modulation of cell-mediated cytotoxicity of lymphocytes by ELF components of weak EM fields (Lyle et al., 1983, 1987) and in the swimming behavior of bacteria elicited by small concentration gradients of an attractant (Aswad and Koshland, 1974; Koshland, 1975).

Stimulus Amplification in Cooperative Systems

Amplification effects in cooperative systems raise questions about thresholds and the minimum size of an effective triggering stimulus. Studies of cooperativity in biological systems have usually focused on the effects of a change in an external parameter on the equilibrium constant of a specific reaction (Blank, 1976; Schwarz, 1975). Although the sharp transition from one highly stable state to another such state can also be achieved by noncooperative means, much larger transition energies would be required and the transition would occur more slowly. Sharp and fast transitions characteristic of many biological systems thus involve cooperative interactions, such as the individually weak forces in a series of hydrogen bonds or in hydrophobic reactions (Engel and Schwarz, 1970; Schwarz, 1975; Schwarz and Balthasar, 1970; Schwarz et al., 1970).

Compared with tissue electric gradients induced by environmental EM fields, the requisite gradients for some known molecular transitions are very large. The helix-coil conformational change in poly(gamma-benzyl L-glutamate) can be induced by a gradient of 260 kV/cm (Schwarz and Seelig, 1968). Long-lasting conformational changes occur in poly (A).2 poly(U) and in ribosomal RNA with pulsed electric fields of 20 kV/cm and with a decay time of 10 usec (Neumann and Katchalsky, 1972). However, these sensitivities for nucleic acid chains in pure solutions contrast sharply with effects of ELF pulsed EM fields on DNA synthesis in cultured mammalian cells, where significant effects occur at field intensities in the range of 10^{-8}-10^{-4}T (Takahashi et al., 1986).

It is therefore clear that observed EM field interactions with cells and tissues based on oscillating ELF tissue gradients between 10^{-7} and 10^{-1} V/cm noted above wouldinvolve degrees of cooperativity many orders of magnitude greater than envisaged in the examples just cited. In part this discrepancy appears to relate to far greater sensitivities of cellular systems to low-frequency EM fields and to RF fields with low-frequency amplitude-modulation than to imposed step functions or DC gradients used in many electrochemical experiments and models to test levels of cooperativity in biological systems (Blank, 1972).

A most important factor in determining this threshold for a low-level coherent oscillation to elicit a cooperative response is the thermal Boltzmann (kT) noise in the system. This is 0.02 eV at room temperature and is the basis of molecular collisional interactions. If modeled on this thermal threshold, the sensing of a gradient of 10^{-7} V/cm would require a cooperative molecular system extending over 300 m. The abundant evidence that extracellular gradients from 10^{-1} V/cm down to this level are biologically significant in systems of cellular dimensions is a salutary reminder of the importance of better understanding molecular and morphological substrates of this transductive coupling. Factors to be considered include temporal entrainment of activity in large systems of random generators by coherent oscillations far weaker than this random activity (Nicolis et al., 1973, 1974). An hypothesis as yet untested suggests that cell surfaces may act as extremely narrow-bandwidth low-pass filters in the transfer of thermal noise long the surface of micron-sized spheres and tubes (Bawin and Adey, 1976), thus enhancing the signal-to-noise ratio of ELF oscillations in the pass band of the filtering system.

Chaotic Models of Biological Nonlinear Oscillating Systems

Self-sustained oscillations in biological systems may be modeled on the requirement for interaction of regular external perturbations with internal oscillations, resulting in synchronization of the system to the external drives (entrainment) (Kaiser, 1984). A sharp frequency response results, with both frequency and intensity windows and rather irregular behavior near the entrainment region.

A further increase in energy of external driving fields, both static and periodic, leads to sequences of period-doubling bifurcations, alternating with quasiperiodic and irregular regions (quasiperiodicity, chaos). As a consequence, a regularly driven self-oscillating system may exhibit intrinsic chaotic behavior, even though the underlying dynamic is strongly deterministic. Finsally, still higher levels of energy input destabilize the system (collapse0, leading to the onset of propagating pulses (solitary waves or _solitons_). A nonlinear temporal structure is thus replaced by a nonlinear spatiotemporal structure.

Kaiser points out that systems having both periodic and chaotic states can exhibit completely different behaviors under periodic driving, and that therein lie additional possibilities for responses of large neuronal systems to imposed fields, as in the genesis of the electroencephalogram in cerebral cortex. Extreme sensitivity of chaotic systems to external driving may carry implications for brain functions in differences between regular and irregular EEG patterns and associated susceptibility to external fields. Circadian pacemaker functions of the suyprachiasmatic nucelus have been modeled as a neural network behaving chaotically for certain functions (Carpenter and Grossberg, 1983; Grossberg, 1983).

Nicolis (1983) has examined the role of chaos in reliable information processing in simple nervous systems, as in the leech or cockroach, where complex behavioral repertoires exist in the absence of elaborate neuronal substrates. This cooperative model is based on the principle that information is produced not only by dissipating degrees of freedom in the system, but also by increasing resolution in systems with few degrees of freedom. For example, certain nonlinear, dissipative systems with just three degrees of freedom can exhibit random behavior which is analogous to that produced by explicit stochastic equations. Instead of creating new degrees of freedom with increased bandwidth or dimensionality of state space, these systems generate iterative self-similar processes which decrease resolution or expand dynamics of trajectoriesin a low-dimensional state space. Sensitivity to small differences in initial conditions gives a probabilistic character to behavior of an otherwise simple deterministic system.

Space-time intermittency is a baffling phenomenon observed in spatially extended systems (Keeler and Farmer, 1986). In general, space-time intermittency occurs when two qualitatively different types of behavior are intermittent in both space and time. A common manifestation occurs in fluid flow, where patches of turbulence are sometimes isolated in space. The interface between laminar and turbulent behavior is dynamic and to the casual observer ppears unpredictable; without changing therameters, turbulence can spread through the entire flow, or disappera, so that the entire flow becomes laminar. Modeled in a one-dimensional lattice of coupled quadratic maps, this system naturally forms spatial domains. Motion of the domain walls causes spatially localized changes from chaotic to almost periodic behavior. The almost periodic phases have signatures close to one, resulting in long-lived laminar bursts with a 1/f low-frequency spectrum.

Dispersive Models of RF Field Interactions at Cell Membranes with Initiation of Solitary Waves

We have discussed above our three-stage model of transductive coupling from membrane surface to cell interior. The model proposes initial interactions over domains of the membrane surface. These interactions are dissipative in nature and highly cooperative in their modulation of Ca binding at sites on terminals of stranded glycoproteins. The ensuing step would be dispersive, and might involve solitary waves (solitons) similar to those proposed by Davydov (1979) moving in sequence down the length of glycoprotein and lipoprotein molecules (Adey and Lawrence, 1984; Adey and Lawrence, 1982).

These solitons may arise in interactions of phonons and excitons along linear molecules that result in nonlinear molecular vibrations. It is proposed that nonlinear interatomic forces (specifically in the hydrogen bond) can lead to robust solitary waves with greatly increased radiative lifetimes (Davydov, 1979; Hyman et al., 1981). Davydov concludes that this would correspondingly increase the tendency for molecular vibrations to be the vehicle for energy transfer over long molecular chains, specifically over the amide "spines" in alpha-helix proteins and DNA with the bond sequence ---HNC=O---HNC=O-, etc. Davydov's nonlinear analysis shows that propagation of amide-I vibrations can couple to nonlinear sound waves in the alpha-helix, and the coupled excitation propagates as a localized and dynamically stable wave. The amide-I vibrations are the source of the longitudinal sound waves which stretch hydrogen bonds along the helix. Soliton formation exhibits a sharp threshold which is a function of energy coupling between the hydrogen bond stretching and the amide-I excitation (Scott, 1981). With a further increase in coupling energy, there is a second threshold above which solitons will not form (Kaiser, 1984). It remains to be determined whether the general properties of these nonlinear mechanisms can account for persisting molecular states in the needed time frame.

Specific Models of Cooperative Actions in Biomolecular Systems

Imposed EM fields have proved unique tools in the developing awareness of the profound importance of the pericellular microenvironment. There is strong evidence that cell membranes are powerful amplifiers of weak electrochemical events in their immediate vicinity; and virtually all these sensitivities appear to involve maximum effects with natural or imposed fields at frequencies below 100 Hz, or with RF fields amplitude-modulated in the same frequency band, a spectral span that we have named the "biological spectrum."

Low-frequency sensitivities have been modeled in terms of Lotka-Volterra (predator-prey) processes involving slow shifts in energy states of coherent populations of fixed chareges on cell surface glycoproteins (Frohlich, 1975); in limit cycle behavior of Ca ions binding to cell surface macromolecules (Kaczmarek, 1976); in chaotic behavior of pseudorhythmic molecular oscillations (Kaiser, 1984), discussed above; and in cyclotron oscillations of Ca ions exhibiting coherent states at the cell membrane surface (Liboff, 1985; Polk, 1984). Also, experimental data aon microwave absorption by red cell membranes (Blinowska et al., 1985) and related enzyme-substrate activity based on quantum cooperative effects suggest cell membranes as a primary site for Frohlich's coherent odcillations and a possible role for local superconductivity (Achimowicz et al., 1977).

Lotka-Volterra Models of Excitations in Cell Membranes

From the observed high sensitivity of biological systems to weak EM
fields, Frohlich has suggested that ELF electric oscillations of the EEG in
brain tissue may relate to a storage mechanism through which a biomolecular
system can store signal energy and, in so doing, overcome thermal noise
(Frohlich, 1977). These electric vibrations cannot arise from a collective
mode based on interactions of various molecular groups, since enormous
tissue volumes would be involved to overcome thermal noise. Instead,
Frohlich has applied a general theory of coherent vibrations in biological
systems (Frohlich, 1968, 1972, 1980, 1986a and b; Bhaumik et al., 1976)
within the framework of the fluid mosaic model.

Collective chemical oscillations that can be represented by the
Lotka-Volterra (predator-prey) equations may occur between an entity formed
of globular proteins and the ions and structured water surrounding them.
These globular proteins would oscillate between a strongly electrically
excited polar state and a weakly polar ground state. A slow chemical
oscillation is thus connected with a corresponding electrical vibration.
With strong electric interactions between the highly polar states and with
considerable damping occurring due to electric currents, limit-cycle
conditions arise that make these oscillations highly sensitive to external
electrical and chemical influences. In this model, expected coherent
oscillations would occur at frequencies around 10^{11}Hz. Applying these
models to the actions of enzymes, there are possibilites for clllective
enzyme reactions, based on long-range selective interactions between enzymes
in excited polar states, other enzymes in the ground state, and substrate
molecules. These collective coherent oscillations at 10^{11}Hz may initiate
chemical oscillations at far lower frequencies in the poulations of
substrate and excited molecules, associated in turn with electrical
oscillations around 10 Hz.

Cyclotron Resonance Models of Ionic Oscillations at Cell Membranes

Polk (1984) noted that free (unhydrated) Ca ions in the earth's
geomagnetic field would exhibit cyclotron resonance frequencies around 10
Hz; and that these cyclotron currents would be as muchas five orders of
magnitude greater than the Faraday currents if the Ca ions exhibited nearest
neighbor coherence. Experimental studies have shown that there are
interactions between the earth's geomagnetic field and a weak imposed low-
frequency EM field (40 V/m peak-to-peak in air, estimated tissue component
10^{-7} V/cm) in determining Ca efflux from chick cerebral tissue (Blackman
et al., 1985b). for example, halving the local geomagnetic field with a
Helmholtz coil rendered ineffective a previously effective 15 Hz field; and
doubling the geomagnetic field caused an ineffective 30 Hz signal to become
effective.

Liboff (1985) notes that for a mean value for the earth's geomagnetic
field of 0.5 G, most of the singly and doubly charged ions of biological
interest have gyrofrequencies in the range 10-100 Hz. Liboff hypothesizes
that imposed EM fields at frequencies close to a given resonance may couple
to the corresponding ionic species in such a way as to selectively transfer
energy to these ions. He proposes that data from the Blackman experiments
cited above may relate to cyclotron resonance in singly ionized K^+, with
secondary effects on Ca efflux.

Local Superconductivity in Cooperative Models of Enzyme Activation

In Frohlich's model of enzymatic activation discussed above, coherent
pumping energy at certain power levels is required to initiate long range
selective forces. Metabolic activity is proposed as a possible energy source

and RF fields may also act as sources of pumping power. Achimowicz et al.,(1977) have proposed an activating mechanism based on quantum cooperative effects that does not require this assumption. Their model proposes that macromolecular interactions destabilize the electron structure of the enzyme-substrate complex. This structure is determined by strong electron (exciton) and phonon-mediated electron-electron interactions. A structural transition occurs, analogous to a Peierls instability in crystal lattices, leading to a softening of the phonon spectrum and a modification of electron levels as electron and phonon systems are conjugated. They hypothesize that renormalization of electron and phonon spectra might lead to high temperature superconductivity; and that phenomenologically, this state would be equivalent to selective intermolecular forces.

CONCLUSIONS

It is at the atomic level that physical, rather than chemical events now appear to shape the flow of signals and the transmission of energy on biomolecular systems. Recent observations have opened doors to new concepts of communication between cells as they whisper together across barriers of cell membranes.

Use of weak amplitude-modulated RF fields as unique tools to study the sequence and energetics of events that couple chemical stimuli from surface receptor sites to the cell interior has identified cell membranes as a primary site of interaction with these fields. Regulation of cell surface chemical events by these fields indicates a major amplification of initial weak triggers associated with binding of hormones, antibodies and neurotransmitters to their specific binding sites. Calcium ions play a key role in this stimulus amplification. The evidence supports nonlinear, nonequilibrium processes at critical steps in transmembrane signal coupling.

Communication between cells through special gap junctions is similarly sensitive to athermal amplitude-modulated RF fields. Cancer promoting substances that act at cell membranes and disrupt gap junction communication are enhanced in their actions by these fields. We hypothesize that cancer promotion with tumor formation may involve dysfunctions at cell membranes, with disruption of inward and outward signal streams. These disorders at cell membranes are distinct from initiating events in cell nuclei that damage genetic stores of DNA. Further research will determine the extent of possible health hazards from concurrent exposures to environmental chemical cancer promoters and EM fields.

Acknowledgments

Studies in this laboratory have been generously supported by the US Department of Energy, the US Environmental Protection Agency, the FDA Bureau of Radiological Health, the US Office of Naval Research, the Southern California Edison Company, the General Motors Medical Research Institute and the US Veterans Administration.

REFERENCES

1. J. Achimowicz, A. Cader, A., L. Pannert, and E. Wojcik, Phys. Lett 60A:383 (1977).
2. W.R. Adey, in: "Functional Linkage in Biomolecular Systems" F.O. Schmitt, D.M. Crothers and D.M. Schneider (eds.) Raven, New York, p. 325 (1975).
3. W.R. Adey, BioSystems 8:163 (1977).
4. W.R. Adey, Physiol. Rev. 61:435 (1981a).
5. W.R. Adey, in: "Biological Effects of Nonionizing Radiation" K.H. Illinger (ed.) Am. Chem. Soc., Washington D.C., p. 271 (1981b).

6. W.R. Adey, in "Synergetics of the Brain" E. Basar, H. Flohr, H. Haken A.J. Mandell (eds) Springer, Berlin, Heidelberg, New York, p. 201 (1983).

7. W.R. Adey, in: "Nonlinear Electrodynamics in Biological Systems" W.R. Adey and A.F. Lawrence (eds.) p.3 (1984).

8. W.R. Adey, Bioelectrochem. Bioenergetics 15:447 (1986).

9. W.R. Adey, in: "Interactions of Biological Systems with Static and ELF Electric and Magnetic Fields". 23rd Hanford Life Sciences Symposium. U.S. Department of Energy, Washington D.C. Symposium Series DOE-60, p. 237 (1988).

10. W.R. Adey, in: "Biophysical Aspects of Cancer" J. Fiala and J. Pokorny (eds.) Charles University, Prague, p.   (1987).

11. W.R. Adey, in: Dynamics of Sensory and Cognitive Processing in the Brain". Second Symposium. E. Basar (ed.) Springer, Heidelberg (1987).

12. W.R. Adey, Neurochem. Res. 13:671 (1988).

13. W.R. Adey, in: "Biological Coherence and Response to External Stimuli" H. Frohlich (ed.) Springer, Heidelberg p. 148 (1988).

14. W.R. Adey and S.M. Bawin, Radio Science 17(5S):149S (1982).

15. W.R. Adey, S.M. Bawin and A.F. Lawrence, Bioelectromagnetics 3:295 (1982).

16. W.R. Adey and A.F. Lawrence (eds.) "Nonlinear Electrodynamics in Biological Systems" Plenum, New York (1984).

17. D. Aswad and D.E. Koshland, J. Bacteriol. 118:640 (1974).

18. E.K. Balcer-Kubiczek and G.H. Harrison, Carcinogenesis 6:859 (1985).

19. V.S. Bannikov, S.M. Bezruchko, E.V. Grishankova, S.B. Kuz'min, Y.A. Mityagin, R.Y. Orlov, S.B. Rozhkov and V.A. Sokolina, Dokl. Akad. Nauk SSR 253:479 (1980); Dokl. Biophys. (Eng. Trans.) 253:119 (1980).

20. P.W.Barber, O.P. Gandhi, M.J. Hagmann and I. Chattergee, IEEE Trans. Biomed. Eng. 26:400 (1979).

21. S.M. Bawin and W.R. Adey, Proc. Nat. Acad. Sci. USA 73:1999 (1976).

22. S.M. Bawin, L.K. Kacmarek and W.R. Adey, Ann. NY Acad. Sci. 247:74 (1975).

23. S.M. Bawin, W.R. Adey and I.M. Sabbot, Proc. Nat. Acad. Sci. USA 75: 6314 (1978a).

24. S.M. Bawin, A.R. Sheppard and W.R. Adey, Bioelectrochem. Bioenergetics 5:67 (1978b).

25. D.E. Beischer and V.R. Reno, Ann. NY Acad. Sci. 247:473 (1975).

26. A.A. Benson, J. Amer. Oil Chem. Soc. 43:265 (1966).

27. D. Bhaumik, K. Bhaumik and B. Dutta-Roy, Phys. Lett 56A:145 (1976).

28. C.F. Blackman, J.A. Elder, C.M. Weil, S.G. Benane, D.C. Eichinger and D.E. House, Radio Sci. 14:93 (1979).

29. C.F. Blackman, S.G. Benane, L.S. Kinney, D.E. House and W.T. Joines, Radiat. Res. 92:510 (1982).

30. C.F. Blackman, S.G. Benane, D.E. House and W.T. Joines, W.T., Bioelectromagnetics 6:1 (1985a).

31. C.F. Blackman, S.G. Benane, J.R. Rabinowitz, D.E. House and W.T. Joines, Bioelectromagnetics 6:327 (1985b).

32. M. Blank, J. Colloid Interface Sci. 41:97 (1972).

33. M. Blank, J. Electrochem. Soc. 123:1653 (1976).

34. K.J. Blinowska, W. Lech and A. Wittlin, Phys. Lett. 109A:124 (1985).

35. C.V. Byus, S. Pieper and W.R. Adey, Carcinogenesis 8:1385 (1987).

36. C.V. Byus, K. Kartun, S. Pieper and W.R. Adey, Submitted to Cancer Research (1988).

37. C.D. Cain, W.R. Adey and R.A. Luben, J. Bone Mineral Res. 2:437 (1987).

38. G.A. Carpenter and S. Grossberg, J. Theoret. Neurobiol. 1:1 (1983).

39. K.S. Cole, Cold Spring Harbor Symp. Quant. Biol. 4:110 (1940).

40. J.S. Coombs, D.R. Curtis and J.C. Eccles, J. Physiol. London 145:505 (1959).

41. M.S. Cooper and N.M. Amer, Phys. Lett. 98A:138 (1983).

42. H. Davson and J.F. Danielli, "The Permeability of Natural Membranes", 2nd Edition, University Press, Cambridge, (1952).
43. A.S. Davydov, Physica Scripta 20:387 (1979).
44. R. Dixey and G. Rein, Nature 296:253 (1981).
45. F. Drissler and R.M. Macfarlane, Phys. Lett. 69A:65 (1978).
46. F. Drissler and L. Santo, in: "Coherent Excitation in Biological Systems" H. Frohlich and F. Kremer (eds.) Springer, p. 6, Berlin (1983).
47. W. Drost-Hansen (ed.) "Cell Associated Water" Academic Press, New York (1978).
48. S.K. Dutta, A. Subramoniam, B. Ghosh and R. Parsad, Bioelectromagnetics 5:71 (1984).
49. J.C. Eccles, "The Neurophysiological Basis of Mind", Clarendon Press, Oxford (1953).
50. G.M. Edelman, I. Yahara and J.L. Wang, Proc Nat. Acad. Sci. USA 70: 1442 (1973).
51. G.S. Edwards, C.C. Davis, M.L. Swicord and J.D. Saffer, Phys. Rev. Lett. 53:1284 (1984).
52. J. Engel and G. Schwarz, Angew. Chem. Int. Ed. 9:389 (1970).
53. R.J. Fitzsimmons, J. Farley, W.R. Adey and D.J. Baylink, Biochim. Biophys. Acta 882:51 (1986).
54. W.H. Fletcher, W.W. Shiu, D.A. Haviland, C.F. Ware and W.R. Adey, Proc. Bioelectromagnetics Soc., 8th Annual Meeting, Madison WI. p. 12 (abstract). (1986).
55. W.H. Fletcher, W.W. Shiu, T.A. Ishida, D.L. Haviland and C.F. Ware, J. Immunol. 139:956 (1987a).
56. W.H. Fletcher, C.V. Byus and D.A. Walsh, Adv. in Exper. Med. Biol. 219:299 (1987b).
57. K.R. Foster and A.W. Guy, Sci. Amer. 255:32 (1986).
58. K.R. Foster and W.F. Pickard, Nature 330:531 (1987).
59. K. R. Foster, M.A. Stuchly, A. Kraszewski and S.S. Stuchly, Biopolymers 23:   (1983).
60. H. Frohlich, Nature 157:478 (1946).
61. H. Frohlich, Int. J. Quant. Chem. 2:641 (1968).
62. H. Frohlich, Phys. Lett. 29A:153 (1972).
63. H. Frohlich, Proc. Nat. Acad. Sci. USA (1975).
64. H. Frohlich, Neurosci. Res. Program Bull. 15:67 (1977).
65. H. Frohlich, IEEE Trans. Microwave Theory Tech. 26:613 (1978).
66. H. Frohlich, Adv. Electronics Electron Phys. 53:85 (1980)
67. H. Frohlich, in: "Modern Bioelectrochemistry", F. Gutmann and F. Keyzer (eds.) Plenum, New York, p. 241 (1986a).
68. H. Frohlich, in: "The Fluctuating Enzyme", G.R. Welch (ed.) Wiley, New York, p. 421 (1986b).
69. L. Furia and O.P. Gandhi, Phys. Lett. 102A:380 (1984).
70. E.J. Furshpan and T. Furikawa, J. Neurophysiol. 25:732 (1962).
71. C. Gabriel, E.H. Grant, R. Tata, P.R. Brown, B. Gestblom and E. Noreland, Nature 328:145 (1987).
72. O.P. Gandhi, K. Sedigh, G.S. Beck and E.L. Hunt, in: "Biological Effects of Electromagnetic Waves", C.C. Johnson and M.L. Shore (eds), Bureau of Radiological Health, Rockville MD. HEW Publication 77-8010, p. 44 (1976).
73. O.P. Gandhi, E.L. Hunt and J.A. d'Andrea, Radio Sci. 12, Suppl.6: 39 (1977).
74. O.P. Gandhi and M.J. Hagmann, in: "The Physical Basis of Electro-magnetic Interactions with Biological Systems", L.S. Taylor and A.Y. Cheung (eds.) University of Maryland, College Park, p. 243 (1977).
75. J.L. Gmitro and L.E. Scriven, in: "Intracellular Transport", K.B. Warren (ed.), Academic Press, New York, p. 221 (1966).

76. E. Grant, in: "The Physical Basis of Electromagnetic Interactions with Biological Systems", L.S. Taylor and A.Y. Cheung (eds.) University of Maryland, College Park, p.113 (1977).
77. I.T. Grodsky, Math. Biosci. 28:191 (1976).
78. S. Grossberg, in: "Synergetics of the Brain", E. Basar, H. Flohr, H. Haken and A.J. Mandell (eds.) Springer, Berlin, Heidelberg, New York, p. 274 (1983).
79. W. Grundler, F. Keilmann and H. Frohlich, Phys. Lett. 62A:463 (1977).
80. M.J. Hagmann, O.P. Gandhi and C.H. Durney, IEEE Microwave Theory Tech. 27:804 (1979).
81. A.L. Hodgkin and A.F. Huxley, J. Physiol. London 117:500 (1952).
82. J.M. Hyman, D.W. McLaughlin and A.C. Scott, Physica D 30:23 (1981).
83. K.H. Illinger, (ed.) "Biological Effects of Nonionizing Radiation", Am. Chem. Soc. Symp. Ser. 157, (1981).
84. C.C. Johnson, C.H. Durney, P.W. Barber, H. Massoudi, S.J. Allen and J.C. Mitchell, Radio Sci. 12, Suppl. 6:57 (1977).
85. L.K. Kaczmarek, Biophys. Chem. 4:249 (1976).
86. L.K. Kaczmarek and W.R. Adey, Brain Res. 63:331 (1973).
87. L.K. Kaczmarek and W.R. Adey, Brain Res. 66:537 (1974).
88. F. Kaiser, in: "Nonlinear Electrodynamics in Biological Systems", W.R. Adey and A.F. Lawrence (eds.) Plenum, New York, p. 393 (1984).
89. A. Katchalsky, Neurosci. Res. Program Bull. 12:30 (1974).
90. A. Katchalsky and P.F. Curran, "Nonequilibrium Thermodynamics in Biophysics", Harvard University Press, Cambridge MA (1965).
91. A. Katchalsky, V. Rowland and R. Blumenthal (eds.) Neurosci. Res. Program Bull. 12:1 (1974).
92. J.D. Keeler and J.D. Farmer, Physica D 23:413 (1986).
93. S. Kinoshita, H. Kuniko and T. Kushida, J. Phys. Soc. Japan 49:314 (1980).
94. M. Kohli, N. Mei, E.W. Prohofsky and L.L. Van Zandt, Biopolymers 20:853 (1981).
95. D.E. Koshland, in: "Functional Linkage in Biomolecular Systems" F.O. Schmitt, D.M. Crothers and D.M. Schneider (eds.) Raven Press, New York, p. 273 (1975).
96. H.N. Kritikos and H.P. Schwan, IEEE Trans. Biomed. Eng. 23:168 (1976).
97. A.F. Lawrence and W.R. Adey, Neurol. Res. 4:115 (1982).
98. A.F. Lawrence, J.C. McDaniel, D.B. Chang and R.R. Birge, Biophys. J. 51:785 (1987).
99. A.R. Liboff, in: "Interactions Between Electromagnetic Fields and Cells" A. Chiabrera, C. Nicolini and H.P. Schwan (eds.), Plenum Press, New York, p. 281 (1985).
100. G.N. Ling, C. Miller and M.M. Ochsenfeld, Ann. NY Acad. Sci. 204:6 (1973).
101. S. Lin-Liu and W.R. Adey, Bioelectromagnetics 3:309 (1982).
102. S. Lin-Liu, W.R. Adey and M.-M. Poo, Biophys. J. 45:1211 (1984).
103. W.A. Little, Neurosci. Res. Program Bull. 15:62 (1977).
104. P.S. Lomdahl, L. MacNeil, A.C. Scott, M.E. Stoneham and S.J. Webb, Phys. Lett. 92A:207 (1982).
105. W.R. Loewenstein and Y. Kanno, Nature 209:1248 (1966).
106. W.R. Loewenstein, Biochim. Biophys. Acta 560:1 (1979).
107. W.R. Loewenstein, Physiol. Rev. 61:829 (1981).
108. R.A. Luben and C.D. Cain, in: "Nonlinear Electrodynamics in Biological Systems", W.R. Adey and A.F. Lawrence (eds.) p. 23 (1984).
109. R.A. Luben, C.D. Cain, M.Y.Chen, D.M. Rosen and W.R. Adey, Proc. Nat. Acad. Sci. USA 79:4180 (1982).
110. D.B. Lyle, P. Schechter, W.R. Adey and R.L. Lundak Bioelectromagnetics 4:281 (1983).
111. D.B. Lyle, R.D. Ayotte, A.R. Sheppard and W.R. Adey, Bioelectromagnetics 9 (in press) (1988).
112. J. Maddox, Nature 324:11 (1986).
113. S. Milham, Environ. Health Perspectives 62:297 (1985)

114. W.H.Moolenaar, R.J. Aerts, L.G.J. Tertoolen and S.W. DeLast, J. Biol. Chem. 261:279 (1986).
115. National Academy of Sciences USA, "Biologic Effects of Electric and Magnetic Fields Associated with Proposed Project Seafarer", Washington DC, 440pp (1977).
116. E. Neumann and A. Katchalsky, Proc. Nat. Acad. Sci. USA 69:993 (1972).
117. P. Newmark, Nature 327:101 (1987).
118. J.G. Nicholls and S.W. Kuffler, J. Neurophysiol. 27:645 (1964).
119. J.S. Nicolis, in: "Synergetics of the Brain", E. Basar, H. Flohr, H. Haken and A. J. Mandell (eds.), Springer, Berlin, Heidelberg, New York, p. 330 (1983).
120. J.S. Nicolis, G. Galanos and E.N. Protonotarios, Internat. J. Control 18:1009 (1973).
121. J.S. Nicolis, E. Protonotarios and E. Lianos, "The Role of Noise in "Self-Organizing" Systems", Univ. of Patras, Greece, Dept. of Electrical Engineering, Technical Report CSB-1, 55 pp (1974).
122. B.O. Nilsson and L.E. Petterson, IEEE Trans. Microwave Theory Tech. 27:616 (1979).
123. Y. Nishizuka, Philos. Trans. R. Soc. London B302:101 (1983).
124. Y. Nishizuka, Nature 308:693 (1984).
125. R.M. Noyes and R.J. Field, Annual Rev. Phys. Chem. 25:95 (1974).
126. H.G. Othmer and L.E. Scriven, Ind. Eng. Chem. 8:302 (1969).
127. H.G. Othmer and L.E. Scriven, J. Theor. Biol. 32:507 (1971).
128. H.G. Othmer and L.E. Scriven, J. Theor. Biol. 43:83 (1974).
129. G. Pasti, B.S. Warren, S.A. Aaronson, and P.M. Blumberg, Nature 324: 375 (1986).
130. A.S. Perlin, Fed. Proc. 36:106 (1977).
131. C. Polk, Proc. Biolectromagnetic Soc., 6th Annual Meeting, p. 77 (abstract) (1984).
132. E.W. Prohofsky, in: "The Mechanisms of Microwave Biological Effects", L.S. Taylor and A.Y. Cheung (eds.) University of Maryland, College Park, p. 7 (1979).
133. M.J. Radeke, T.P. Misko, C. Hsu, L.A. Herzenberg and M. Shooter, Nature 325:393 (1987).
134. L. Rayleigh, Philos. Mag. 32:529 (1917).
135. H. Riedel, J. Schlessinger and A. Ullrich, Science 236:197 (1986).
136. J.D. Robertson, J. Cell. Biol. 19:201 (1963).
137. M. Rodbell, M.C. Lin and Y. Salomon, J. Biol Chem. 249:59 (1974).
138. D.A. Savitz, H. Wachtel and F. Barnes, National Contractors' Review, US Dept of Energy, Office of Energy Storage and Distribution, Washington DC 20585, and Electric Power Research Institute Health Studies Program, Palo Alto, Calif. 94303. Proceedings, November, 1986.
139. F.O. Schmitt, D.M. Schneider and D.M. Crothers (eds.), "Functional Linkage in Biomolecular Systems", Raven Press, New York (1975).
140. H.P. Schwan, Adv. Biol. Med. Phys. 5:147 (1957).
141. H.P. Schwan, Proc. IRE 47:1841 (1959).
142. H.P. Schwan, in: "Medical Physics", O. Glasser (ed.), Yearbook Publishers, Chicago, p. 1, (1960).
143. H.P. Schwan, J. Cell Comp. Physiol. 66,Suppl.:5 (1965a).
144. H.P. Schwan, Ann. NY Acad. Sci. 125:344 (1965b).
145. H.P. Schwan, in: "Biological Effects and Health Hazards of Microwave Radiation", P. Czerski (ed.) Polish Med. Publ., Warsaw, p. 152 (1974).
146. H.P. Schwan, Neurosci. Res. Program Bull. 15:88 (1977).
147. H.P. Schwan, R.J.Sheppard and E.H. Grant, J. Chem. Phys. 64:2257 (1976).
148. G. Schwarz, in "Functional Linkage in Biomolecular Systems", F.O. Schmitt, D.M. Schneider and D.M. Crothers (eds.), Raven Press, New York, p. 32, (1975).

149. G. Schwarz and W. Balthasar, Eur. J. Biochem. 12:461 (1970).
150. G. Schwarz, S. Klose and W. Balthasar, Eur. J. Biochem. 12:454 (1970).
151. G. Schwarz and J. Seelig, Biopolymers 6:1263 (1968).
152. A.C. Scott, Phys. Lett. A86:60-62 (1981).
753. A.C. Scott, Phys. Rev. A31:3518 (1985).
154. A.R. Sheppard, W.F. Pickard and S.M. Bawin, Bioelectromagnetics Society, Proc. 6th Annual Meeting, p. 50 (1984).
155. S.J. Singer and G.L. Nicholson, Science 175:720 (1972).
156. K.D. Straub, in: "The Physical Basis of Electromagnetic Interactions with Biological Systems", L.S. Taylor and A.Y. Cheung (eds.) University of Maryland, College Park, p. 35 (1977).
157. M.L.Swicord, G.S. Edwards and C.C. Davis, in: "Nonlinear Electrodynamics in Biological Systems", W.R. Adey and A.F. Lawrence (eds.) Plenum, New York, p. 35 (1984).
158. M.L. Swicord, G.S. Edwards, J.L. Sagrapanti and C.C. Davis, Biopolymers 22:2513 (1983).
159. K.Takahashi, I. Kaneko, M. Date and E. Fukada, Experientia 42:185 (1986).
160. T.L. Thomas, P.D. Stolley, A. Stemhagen, E.T.H. Fontham, M.L. Bleeker, P.A. Stewart and R.N. Hoover, J. Nat. Cancer Inst. 79:233 (1987).
161. J.E. Trosko, Eur. J. Cancer Clin. Oncol. 23:599 (1987).
162. J.E. Trosko and C.C. Chang, in: Genetic Toxicology of Environmental Chemicals, Part B: Genetic Effects and Applied Mutagenesis", Liss, New York, p. 21 (1986).
163. A.M. Turing, Philos. Trans. Roy. Soc. London B237:37 (1952).
164. A. Ullrich, L. Coussens, J.S. Hayflick, T.J. Dull, A. Gray, A.W. Tam J. Lee, Y. Yarden, T.A. Libermann, J. Schlessinger, J. Downward, E.L.V. Mayes, N. Whittle, M.D. Waterfield and P.H. Seeburg, Nature 309:428 (1985).
165. J.H. Van Vleck and V.F. Weisskopf, Rev. Mod. Phys. 17:27 (1945).
166. L.L. Van Zandt, M. Kohli and E.W. Prohofsky, Biopolymers 21:1465 (1982).
167. S.J. Webb, Phys. Rep. 60:201 (1980).
168. S.J. Webb and D.D. Dodds, Nature 218:374 (1968).
169. N. Wertheimer and E. Leeper, Am. J. Epidemiol. 109:273 (1979).
170. J. Wyman, Adv. Protein Chem. 4:407 (1948).
171. J. Wyman and D.W. Allen, J. Polymer Sci. 7499 (1951).
172. H. Yamasaki, in: "Nongenotoxic Mechanisms in Carcinogenesis", B.E. Butterworth and T.J. Slaga (eds.) Cold Spring Harbor Laboratory, Banbury Report 25, p. 297 (1987).
173. L.P. Yotti, C.C. Chang and J.E. Trosko, Science 206:1089 (1979).

BIOLOGICAL RESPONSES TO MICROWAVE RADIATION:

REPRODUCTION, DEVELOPMENT AND IMMUNOLOGY

Huai Chiang and Binjie Shao

Microwave Research Lab
Zhejiang Medical University
Hangzhou, 310006, China

INTRODUCTION

Microwave (MW) exposure has been reported to affect reproduction, development and immune responses through a variety of combinations of exposure frequency, duration and power density.

Reproductive efficiency concerns the capacity of the dam or sire to effect a conception, and bear and rear offspring.Changes in this capacity might be due to alterations in behavior, physiology, or morphology. This paper will confine its review mostly to recent results on the male. It has been shown that spermatogenesis can be inhibited after testis are heated locally to above 40°C by microwave irradiation (Zhou, et al, 1980, Rou, et al, 1986). Moreover, it has been suggested that spermatocyte is the most sensitive cell to microwave radiation; followed by spermatid, spermatogonium stem cell and interstitial cell (Saunders, 1981; Chang, et al, 1984; Lebovitz, et al, 1987). It should be noted that in recent years, the application of microwave contraception has been extended from animal experiments to clinical applications in China (Fang, et al, 1982). However, the increase in chromosomal aberration of spermatogenic cells, in incidence of sperm abnormallity and in dominant lethal mutation rate of exposed animals after microwave radiation indicated that there might exist a potential risk of genetic effect in microwave contraception.

There has been a growing concern about teratogenic effects of microwave radiation (Lary and Conover, 1987). It is recognized that high power density exposure increases death rate, gross malformations, and retards fetal growth. Hyperthermia is usually considered the cause of these phenomena with a threshold of about 41°C. Note that the same temperature threshold for both birth defects and prenatal death was reported by Lary, et al (1986). In general, the developmental effects occurring in the absence of a measurable temperature increment in experimental animals are relatively weak, and most of the functional changes are reversible in later life.

The immunological systems of humans and experimental animals are particularly susceptible to MW irradiation. Studies in animals have shown inconsistent results in various elements of the system. In some instances a thermal burden to the exposed animal has been credited, while in others a nonthermal or direct interaction of MW with the system has been suggested. One of the most consistent findings of MW-induced changes is hematopoietic system including increased lymphocyte formation and its higher activity following exposure of several species to various frequencies of MWs (Baranski 1971, 1972).

Under defined conditions, MW fields can stimulate immune system functions for short periods, such as diathermy and some cancer treatment in the clinic or laboratory using moderate levels of MW fields. If the exposure intensities of MW are over the physilolgical burden in an animal (or a human), the immune function would be suppressed, particularly when the exposure duration is longer. There are also evidences that long-term exposure of animals to low-level MW fields may lead to transient suppression of certain immunologic reactions, which suggests existence of cumulative effect of "subclinical damage" on the immune system under prolonged low power MW irradiation. During the course of a long-term MW exposure, earlier immune suppression might be found associated with higher power MW irradiation and later immune suppression with lower power irradiation.

EFFECTS ON REPRODUCTION

Effects of microwave on reproduction should be included the effects on testis and female genitals. This review will be limited to the former, as there are only a few literatures of the latter published.

Spermatogenesis Inhibition

Spermatogenesis can be inhibited after testes are heated locally to high temperature. Zou et al(1980) reported that as the testes of adult rabbits were heated to 41 - 43°C for 15 minuts by 2450 MHz microwave radiation. The mean sperm count declined from 414*10^9/L to 39*10^9/L 13 - 18 days after the treatment, and maintained low level when the treatment was repeated every two months. It began to rise two months after the termination of the treatment. When testes of dogs were irradiated with microwave of the same frequency and the local temperature was raised to 42$\pm$1° C for 20 minutes, the mean sperm density of the dogs was reduced from 146.6*10^9 /L to 31.2*10^9/L one month after treatment; the survival rate of sperms from 92.7% to 60.2%, and the motility of sperms decreased gradually from 7 - 8 cm per 8 hours to 1 cm or even to zero per 8 hours. These changes may last for 3 - 4 months (Rou,et al, 1986).

Lebovitz and Johnson (1983) failed to demonstrate any significant effect of microwave radiation on testicular function (daily sperm production per testis and per gram of testis) at a thermally significant dose rate of 6.3 mW/g. In the experiment, male rats were exposed for 6 hours per day for nine days to pulse-modulated microwave radiation (1.3 GHz, at 1 us pulse width, 600 pulses per second). Exposures were carried out in cylindrical waveguide sections. The mean elevation in core temperature of the rats was on the order of 1.5^{o}C to 2°C, the core temperature being less than 40°C, at 6.3mW/g. However, when the rats were exposed to micro-wave radiation with the same exposure equipment and a whole body dose rate of 10.5 mW/g for 30 minutes followed by 7.7mW/g for 60 minutes, the testis temperature reached 40°C, then a decrease in daily sperm production was observed(Lebovitz, et al, 1987).

As to humans, hundreds of male volunteers have received microwave irradiation as a modality of contraception.Fang et al (1982) investigated 53 volunteers who were exposed to 2450 MHz microwave radiation as a male contraception. They found that if testes were heated to 40° - 42°C for 30 minutes once a month, then the sperm counts were reduced to a level of dysspermatogenic sterility seven weeks after the first irradiation, i. e. less than $5*10^9$/L, and persisted at low level continuously. Thereafter stabilizing for 3 - 4 weeks the contraception can be considered effectiveness and other contraceptions could be withheld. No changes in sexual function and no side effects on their health or labour capacity were observed in the volunteers who received the microwave contraception for nine months to over two years, and some of them have been followed up for 5 years.

The mechanism of tissue temperature elevation by microwave irradiation is internal generation of heat, so it is more effective than that induced by hot-water and infre-red heating. The applicators of microwave equipment used in animal experiment and volunteers mentioned above are microwave antennas. Testes were irradiated in near field, and so it was difficult to control the intensity and the even distribution of irradiation, thus contraception failed on occasion. Recently, a specialized resonant cavity applicator suitable for testes exposure has been developed in China, which makes the energy well-distributed, and lessens enormously exposure of the whole-body to microwave except testes.

With regard to the harmful effects to workers with long-term exposure to microwaves, Lancranjan et al (1975) reported a high frequency of libido paucity and various alterations in spermatogenesis in 74% of the subjects observed. The alterations consisted of asthenospermia, hypospermia and / or teratospermia. However, there were no similar reports from other laboratories.

Histological and Cytological Studies

The histological and cytological studies of testes following microwave irradiation showed unanimous results. Observations on cytology and autoradiographic kinetics were carried out in rats exposed to 2450 MHz microwave radiation for 30 minutes to elucidate the effects on spermatogenocyte of different types and at various stages (Liu, et al, 1981.) The results revealed as follows. First, a lot of primary spermatocytes in the pachytene stage began to degenerate and necrotize two hours after irradiation. Pyknotic adhesion of the chromosomes with deeper stain appeared; then separation of the nucleus from the cytoplasm, degenerated cell swelling and cytoplasmic vacuoles occurred; furthermore, disappearance of nuclear membrane, karyolysis and cell necrosis were found. Secondly. early-phase spermatids were also sensitive to microwave irradiation. Cellular swelling, scaling of acrosomal granules, displacement of the nuclear cap, vacuolization of the nucleus appeared, and fussion of the degenerated spermatids resulted in the formation of multinuclear giant cells, which cast off into the tubular lumina. The spermatogenic cells especially the seminifeocus stem cells remained unaffected while spermatocytes and spermatids were damaged. Degeneration and necrosis of interstitial cells took place only when the temperature shot up to 43°- 45° C. It is considered that cells in the phase of luxurious meiotic division are most vulnerable to microwave irradiation; the refractory type of spermatogonia and those cells of well differentiation revealed high tolerance, such as supporting cells, interstitial cells and matured sperms. Many experiments demonstrated that the primary spermatocyte are most sensitive to microwave exposure, and next the spermatids (Saunders, 1981; Chiang,et al, 1984; Lebovitz, et al, 1987). Hall et al (1982) reported the lack of influence of microwave radiation on turkey sperms at 10 or 50 mW/g in vitro. The percentage of viability, percentage of abnormal sperms and the release of LDH and glutamic oxalic transaminase were examined before and after microwave exposure, no changes was observed.

In another experiment (Hall, et al, 1983), the irradiated turkey sperms were used to inseminate virgin turkey hens artificially. The mean number of eggs, percentage of fertile eggs, percentage of early and late deaths, etc. were assessed, There was also no effect on fertilizing capacity of the sperms. These experiments demonstrated that matured sperms are not damaged by microwave radiation at 50 mW/g.

Li, et al (1982, 1986, 1987) reported the changes in ultrastructure after testes were exposed to microwaves. Ultrastructural observation on the convoluted epithelium showed that changes in primary spermatocytes, early spermatids and spermatogonia were all similar, including dilatation of the smooth endoplasmic reticulum and perinuclear space, mitochondrial swelling, local rupture of the nuclear envelope, and cell disruption. Damages in ultrastructure of the primary spermatocytes and spermatids were more serious than those in spermatogonia (1982). The main alteration in principal cells of epididymal epithelium were swelling of microvilli, decreasing of micropinocytosis, dilatation of endoplasmic reticulum and Golgi complex, and swelling of mitochondria. In addition, there were many myelin figures in the microvilli zone and cytoplasm (1986). Thickening and infolding of the inner most non-cellular layer and degeneration of the myoid cells in peritubular tissues of convoluted tubules were detected (1987).

Genetic Effects

One of the problems most concerned is the genetic effects resulting from microwave irradiation of testes. The experiments on chromosome aberration of spermatogenic cells, sperm malformation, and dominant lethal mutation were carried out in recent years.

Chiang and Lou (1983) reported that male mice were exposed to 2450 MHz microwave at 37 mW/cm^2, and the surface temperature of testis rose to about 40°C for 15 minutes. Premature separations of sex chromosomes were remarkably increased in primary spermatocyte at the first meiosis in metaphase. An increase in chromosome aberration rate, including chromatid break, fragmentation, and chromosome pulverization in a few cells was observed. Manikowska-Czerska(1982) reported that an increase of univalents and translocations was observed at meiotic metaphases I after the male CBA mice were irradiated with 2450 MHz waveguide facility for 30 minutes daily, 6 days per week for two weeks at the whole body dose rate of 10 or 20 mW/g. The translocations were also observed at either 915 MHz or 9.4 GHz frequencies and at either dose rates of 1 and 20 mW/g (1983). Spermatogonic cells can keep alive due to its high resistance while other spermatogenic cells could not develop into sperms or become exhausted. So the damages on spermatogonia due to microwave irradiation may bring about genetic risk. The experiments on chromosome aberration of spermatogonia were carried out in rats and mice exposed to microwave radiation at high intensities by Liu et al(1984) and Chiang et al(1983). The temperature of the testes reached about 42°C. The results showed high aberration rates in exposed animals as compared with the controls. The main types of chromosome aberration were chromatid break, fragmentation and ringforms. Chromosome pulverlization was found in a single cell. Although microwave has not sufficient energy such that one photon can break molecules, breakage will occur as a consequence of the combined action of many photons acting simultaneously (Leonard et al, 1983).

The formation of deformed sperm was determined by combined factors of heterochromosomes and euchromosomes. Goud et al (1982) reported a high proportion of sperm shape abnormalities occurred in male mice exposed to 2450 MHz microwave at a power density of 170 mW/cm^2 for 70 sec, on the 35th day after irradiation. Chiang and Wang (1984) studied the incidence

144

of sperm abnormality to analyze the potential genetic risk on spermatogenic cells of different stages. Having been irradiated with 2450 MHz microwave at 32 mW/cm^2 for 15 minutes, mice were sacrificed and sperms were examined at the 1st, 3rd, 5th and 9th weeks after exposure. Maximum incidence of sperm abnormality occurred at the 5th week after exposure with the majority of "amorphous type" sperms. The incidence of abnormalities at one week following exposure was also higher than that of the control, and was characterized as high proportion of "plump head" type with loose and enlarged head. These results also demonstrated that the primary spermatocytes were most easily damaged by microwave radiation and the next was spermatid cells. By the 9th week after exposure, the incidence of abnormal sperm returned to the control level. But the deformed sperms lasted for a much longer period of time in human reported by Liu et al (1986). They found that the incidence of deformed sperms and cast-off cells were still higher in seminal fluid of 16 microwave contraception volunteers than that in 6 controls after suspending irradiation. The spermatid accounted for the majority of exfoliated cells, and primary spermatocyte being the next. These results indicated that the phenomenon of degeneration did not subside completely one year after the termination of irradiation. As a few spermatogonia stem cells were potentially damaged during exposure, they maintained the abilities of division and differentiation, the new cells were produced and differentiated continously, and then the degenerated cells appeared in the semen. However, the authors pointed out that a local over heating, because of the inhomogeneous distribution of power density caused by the antenna applicator, may result in damage to the spermatogonia in the local area.

Studies on dominant lethal testing after exposure of male mice to microwave were carried out by many authors. Goud (1982) reported that the pre-implantation and total losses were increased after male mice were exposed to 2450 MHz at 170 mW/cm^2. Saunders et al (1981) reported no significant reduction in post-implantation survival after the lower halves of mice bodies exposed to 2450 MHz microwave for 30 minutes at the half body dose rate of 43 W/kg and mated at intervals to females over the following 8 - 10 weeks. Szerski et al (1985) also reported the effects of microwave exposure on post-implantation loss of mice. Male ICR mice were exposed 30 minutes daily, six day per week for two consecutive weeks at the dose rates of 0, 1, and 10 mW/g in a 2450 CW waveguide system. After exposure, each male animal was caged for 3 days with females, The mating procedure was repeated over 10 weeks. The results showed that the post-implantation loss was increased in litters sired by exposed males; stem cells and early spermatogonia were resistant to the treatment, and the spermatocytes I were the most sensitive stage. Wu and Chen (1985) also reported that the dominant lethal rate in male mice mating 5th week after microwave exposure was the highest, and most embryos died at early and middle stages. Manikowska-Czerska et al (1983) exposed male mice to 915 MHz microwave at the dose rates of 1 and 20 mW/g 30 minutes daily, six days per week for two weeks. Three weeks after the end of exposure, males were mated with females. The post-implantation loss, about 20 % in both the 1 and 20 mW/g groups, was more than 4.9% loss in the sham exposed group. Karyotypic analysis was also performed on fetal liver cells and the chromosomal translocations were found, The authors concluded that the appearance of translocations in offspring confirmed the existence of chromosome damage detected by direct examination in meiotic cells.

Comments

Spermatogenesis can be inhibited after testes are heated locally up to over 40°C by microwave irradiation. In recent years, the application of microwave contraception has been extended from animal experiments to clinical applications, and a new specialized resonant cavity applicator

has been developed in China. It is demonstrated that spermatocyte is
most sensitive to microwave radiation; next is spermatid, spermatogonium
stem cell, interstitial cell and matured sperm are usually not damaged by
microwave radiation.

The increase in chromosomal aberration of spermatogenic cells, in
incidence of sperm abnormality and in dominant lethal mutation rate of
exposed animals after microwave radiation indicated that there might
exist a potential risk of genetic effect in microwave contraception.
Nevertheless, the adverse effects might be reversible. It is necessary to
study further the long-term effects including dominant lethality and
morphologic and functional teratogenesis of offspring in order to certify
the safety of microwave as a male contraceptive measure.

EFFECTS ON DEVELOPMENT

Conventional Teratogenic Studies

There has been a growing concern about teratogenic effects of microwave radiation. In a recent review of the teratogenic effect of RF radiation, Lary and Conover (1987) arranged the literatures in order of the
maternal colonic or rectal temperature induced by RF exposure, because of
the conspicuous correlation between RF-induced hyperthermia and teratogenic / embryotoxic effects. It is recognized that exposure to high power
density energy increases death rate, gross malformations, and retards
fetal growth. Hyperthermia is usually considered the cause of these
phenomena with a threshold of about $41^{\circ}C$. Conover et al (1980) reported a
thermal threshold of $41.5^{\circ}C$ for the occurrence of teratogenic and embryotoxic effects. The same temperature threshold for both birth defects and
prenatal death was reported by Lary et al (1986).

Nawrot (1981,1985) reported that a power density of 30 mW/cm^2 (SAR,
40.2mW/g) for mice was near the teratogenic threshold level. Berman et al
(1982) reported that the rectal temperatures of pregnant hamsters after
exposure to 30 mW/cm^2 (SAR, 9 mW/g) 2450 MHz for 100 minutes daily on
days 6-14 of gestation were increased by $1.6^{\circ}C$ over sham-irradiated dams,
and significantly increased fetal resorptions, decreased fetal body
weight, and decreased skeletal maturity were observed. The authors suggested that the hamster fetus may be more susceptible to microwave radiation.

With mice exposed to 2450 MHz microwave at 28 mW/cm^2 (SAR, 16.5
W/kg), Berman et al (1984) found that there was a persistant delay in
postnatal development of the brain in the neonates which had been exposed
during gestation. Schmidt et al (1984) reported that pregnant Sprague-
Dawley rats were exposed to a low level, a SAR of 0.4W/kg, 2450 MHz
microwave 24 hours per day. No significant differences were noted in the
number of survivors, fetal lengths and weights, skeletal or other abnoma-
lities. The authors concluded that microwave exposure does not affect
fetal development in the absence of maternal thermal stress. However,
Tofani et al (1986) found the adverse effect of continuous low level
exposure to RF radiation on intra-uterine development in rats. Pregnant
Sprague-Dawley rats were exposed to 27.12 MHz wave at a power density of
0.1 mW/cm^2, the upper SAR was 0.00011 W/kg. The results showed that
exposure to RF radiation caused a great number of dams with total resorp-
tion and high post-implantation losses; an incomplete ossification of
cranial bones was also found in the fetuses of the exposed groups. The
authors suggested that some of the effects be ascribed to a specific
action of RF occurring independently of the rise in temperature.

Behavioral Teratology

Since the methodology typically employed to assess structural defects requires removal and necropsy of fetuses prior to normal parturitions and performance deficits may occur in the absence of gross teratogenic anomalies, behavioral teratogenesis has been studied in this field.

Jensh (1983) reported the effects of protracted prenatal exposures to a 20 mW/cm^2 2450 MHz microwave on rat behavior. Behavioral tests begun at 60 days of age included conditioned avoidance response, water T-maze, open field, 24-hour activity wheel, forelimb hanging, and swimming. Results of open field and activity wheel tests indicated that female offspring might be preferentially sensitive to irradiation. Irradiated males appeared less active than females did in the activity wheel but more active than control males. No significant alterations were observed in other behavioral tests. The postnatal psychophysiologic evaluations were carried out in another study (Jensh, 1984). Significant differences between prenatal exposure and sham exposure rat groups were observed for eye opening, water T-maze and open field test. The author indicated that exposure at the power density of 35 mW/cm^2 of 6000 MHz microwave may result in subtle long - term neurophysiologic alterations not detectable at term using conventional morphologic procedures. Galvin et al (1986) reported that rats exposed to 2450 MHz at 10 mW/cm^2 prenatally (on days 5-20 of gestation) or perinatally (plus days 2 through 20 postnatally) were examined by a neurobehavioral test battery on postnatal days 30 and 100. Body mass, locomotor activity, startle to acoustic and air-puff stimuli, fore-hindlimb grip strength, negative geotaxis, reaction to thermal stimulation, and swimming endurance were assessed. The results showed that the prenatally and the perinatally exposed rats weighed more than sham-exposed rats at 30, but not at 100, days of age. In additions, the perinatally exposed animals had less swimming endurance at 30, but not at 100 ,days of age relative to sham exposed rats. For the other measures, only the air-puff startle response was altered with a longer latency and was limited to the prenatally exposed female pups. The essentially same results appeared again in their second experiment. Kaplan et al (1982) reported a lack of biological and behavioral effects of prenatal exposure to 2450 MHz microwave at the dose rates of 0.034, 0.34 or 3.4 W/kg.

Lovely et al (1983) reported that pregnant rats were exposed for 20 hours per day for 19 days of gestation to microwave at 0.5 mW/cm^2 (SAR, 0.3 W/kg). No changes in core temperature were observed as a consequence of subchronic exposure to the microwave. They found that the prenatally exposed female rats had greater changes in skin temperature when taken from their foster mother for 2 hours on day 4 of life; other fostered groups had no such changes. Female progeny developed higher temperatures as young adults and later in life than control rats. While male progeny were different in their ability to retain core heat in a four-hour cold stress test conducted at 90 days of age. The authors suggested that the various effects of prenatal exposure appeared to be sex-specific. O'Conner et al (1983) reported the behavioral evaluation of rats exposed in utero to 30 mW/cm^2 of 2450 MHz microwave radiation. The rats did not show any difference in avoidence learning and other behavioral tests. But the exposed rats when placed in a runway heated to form a gradient from 17°C to 38°C spent more time in the cooler sections of the runway. The authors indicated that the exposed animals evidenced postnatal sensitization solely to thermal stimuli.

Other effects on Development

Albert et al (1981a) reported the effects of prenatal and / or postnatal microwave exposure at a dose rate of 2 mW/g or 2.77 mW/g on the Purkinje cells in rat cerebellum. Quantitative studies of Purkinje cells showed a significant and irreversible decrease in rats irradiated during fetal or fetal and early postnatal life. In animals exposed postnatally, and euthanized immediately after irradiation, significant decrease in the relative number of Purkinje cells was apparent, restoration occurred after forty days of recovery. But no significant effect was found in the brain of squirrel monkey prenatally exposed to 2450 MHz microwaves at 3.4 W/kg as compared to sham-irradiated controls (Albert, et al, 1981b). A slight retardation in the development of cerebellar cortex in Japanese quail due to microwave prenatal exposure at 5 mW/cm^2 SAR, 4.03 mW/g) was reported by Inouye et al (1982). Decline of the external granular layer, growth of the molecular layer, differentiation and the alignment of Purkinje cells and the accumulation of granule cells were temporary altered and turned to normal at 8 weeks of age.

Nakas (1981) reported the effects of repeated 2450 MHz microwave radiation at a non-thermal dose (6.54 J/g) on the acetylcholinesterase (AChE) activity in developing brain. Rats were subjected to daily exposure starting on day 7 after birth for 8 days. The AChE activities in the brain of rats were determined at 15 days of age. A significant depression of AChE activity was observed in the exposed rats. Chernovetz et al (1977) reported that pregnant rats exposed to microwave energy at the dose rate of 31 mW/g had a significant decrease in brain norepinephrine of the progeny and that a decrease in dopamine was indicated. Histochemical analyses of catecholamine (CA), mainly norepinephrine and dopamine, and monoamine oxidase (MAO) in mouse hypothalamus were performed with a microspectrophotometer by Chiang and Yao (1987) in studying the effects of pre-natal and post-natal microwave radiation on developing mouse brain. Pregnant mice were irradiated for 5 hours daily throughout the pregnancy with pulsed microwave at an incident average power density of 8 mW/cm^2(SAR 3-3.5 mW/g). After birth, from day 3 to 20, half the offspring were irradiated at 1 mW/cm^2. The results showed that CA in the hypothalamus was slightly decreased after prenatal exposure, but decreased much more in these supplemented with postnatal exposure. The changes found in MAO activity in the hypothalamus were similar. Since MAO is the main catabolic enzyme of CA, the finding of decreased CA may have resulted from less synthesis rather than enhanced catabolism.

Chiang and Yao (1987) reported that either prenatal or postnatal exposure can decrease SDH in the hypothalamus of mice in the experiment aforementioned. The authors indicated that the activity of SDH, a marker enzyme located on brain mitochondria, is non-specific indicator. But it might be a more sensitive and more reliable indicator in evaluation the effects of prenatal microwave exposure than the developmental behavioral tests which are usually used. Jamakosmanovic (1981) reported changes in ATP, ADP and AMP in rat brain during postnatal development following in utero microwave exposure on day 10 of gestation at 8 mW/cm^2. The author suggested that the central nervous system is very sensitive to non-thermal doses of absorbed microwave energy during a single prenatal exposure.

McRee et al (1983) reported the reproductive effect of microwave exposure during embryogenesis. Japanese quail embryos were exposed continuously to 2450 MHz microwave radiation during the first 12 days of embryogenesis with an incident power density of 5 MW/cm^2(SAR, 4.03 mW/g). At 23 weeks of age an assessment of the reproductive capacity of the males was performed. It was found that the spermatozoal numbers and motility in semen samples were reduced significantly, and the percentage of fertile eggs was significantly reduced when males exposed during their embryogenesis were mated to sham control females.

Galvin et al (1983a) reported the effects of microwave radiation during gestation period on the postnatal hematological and immunological changes in animals. Pregnant rats were exposed to 2450 MHz microwave radiation at 10.3 mW/cm^2 for 3 hours per day from day 3 through day 20 of gestation. The exposure did not result in any increase in maternal core temperature. The white blood cell count in pups at 10 days of age was lower in exposed group; the exposed female pups exhibited significantely lower percentage of lymphocyte and higher percentage of neutrophil. At day 30 postpartum, these differences were no longer noted between exposed and sham-exposed pups. The authors suggested that there might be a apparent recovery from prenatal microwave exposure. Galvin et al (1983b) reported further that microwave exposure during embryogeny may produce latent hematological effects which would appear in adult quail. Fertile Japanese quail eggs were exposed to 2450 MHz microwave for the first 12 days of embryogeny to 5 mW/cm^2(SAR, 4.03 mW/g), RBC numbers, packed cell volumes, total hemoglobin levels, and mean cell volumes were greater in the exposed group for both 6 and 12 week old quails. Reduced delayed hypersensitivity to PHA and together with leucocytosis was observed in exposed females at 12 and 22 weeks of age, but not at 6 weeks of age. Smialowicz et al (1979) reported that in the rats exposed from day 6 of gestation through 41 days after birth to 2450 MHz microwave at 5 mW/cm^2(SAR, 1-5 mW/g), their lymphocytes responded to a significantly greater extent than those of control animals following stimulation in vitro with T- or B- cell mitogens. Four other experiments were conducted with 425 MHz microwave (Smialowiez, et al, 1982). A significantly enhanced response of lymph-node but not of blood lymphocytes from irradiated rats following stimulation with mitogens was demonstrated in two of four experiments. However, in another experiment, the pups which not subsequently irradiated, showed a similar increase response of node but not of blood lymphocytes to T-cell mitogens at 42 days of age.

Shandala and Vinogradov (1982) studied the effect of autoantibodies on embryogenesis. Pregnant rats were immunized with serum from irradiated or intact donor rats. Immunization with serum from intact rats did not effect the course of pregnancy, while immunization with serum from irradiated rats resulted in an increased embryolethality and a decreased viability in offsprings. Shandala (1987) speculated that the study of autoallergic processes at the fetus and the postnatal stage would provide a promising modality to assess the detrimental effect of nonionizing radiation on immune system. Another experiment was performed with similar results. The author concluded that microwave radiation at 0.5 mW/cm^2 promotes the formation of antibrain antibodies in organism, which are of hazard effect. The stimulation of autoimmune processes influenced by microwave might be regarded as a compensation contributing to support the immune homostasis.

A interesting finding that microwave radiation enhances teratogenic effect of cytosine arabinoside (ara-c) in mice was reported by Szmigielski et al (1984). Pregnant mice were irradiated in microwave field at 10 or 40 mW/cm^2 for two hours daily. On the 9th day of pregnancy, all dams were injected intraperitoneally with 10 mg/kg of ara-c, a known teratogen. It was found that in mice treated with both ara-c and microwave exposure the frequency of resorptions and fetal malformations, including cleft palate, cleft lip and palate, and other defects, was significantly higher than in mice treated with ara-c or with microwaves exposure alone. A significant enhancement of the teratogenic potency of ara-c after combined exposure to both ara-c and microwaves during pregnancy was also reported (Marcickiewicz, et al, 1986). They concluded that the relative weak embryotoxic effects of microwaves may be revealed and markedly enhanced after combination of radiation with other teratogenic factors.

Comments

Many of the authors who reported structural malformations have employed highly intense irradiation which will impose a conspicuous thermal insult to the irradiated subject. There are also growing investigations reporting developmental functional effects in the absence of a measurable temperature increment. Because of the highly nonuniform absorption of RF energy in exposed animals, the developmental effects that occur in the absence of a measurable increase in rectal or colonic temperature cannot necessarily be attributed to a nonthermal causative mechanism. Nevertheless, the nonthermal effects on prenatal development cannot be excluded, particularly on a developing central nervous system, which is most sensitive to RF exposure.

In general, the developmental effects occurring in the absence of a measurable temperature increment in experimental animals are relatively weak, and most of the functional changes are reversible in later life. There is no evidence to suggest that exposure to environmental RF field have any injurious effects on development in human beings at present.

High intensity exposure may occur from medical exposure to RF diathermy equipment and from close-proximity exposure to high intensity antennas. A cohort epidemiology study on the effects of shortwave and microwave equipment used during pregnancy was conducted by Kallen (1987). Within the cohort, infants born with serious malformations and perinatally dead infants were examined as specific cases. The results showed a marginally statistically significant effect of occupational exposure to shortwave equipment. No other epidemiological studies on the developmental effects of RF field during pregnancy were reported. However, it is advisable that pregnant woman should avoid any possible exposure to intense RF field either at work or for theraputic purpose.

For developmental effects, especially the functional teratological effects, there is an obvious need to study further the problem of long-term exposure to a relatively weak field. Studies are also needed to elucidate the possibility of co-teratogenic effect of RF field.

EFFECTS ON IMMUNOLOGY

The immunological systems of humans and experimental animals are particularly susceptible to microwave (MW) irradiation. Studies in animals have inconsistent results in various elements of the system. In some instances a thermal burden to the exposed animal has been credited, while in others a nonthermal or direct interaction of MW with the system has been suggested. This paper reviews the effects of MW on the immunological system of humans and laboratory animals covering mainly the recent articles. For convenience, the review is divided into in vivo and in vitro studies, tumor treatment, action mechanisms and comments.

In Vivo Studies

Many species of animals including mice, rabbits, Chinese hamsters, and others have been used in studies of immune effects of MW. One of the most consistent findings of MW-induced changes in the hematopoieptic system is increased lymphocyte formation and its higher activity following exposure of several species to various frequencies of MWs. Baranski (1971,1972) exposed guinea pigs and rabbits to continuous or pulse-modulated 3000 MHz MW at an average power sensity of 3.5 mW/cm^2 three hours daily for three months. At this power level the body temperature of the animals was not elevated. Increases in absolute lymphocyte counts in peripheral blood, abnormalities in nuclear structure, and mitosis in the erythroblastic cell series in the bone marrow and in lymphoid cells in lymph nodes and spleen were observed. In a study by Czerski (1975) mice were exposed for six hours daily to 2950 MHz pulse-modulated MWs at 0.5 mW/cm^2 for 6 or 12 weeks. After six weeks the relative number of lymphoblasts in the lymph nodes of exposed mice increased considerably. In another experiment (Czerski,1975) rabbits were exposed 2 h daily, six days weekly for six months to 2950 MHz pulsed MW at 5 mW/cm^2. Peripheral blood lymphocytes from these animals when cultured for seven days in vitro underwent increased "spontaneous lymphoblastoid transformation". Maximum increases occurred after one or two months of exposure, returned to base line, and rose again one month after irradiation had been terminated. Chiang et al (1982) exposed mice to 3.0 GHz (PW) MWs at 5 mW/cm^2 (estimated SAR, 4.5 mW/g) for 3 h daily for 37 days. Following the exposure period peripheral blood lymphocytes were markedly increased. Shao et al (1986) reported that Wistar rats were exposed to 3.0 GHz (PW) MWs at 7 mW/cm^2 (estimated SAR = 1.4 mW/g) for 60 min daily for 3 successive days. After the last exposure the absolute lymphocyte counts were increased significantly. Miro et al (cited from Smialowicz, 1979) exposed mice to 3105 MHz pulsed MWs continuously over a 145-hour period at a power density of 2 mW/cm^2. The lymphoblastic cells in the spleen and lymphoid areas of exposed mice increased. Smialowicz et al (1982a) exposed rats pre- and postnatally to 425 MHz CW fields at a forward power of 20 W and found an increased responsiveness of their lymphocytes to mitogens.

The particular susceptibilities of lymphocytes to MW described above have led to examination of the effects of MW radiation on the immune system. For example, Czerski (1975) reported that mice exposed for six weeks to 2950 MHz pulsed MWs at 0.5 mW/cm^2 had significantly greater numbers of antibody-producing cells and higher serum antibody titers following immunization with sheep red blood cells (SRBC). Mice exposed for 12 weeks did not show this increased responsiveness. Shao and Wang (1985) exposed Kunming mice to 3 GHz MW (937 Hz pulse reptition and 1.2 us pulse width) for 1 h once daily for two successive days at power densities of 1,5,7 and 12 mW/cm^2 (estimated SAR, 0.9, 4.5, 6.3 and 10.8 mW/g respectively). SRBC injected into the peritoneal cavity of the subjects on the 3rd day and hemagglutination titer (HAT) tests were conducted on the 8th day after MW irraiation. As compared with sham-irradiated controls no significant difference was observed in HAT level in sera of exposed mice at a power density of 1 mW/cm^2, but there were marked increases at power densities of 5, 7 and 12 mW/cm^2 (p<0.005, 0.001 and 0.005 respectively). The power density of 7 mW/cm^2 (estimated SAR, 6.3 mW/g) that elicited a maximum increase in HAT might be regarded as optimal stimulus for the immune system. However, Smialowicz et al (1979, abstract) evaluated the primary immune response of mice to SRBC after exposure to 425 MHz fields in a TEM transmission line for 1 h on each of 5 consecutive days. Exposures to CW fields were made at power densities of 39, 10 and 2.5 mW/cm^2 (SAR/unit power density = 0.11 mW/g/mW/cm^2) and to pulsed fields at 9, 2.5 and 0.63 mW/cm^2 (1 ms pulses at 250 pps). There was no difference in primary immune response between exposed and sham-exposed mice or between mice exposed to CW and pulsed fields.

Wiktor-Jedrzejczak et al (1977a,b,c) exposed mice in a rectangular wave guide to 2450 MHz MWs for 30 min at an average dose rate near 14 mW/g. At 3, 6 and 12 days following single or multiple exposure, mice were tested for: the relative frequency of T- and B-cells in spleen, the functional capacity of spleen cells to T- and B-cell specific mitogens, and ability to respond to SRBC or dinitrophenyllysine-Ficoll. A single 30 min exposure induced a significant increase in the proportion of complement receptor positive lymphocytes (CR⁺L) in the spleen of mice which peaked six days following exposure. This effect was further enhanced by repeated (three times) exposures which also produced a significant increase in the proportion of immunoglobulin positive (Ig+) spleen cells (1977b). A significant increase in the proportion of Fc receptor positive (PcR+) cells in the spleen was observed seven days following a single exposure for 30 min. The type and combination of surface receptors (CR, Ig, Fc) expressed on splenic B-cells represent different maturational stages in B-cell development. Wiktor-Jedrzejczak et al (1977a,b,c) were unable to demonstrate any change in the total number of theta positive (θ+) T-cells in the spleen of mice following a single or multiple exposure to 2450 MHz MWs nor change in in vitro spleen cell response to stimulation by the T-cell specific mitogens PHA and conconavalin A (Con A) (1977a). The response of spleen cells to pokeweed mitogen (PWM), which stimulates both T- and B-cells, was also unchanged. The response to the B-cell specific mitogens lipopolysaccharide (LPS) , polyinosinic acid (Poly I.C), and purified protein derivative of tuberculin (PPD), however, significantly increased over controls following a single exposure (1977a). These results agree with the observed changes in the proportion of cells bearing different surface markers. Wiktor-Jedrzejczak et al (1977a) noted that MW irradiation did not stimulate lymphoid cell proliferation ₚer se, but appeared to act as a polyclonal B-cell activator, which led early maturation of noncommitted B-cells. But Smialowicz (1979) couldn't repeat the results (e.i. increased CR+L in splenic cells after MW exposure) of the study by Wiktor-Jedrzejczak et al. The differences in results were initially suspected to be due to differences in exposure system: a wave-guide system in the studies by Wiktor-Jedrzejcak et al, and exposure under far field conditions in an anechoic chamber by Smialowicz et al. Subsequent observations showed that the MW-induced effect (increased CR+L in splenic cells) observed by Wiktor-Jedrzejczak et al was mouse strain specific: the effect could be induced (by Wiktor-Jedrzejczak et al) in CBA/J mice but not in BALB/c. The reason for this strain specificity is not clear yet at this time. In another experiment, however, Smialowicz et al (1981) reported that two groups of mice (one was young adult, 10-12 week old; another was adult, 16 week old) were exposed to 2450 MHz MW at 40 mW/cm² (SAR, 28 mW/g) once for 30 min. They found that after exposure the CR+ spleen cells increased significicntly, accompanied by a significant decrease in the number of nucleated cells in the spleen of these mice, but in the young adult one, such changes were absent. The authors concluded that the age of the mouse, like the strain of the mouse, the MW exposure characteristics and the environmental conditions, may also be a source of variation that affect some elements of immune response to MW. A variation, related to the age of the mouse, that affects immune response to MW was also observed in another experiment. Shao and Wang (1985) exposed young (4-5 week old), young adult (9-10 week old) and adult (over 16 week) Kunming mice to 3 GHz (PW) MW irradiation for 1 h once daily for two successive days with three sham-exposed groups at corresponding ages each, and immunized these animals with SRBC by a single peritoneal injection on the 3rd day after exposure, and assayed HAT levels in sera of animals on the 8th day after exposure. They found that there were significant increases of HAT level in exposed young adult and adult mice, but no difference in exposed young ones as compared with sham-exposed animals respectively.

There are a few reports dealing with the effects of MW on macrophages, which play an important role in the immune system. Huang and Mold (1980) observed significant modulation of responsiveness of spleen lymphocytes to mitogens in mice exposed to MWs (2.45 GHz) at power densities of 5-15 mW/cm² . In addition, peritoneal macrophages from irradiated animals significantly suppressed B lymphocyte responsiveness to lipopolysaccharide in co-culture experiments. Szmigielski et al (cited from

Smialowicz, 1979) reported in vitro and in vivo inhibition of virus multiplication by MW hyperthermia, an observation that is consistent with the results reported by Rama Rao et al (1983). They exposed hamsters to MW field (2.45 GHz; 25 mW/cm^2 for 1 h), which resulted in activation of peritoneal macrophages that were significantly more viricidal to vaccinia virus as compared to sham-exposed or normal controls. Macrophages from microwave-exposed hamsters became activated as early as 6 h after exposure and remained activated for up to 12 days. The activation of macrophages by MW exposure paralleled the macrophage activation after vaccinia virus immunization. Activated macrophages from vaccinia-immunized hamsters did not differ in their viricidal activity when the hamsters were MW-exposed or sham-exposed. Exposure for 1 h at 15 mW/cm^2 did not activate macrophages, while 40 mW/cm^2 exposure was harmful to some hamsters. Average maximum core temperatures in the exposed (25 mW/cm^2) and sham groups were 40.5°C and 38.4°C respectively. In vitro heating of macrophages to 40.5°C was not as effective as in vivo microwave exposure in activating macrophages to the viricidal state.

It is evident that early studies of MW exposure of experimental animals were usually at high power densities for short periods; while more recent studies were at lower power densities and involving longer duration of exposure, which are more pertinent in extrapolation of potential MW hazards arising from chronic exposure. Rencently, Shao et al (1988) reported that Wistar rats and Kunming mice were exposed to pulse MW irradiation with 3 GHz at power densities of 1.0 and 0.1 mW/cm^2 (estimated SAR, 0.32-0.2 mW/g, rats) and 0.5 and 0.05 mW/cm^2 (estimated SAR, 0.6-0.45 and 0.06-0.045 mW/g) in an anechoic chamber, 5 h daily, 6 days per week for 11 months. The results showed that as compared with sham-exposed groups, the thymus weight and substantial tissue, the number of its cells, and HAT level in irradiated rats were decreased after the end of exposure; neutrophil phagocytosis in irradiated rats was suppressed from the fourth month through the end of exposure; and absolute lymphocyte count and HAT level were enhanced in 0.05 mW/cm^2 (estimated SAR, 0.06-0.045 mW/g) irradiated mouse group at the termination of exposure. The results suggest the existence of accumulative effects of "subclinical damage" on immune system under prolonged low power MW irradiation. Earlier immune suppression was found associated with higher MW irradiation and later immune suppression with lower irradiation or even with temporary increase of immune response.

In Vitro Studies

In in vitro experimental protocols, it is easier to determine and to reproduce the amount of absorbed MW energy, more exactly to examine immunocompetent cells in an unstimulated state, more easily to control study conditions, and to allow analysis of mechanisms for effects. However, there also exists a series of disadvantages with regard to eventual formulation of guidelines for human exposure limits in vivo, in which other physiological systems and interactions would have to be considered. Several studies have attempted to determine whether in vitro exposure of lymphocytes with MW leads to direct changes in their metabolic or functional states. In an early study, Stodolnik-Baranska (cited from Smialowicz, 1979) exposed human lymphocytes in culture to 3000 MHz pulsed MWs at 7 or 14 mW/cm^2. Lymphocytes were irradiated for 4 h daily at 7 mW/cm^2 for three to five days, while those exposed to 14 mW/cm^2 were irradiated for 15 min for three to five days. After five days in culture, the MW-exposed cells had undergone a fivefold increase in blast transformation compared to controls. Baranski and Czerski (1976) reported that exposure of human lymphocytes to 10000 MHz at power densities between 5 and 15 mW/cm^2 could induce lymphoblastoid transformation. At power densities below 5 mW/cm^2 this effect was not observed, while at power levels above 20 mW/cm^2 cell viability decreased. The induction of blast transformation depended upon the stopping of the exposure (5 to 15 mW/cm^2) at the moment when the temperature of the medium reached 38°C. These results suggest that the MW-induced blast transformation is actually due to a thermal effect.

Similar increases in the lymphoproliferative response of cells exposed to temperatures greater than 37°C have been reported. Human lymphocytes, when cultured at 39°C with the mitogens PHA or Con A, showed an enhancement and earlier onset of ^{3}H-thymidine incorporation compared with cultures incubated at 37°C (Ashman and Nahmais, 1978). In a similar study, Roberts and Steigbigel (1977) reported that the in vitro human lymphocyte response to PHA and the common antigen streptokinase-streptodornase was enhanced at 38.5^{O}C as compared to 37°C. Smith et al (1978) reported that the in vitro response of human lymphocytes to PHA, Con A, PWM, and allogeneic lymphocytes in mixed lymphocyte (MLS) was markedly enhanced by culture at 40°C. These studies demonstrate the need to monitor and to control the temperaure exposed to MW. Without adequate temperture control, it is virtually impossible to accept the in vitro effects as due to MW itself.

Smialowicz (1979) reported that murine splenic lymphocytes were exposed to 2450 MHz (CW) MWs for one,two, or four h at 37°C at an absorbed dose rate of approximately 19 mW/g. Following irradiation, the temperature of the exposed cultures did not differ significantly from controls and cell viability was unchanged. Following irradiation, cells were cultured for 72 h in the presence of T- or B-cell mitogens and the proliferative response was measured by ^{3}H-thymidine incorporation. No difference was found in the blastogenic response of MW-exposed and sham-exposed spleen cells to any of the mitogens employed. In a similar experiment, Hamrik and Fox (1977) exposed rat lymphocytes to 2450 MHz (CW) MWs for 4, 24, or 44 h at 5, 10, or 20 mW/cm^2 (SAR, 0.7, 1.4 and 2.8 mW/g respectively). Unlike the previous studies, Hamrick and Fox (1977) exposed whole blood preparations to MWs. Transformation of unstimulated or PHA-stimulated lymphocytes was measured using ^{3}H-thymidine. No significant difference was found in the proliferative capacity of lymphocytes from exposed and control cultures. The effects of 2450 MHz (CW) MW irradiation on the growth and viability of cultured human lymphoblasts was studied by Lin and Peterson (1977). Human lymphoblasts were exposed to 2450 MHz (CW) MWs in a waveguide for 15 min at incident power densities of 10 to 500 mW/cm^2. The corresponding rates of energy absorption were up to 1200 mW/g. No temperature increase was found, even at the highest power density in the capillary tube which held the cell suspension in the wave guide. No change was observed in the viability or growth of MW-exposed lymphoblasts compared to controls -- further evidence that in the absence of heating, no change in lymphocyte activity occured following MW exposure in vitro. In another experiment Lin et al (1979) studied the effects of 2450 MHz (CW) MWs on the ability of mouse bone-marrow cells exposed in vitro to form granulocyte/macrophage colonies when grown in a methylcellulose-culture system. The bone-marrow cells were exposed for 15 min at 30 to 1000 mW/cm^2 (SAR, 60 to 200 mW/g). At 30 mW/cm^2, there was no effect on the ability of hematopoietic stem cells to form colonies. As the exposure was increased in steps to 1000 mW/cm^2, there was a progressive reduction in the number of colonies formed by MW-exposed cells as compared with sham-exposed cells.

In vitro exposure of macrophages to 2450 MHz has been reported to depress phagocytosis by Mayers and Habershaw (1973). Monolayer cultures of mouse peritoneal macrophages were perfused with suspensions of human erythrocytes while simultaneously exposed to 2450 MHz MWs at 50 mW/cm^2. The phagocytic index of exposed cultures was significantly lower than control after a 30 min exposure. Macrophage phagocytic activity was restored to normal if the MW irradiation was discontinued. When the MWs were on a 2.5°C temperature increase was observed, but the final temperature in any given experiment did not exceed 36.2°C. These investigators concluded that the observed depression of phagocytosis in irradiated cultures was not thermally induced and that the 2.5°C rise in temperature during irradiation would have been expected to enhance rather than depress phagocytosis because optimal phagocytosis occurs at a temperature of 38.5°C. The mechanism by which this effect is induced is not known yet, but heating effects are difficult to dismiss at such a high power density. While the temperature of the suspension medium did not exceed 36.2°C, thermal gradients of much higher temperature would be expected at the macrophage-glass interface.

A MW-induced effect on granulocyte integrity and viability was reported by Szmigielski (1975), who exposed rabbit granulocytes in vitro to 3000 MHz (CW) MWs at 1 or 5 mW/cm^2 for 15, 30 or 60 min. Cultures exposed at 5 mW/cm^2 for 30 or 60 min had increased the proportion of cell death as demonstrated by an increase in nigrosine staining and enhanced liberation of lysosomal enzymes. Exposure to 1 mW/cm^2 fields did not cause the increase of cell death but led to a partial liberation of hydrolases. No temperature change was observed in the MW-exposed cultures. The liberation of granulocyte acid phosphatase and lysozyme was observed in cell suspensions exposed to either 1 or 5 mW/cm^2 and both exhibited a time- and dose-dependent relation. Szmigielski (1975) suggested that low-level MWs may affect the cellular membrane. The possible production of thermal gradients produced in the culture vessels by MW might explain these effects.

Tumor Treatment

Over the past several decades increasing evidence for the beneficial effects of partial or whole-body MW-induced hyperthermia has accumulated. Several studies have demonstrated that MW-induced hyperthermia may benefit a variety of diseases, including cancer. Szmigielski et al (cited from Zmialowicz, 1979) reported that local heating (43°C) of the Gu'erin epithelioma in Wistar rats by 2450 MHz (CW) MWs both inhibited tumors and stimulated the immune reaction against the tumor. Nonspecific immune reactions stimulated by this treatment were the antibody response to bovine serum albumin (BSA), high reactivity of spleen lymphocytes to the mitogen PHA, and increased serum lysozyme levels as a measure of macrophage activity. Tumor-specific reactions observed were increased cytotoxicity of spleen cells and peritoneal macrophages to cultured tumor cells. Similar results were reported by Marmor et al (1977), who exposed tumors in mice to local 1356 MHz irradiation. EMT-6 tumors were found to be highly sensitive to MW heating. The cure rate was a function of temperature and duration of exposure: a five-min exposure at 44°C cured almost 50% of the tumors; a 20-min exposure at 44°C cured 100% of the tumors. To determine the effectiveness of MW heating on tumor regression, tumor-cell-survival studies were done on EMT-6 tumors treated in situ. Cell inactivation by MW heating was similar to that for hot water bath heating. The results indicated that direct cell killing could not account for the observed cures, and these investigators suggested that hyperthermia may stimulate tumor-directed immune response.

Szmigielski et al (1977) exposed mice bearing transplanted sarcoma-180 tumors for 2 h daily on the first through 14th day after transplantation to 3000 MHz MWs at 40 mW/cm^2. This exposure led to a 3 to 4°C increase in rectal temperature, and resulted in a reduction of tumor mass by approximately 40%, a reduction enhanced when MW hyperthermia was combined with Colcemide, Streptolysin S or both. Colcemide enhances the

inhibiting effect of MWs on proliferation of cells in vitro and Streptolysin S is an antineoplastic substance. Szmigilski et al (cited from
Smialowicz, 1979) suggested that immunostimulation is important in the
complex inhibition of tumor growth by increased temperatures.

Janiak and Szmigielski (1982) have found that whole-body MW hyperthermia led to inhibition of various immune reactions, whereas local
heating of tumours stimulated the function of macrophages and/or lymphocytes in tumor-bearing hosts. In most of their studies daily sessions of
2450 or 3000 MHz whole-body MW irradiation at intensities in the range
of 40-50 mW/cm^2 led to elevation of rectal temperature by 3-4°C and
resulted in temporary inhibition of growth of Sarcoma-180 and Sarcoma-L1
in mice. Additional treatment of tumor-bearing hosts with immunomodulatory
agents, such as Streptolysin S, Corynebacterium parvum, interferon or
poly I:C enhanced the tumour-inhibitory effect of MW hyperthermia (Szmigielski et al, 1978). These earlier studies, however, did not provide the
direct conclusions as to the possible involvement of the immune system in
the above effect.

In recent studies Roszkowski et al (1980) found that whole-body
MW hyperthermia, leading to temporary regression of transplantable Sarco-
ma-L1 in mice, resulted at the same time in the elevation of the number
of lung metastases in these animals. In addition, normal animals exposed
to MW hyperthermia prior to implantation of neoplastic cells exhibited
significantly accelerated tumor growth and an increased number of spontaneous metastases to the lungs, as compared to untreated, control mice.
Also the number of lung colonies developing after i.v. injection of tumor
cells was significantly higher in hyperthermia-treated normal and tumor-
bearing mice, as compared to their untreated, control littermates. The
development of lung metastases in tumor-bearing animals or in animals
injected with tumor cells is thought to be controlled by nonspecific
activity of cytotoxic macrophages and/or natural killer (NK) lymphocytes.
Szmigielski and Janiak (1978) reported that when tumor-bearing rats were
exposed to local MW heating of neoplastic tissue completely different
results were obtained. In these studies MW hyperthermia was applied using
a round field electrode connected with a typical 2450 MHz medical diathermy
machine, while shielding the parts of the body other than the exposed
tumor area. This resulted in an increase of intratumor temperature up to
43°C, which remained at this level for 1 h of daily exposure. MW hyperthermia was applied 3 times per week at three differnt schedules. All
schedules resulted in inhibition of tumor growth but no total cures
observed. This inhibitory effect was even more pronounced when MW irra-
diations were combined with injections of bacterial immunostimulators
like Streptolysin S or C Parvum. The regression of tumors resulting from
the combined treatment with local MW hyperthermia and immunopotentiators
was accompanied by stimulation of cell-mediated immunity. This was mani-
fested by the enhanced synthesis of bovine serum albumin antibodies; high
reactivity of splenocyts to PHA and to mitomycin-inhibited tumor cells,
increased cytotoxicity of spleen lymphocytes and peritoneal macrophages
for cultured neoplastic cells, and elevated levels of serum lysozyme as
an indicator of granulocyte and macrophage turnover. The results strongly
suggest the involvement of both specific and non-specific cellular immunity
in hyperthermia-induced tumor inhibition. Nevertheless, it remains to be
evaluated whether the above stimulation is a primary or a secondary
event, and hypotheses of these results need experimental confirmation.

In general, there exist discrepancies in the response of various
species with different neoplasmas to MW-induced hyperthermia, which indicates that more work in this area is needed. It is not known whether MW-
induced hyperthermia affords the host more immunologic benefit than
conventional heating of tissue. What is certain, however, is that MW-

generating devices may contribute to cancer treatment by providing a means to generate intense heat in a localized defined area of tissue.

Action Mechanisms

It is apparent that partial or whole-body exposure animal to MW may lead to a variety of changes in their immunological system, but the action mechanisms are still not very clear, i.e. the significance of the changes caused by MW is difficult to interpret. While some studies indicate that MW increases responsiveness of lymphocytes and potentiates the immune response to antigen, others indicate depressed responsiveness. In most cases, however, these alterations can be attributed to stress-type responses. Liburdy (1979,1980) has investigated the relationship between MW exposure and steroid release associated with thermal stress. He noted that induction of a transient neutrophilia and lymphopenia by MW exposure was similar to the changes induced by endogenous release or exogenous administration of adrenocorticotrophic hormones. The differnce between MW heating and warm air heating might be due to the fact that rapid MW heating may be a stress factor for a greater proportion of the exposure time than the more slowly acting warm air treatment.

Liburdy (1980) characterized the effects of MW exposure on lymphocyte circulation in vivo and the relationship of changes in circulation to steroid release associated with thermal stress and the process of thermoregulation. Mice (C57Bl/6 and BALB/c) were exposed to 26000 MHz (CW) at 25 or 5 mW/cm^2 (SAR, 19 or 3.8 mW/g) for one hour. Immediately before exposure, mice were injected with ^{51}Cr-labelled splenic lymphocytes. One, 6 and 24 h after cell injection, the lung, liver, spleen, and lower extremity long bones (bone marrow) were removed and assayed for proportions of labelled cells. Sham-exposed, warm air-exposed (63°C oven) and steroid-treated groups (4 mg/kg methyl prednisolone) were also assayed. Both exposure to 25 mW/cm^2 MW and exposure to warm air led to an increase in core (colonic) temperature (2°C) within 15 min, followed by stable hyperthermia due to adjusted thermoregulation. Liburdy called attention to the fact that, although the absolute increases in temperature for the 25 mW/cm^2 MW and warm air exposures were identical, however, the rates of core temperature increase for the treatments were significantly different, with a more rapid increase noted in the MW-exposed animals. Steroid output in the MW-heated mice was threefold greater than that in warm air-heated mice, despite identical final core temperatures.

Normally, administered splenic lymphocyte traffic consists of a rapid influx of injected cells into the lungs, followed by migration of most cells ("homing") to the spleen with limited migration to the liver and little, if any, migration to the bone marrow. In Liburdy's studies, the hyperthermic (25 mW/cm^2) MW exposure led to a 37% decrease in lymphocyte migration from the lung to the spleen, and a threefold increase in lymphocyte migration to the bone marrow. Similar changes were noted with administration of exogenous steroids. Lymphocyte traffic to the liver was unchanged by the MW exposure or other treatments. Neither exposure to MW (5 mW/cm^2) without a detectable increase in core temperature nor thermogenic warm air exposure resulted in altered lymphocyte traffic.

Liburdy suggested that such whole-body MW exposure could be perceived as a heat stress that stimulates the hypothalamic-hypophyseal-adrenal axis to trigger the release of adrenal steroids into the blood. MW exposure without a detectable increase in core temperature in these experiments may have constituted a thermal stress, but one with different effects on the hypothamic-hypophyseal-adrenal axis.

An action mechanism of macrophage resistance to vesicular stomatitis virus infection after MW exposure to hamsters was recently reported by Rama Rao et al (1984). Exposure of hamster to MW energy (2450 MHz, 25 mW/cm^2, 1 h) resulted in activation of peritoneal macrophages (PM) to a viricidal state restricting the replication of vesicular stomatitis virus (VSV). The PM from MW-exposed hamsters were viricidal as early as 1 day after exposure and remained active for 5 days. Immunization of hamsters with vaccinia virus induced viricidal PM by 3 to 4 days and they remained active for 7 days. To test the hypothesis that thermogenic MW exposure results in the release of endotoxin across the intestinal epithelium which subsquently activates PM, hamster were injected with lipopolysaccharide (LPS) and their viricidal activity was studied. LPS in vitro (0.2 ug) and in vivo (0.5 ug) activated macrophages to a viricidal state. When administered in vivo, LPS (0.5 ug) activated macrophages as early as 1 day and the activity remained for 3 days. While MW exposure of PM in vitro failed to induce viricidal activity, exposure of PM to LPS in vitro induced strong viricidal activity. This suggests that in vitro response of PM to MW is an indirect one, which is consistent with the hypothesis that MW-induced PM viricidal activity may be mediated via LPS. In preliminary experiments MW exposure resulted in extended survival time for hamsters challenged with a lethal dose of vesicular stomatitis virus supporting the concept that MW-activated PM may be a useful therapeutic modality.

However, convincing evidence of action mechanisms for a nonthermal or direct interaction of MW with hematopoietic cells, particularly with immunocompetent cells, in vitro or in vivo is not available. Evidence is increasing that MW fields may interact and cause alterations at the membrane level of organization in nervous tissue (Bawin et al, 1975), but no such clear evidence is available for blood cells. This does not exclude the possibility of such interactions, and continued research in this area of action mechanism is necessary. What appears to be evident is that the immunologic system is sensitive to MW fields needs to be determined so that the potential health risk (or benefit) to humans can be better evaluated.

Comments

The data cited above indicate that interactions of MW with the immunologic system are complex and difficult to define. In most reports, the thermal influence of MW on the observed alterations is obvious, and others, however, in which the thermal influence is not obvious, and the observed changes are more difficult to explain on thermal-stress, although mere lack of rectal temperature increase following exposure to MW does not exclude a possible thermal interaction which the animal can compensate and control.

In conclusion, some regularities of MW effects on immune system might be established, i.e. under defined conditions, MW fields can stimulate functions of the immune system for short periods, such as diathermy and some cancer treatment in clinic or laboratory using reasonable level of MW fields during a limited course; if the exposure intensities of MW are over the physiological burden in an animal (or a human), the immune function would be suppressed, particularly when the exposure duration is longer; and there are also evidences that long-term exposure of animals (mice) to low-level MW fields may lead to transient suppression of certain immunologic reactions, which suggests existence of accumulative effect of "subclinical damage" on immune system under prolonged low power MW irradiation. During the course of a long-term MW expoosure, earlier immune suppression might be found associated with higher power MW irradiation and later immune suppression with lower power irradiation.

REFERENCES

Albert E. N., Sherif M. F., and Papadopoulos N. J., 1981a, Effect of
    nonionizing radiation on the Purkinje cells of the uvula in squir-
    rel monkey cerebellum, Bioelectromagnetics, 2:241.
Albert E. N., Sherif M. F., and Papadopoulos N. J., 1981b, Effect of
    nonionizing radiation on the Purkinje cells of the rat cerebellum,
    Bioelectromagnetics,2:247.
Ashman R. B. and Nahmias A. J., 1978, Effect of incubation temperature on
    mitogen responses of lymphocytes from adult peripheral blood and
    from cord blood, Clin. Exp. Immunol., 33:319.
Baranski S., 1971, Effect of chronic MW irradiation on the blood forming
    system of guinea pigs and rabbits, Aerospace Med., 42:1196.
Baranski S., 1972, Effect of microwaves on the reactions of the white
    blood cells system, Acta Physiol. Polonica, 23:685.
Baranski S. and Czerski P., 1976, Biological effects of microwaves,
    experimental data, In:"Biological Effects of Microwaves", Strods-
    sburg, Pa, Dowden, Hutchinson and Ross, P. 133.
Bawin S. M., Kaczmarek L. M. and Adey W. R., 1975, Effects of modulated
    VHF fields on the central nervous system, Ann. N. Y. Acad. Sci.,
    247:74.
Berman E., Carter H. B., and House D. E.,1982, Observation of Syrian
    hamster fetuses after exposure to 2450 MHz microwave, J. Microwave
    Power, 17:107
Berman E., Carter H. B. and House D. E. ,1984, Growth and development of
    mice offspring after irradiation in utero with 2450 MHz microwave,
    Teratology, 30:393.
Chernavetz M. E. ,Justesen D. R. and Oke A. F. ,1977, A teratologic study
    of the rat:microwave and infrared radiations compared, Radio Sci.,
    12(6s):191.
Chiang H., Shao B. J., wang X. H. and Lou Y., 1982, A preliminary obser-
    vation on the effects of MW on immune function in mice, J. Zhejiang
    Med. Univ., 11:221. (Chinese)
Chiang H. and Lou Y., 1983, Effects of microwave radiation on chromosome
    of testes in mice, Ratiation Protection, 3:33. (Chinese)
Chiang H. and Wang M., 1984, Morphological change in mouse sperm fol-
    lowing microwave exposure,J. Bioelectricity, 3:367.
Chiang H. and Yao G. D. , 1987, Effects of pulsed microwave radiation
    pre- and post-natally on the developing brain in mice, J. Bioelec-
    tricity, 6:197.
Conover D. L., Lary J. M. and Hanser P. L. , 1980, Thermal threshold for
    teratogenic response in rats irradiated at 27.12 MHz,Bioelectromag-
    netics, 1:204.
Czerski P., 1975, Microwave effects on the blood-forming system with
    particular reference to the lymphocyte, Ann. N. Y. Acad. Sci.,
    247:232.
Czerski P., Gloser Z. R. ,Drop B. A., Silverman P. M. and Manikowska-
    Czerska E., 1985, Dominant lethal test following 2450 MHz CW expo-
    sure of male ICR mice, BEMS 7th Annual Meeting Abstracts, p. 49.
Fang B. R., Lu Q., Li F. B., Zou R. B. and Liu Y. H., 1982, Application
    of microwave in male contraception, Chin. J. Urology, 3:75. (Chinese)
Galvin H. J. , MacNichols G. and McRee D. I., 1983a, Effects of 2450 MHz
    microwave radiation during the gestational period on the postnatal
    hematology of rats, Cell Biophys., 5:33.
Galvin M. J. , McRee D. I. and Thaxton J. P., 1983b, Physiological abnor-
    malities in juvenile and Japanese quail exposed to MW radiation
    during embrygeny, BEMS 5th Annual Scientific Session Abstracts, p.
    47.
Galvin M. J., Tilson H. A., Mitchell C. L., Peterson J. and McRee D. I.,
    1986, Influence of pre- and postnatal exposure of rats to 2.45 GHz

microwave radiation on neurobehavioral function, <u>Bioelectromagne-tics</u>, 7:57.

Goud S. N.,Usharani M. V.,Reddy P. P., Reddi O. S., Rao M. S. and Saxena V. K., 1982, Genetic effects of microwave radiation in mice, <u>Muta-tion Research</u>, 103:39.

Hall C. A., Galvin M. J.,Thaxton J. P. and McRee D. I., 1982, Interaction of microwave radiation with turkey sperm, <u>Radiat. Environ. Biophys.</u>, 20:145.

Hall C. A., McRee D. I.,Galvin M. J., White N. B., Thaxton J. P. and Christensen V. L., 1983, Influence of in vitro microwave radiation on the fertilizing capacity of turkey sperm, <u>Bioelectromagnetics</u> 4:43.

Hamrick P. E. and Fox S. S., 1977, Rat lymphocytes in cell culture exposed to 2450 MHz (CW) microwave radiation, <u>J. Microwave Power</u>, 12:125.

Huang A. T. F. and Mold N. G.,1980, Immunologic and hematopoietic alter-ations by 2450 MHz electromagnetic radiation, <u>Bioelectromagnetics</u>, 1:77.

Inouye M., Galvin M. J. and McRee D. I., 1982, Effect of 2.45 GHz micro-wave radiation on the development of Japanese quail cerebellum, <u>Teratology</u>, 25:115.

Jamakosmanovic A., 1981, The levels of ATP, ADP and AMP in rat brain during postnatal development following in utero 2450 MHz microwave irradiation, <u>Period. Biol.</u>, 81:151.

Janik M. and Szmigielski S., 1982, Alteration of the immune reaction by whole-body and local MW hyperthermia in normal and tumor-bearing animals, Review of own 1976-1980 experiments,<u>Br. J. Cancer</u>, 45 Suppl V:122.

Jensh R. P., Vogel W. H. and Brent R. L., 1983, II. Postnatal psychophy-siologic analylsis, <u>J. Toxicol.Environ. Health</u>, 11:37.

Jensh R. P., 1984, Studies of the teratogenic potential of exposure of rats to 6000 MHz microwave radiation, II. Postnatal psychophysiolo-gic evaluations, <u>Ratiat. Res.</u>, 97:282.

Kallen B., 1987, Search for teratogenic risks with the aid of malforma-tion registries, <u>Teratology</u>, 35:47

Kaplan P., Rebert C., Luman K. and Gage M, 1982, Biological and behavioral effects of prenatal and postnatal exposure to 2450 MHz electromag-netic radiation in the Squirrel monkey, <u>Radio Science</u>, 17(5s):135s.

Lancranjan L., Marcanescu M., Rafaila E., Klepsch I. and Popescu H.I., 1975, Gonadic function in workmen with long term exposure to micro-waves, <u>Health Phys.</u>, 29:381.

Lary J. M., Conover D. I., Johnson P. H. and Hornung R. W., 1986, Dose-response relationship between body temperature and birth defects in radiofrequency-irradiated rats, <u>Bioelectromagnetics</u>, 7:41.

Lary J. M. and Conover D. L., 1987, Teratogenic effects of radiofrequency radiation, <u>IEEE Engin. in Med. and Bio.</u>, 6:42.

Lebovitz R. M. and Johnson L., 1983, Testicular function of rats fol-lowing exposure to microwave radiataion, <u>Bioelectromagnetcs</u>, 4:107.

Lebovitz R. M., Johnson L. and Samson W. K., 1987,Effects pulse-modulated microwave radiation and conventional heating on sperm production, <u>J. Appl. Physiol.</u>, 62:245.

Leonard A., Berteaud A. J. and Bruyere A., 1983, An evaluation of the mutagenic, carcinogenic and teratogenic potential of microwave, <u>Mutation Research</u>, 123:31.

Li W. X., Ding Z. L. and Bao D. Q.,1982 Ultrastructural study of effects of microwave irradiation on seminiferous epithelium in rabbits, <u>Reproduction & Contraception</u>, 2:41. (Chinese)

Li W. X., Bao D. Q., Wang Y. P. and Tang Y., 1986, Ultrastructural study of microwave irradiation effect on epididymal epithelium in rabbits, <u>Reproduction & Contraception</u>, 6:22. (Chinese)

Li W. X.,Bao Q. D., Wang Y. P. and Tang Y., 1987, Ultrastructural and alkaline phosphotase cytochemical effects of microwave irradiation on peritubular tissues of seminiferous tubules in rabbit, Reproduction & Contraception, 7:35. (Chinese)

Liburdy R. P., 1977, Effects of radiofrequency radiation on inflamation, Radio Sci., 12(Suppl):179.

Liburdy R. P., 1979, Radiofrequency radiation alters the immune system: Modulation of T- and B-lymphocyte levels and cell-mediated immunocompetence by hyperthermic radiation, Radiat. Res., 77:34.

Liburdy R. P., 1980, Radiofrequency radiation alters the immune system: II. Modulation of in vivo lymphocyte circulation, Radiat. Res., 83:66.

Lin J. C. and Peterson W. D. 1977, Cytological effects of 2450 MHz CW microwave radiation, J. Bioeng., 1:471.

Lin J. C., Ottenbreit M. J., Wang S., Inone S., Bollinger R. O. and Fracassa M., 1979, Microwave effects on granulocyte and macrophage precusor cells of mice in vitro, Radiat. Res., 80:392.

Liu Y. H., Ou D. Y., Chen Z. C. and Zou R. B., 1981, Cytological and autoradiographic observation on microwave inhibition of white rats, Bulleting of Hunan Teachers University, 1:59. (Chinese)

Liu Y. H.,Yang X. Z., Ren W. H. and Peng P., 1984, Effects of microwave irradiation on chromosome in spermatogenic cells of rats, Reproduction & Contraception, 4:19. (Chinese)

Liu Y. H., 1986, A study of the long-term effects of human spermatogenesis after microwave treatments of testes, In:"Abstracts of the Fifth National Biophysics Conference", Hangzhou, China, p. 289. (Chinese)

Lovely R. H., Mizumori S. I. Y., Johnson R. B. and Guy A. W., 1983, Subtle consequences of exposure to week MW fields: Are there nonthermal effects? In:"Microwave and Thermoregulation" (p. 401-429) Adair E. R., ed., Acadamic Press, N. Y.

Manikowska-Czerska E., Czerski P. and Leach W. M., 1982, Changes in metaphase counts, meiotic translocations and univalency following 2.45 GHz exposure of male CBA mice BEMS 4th Annual Scientific Session Abstracts

Manicowska-Czerska E. Czerski P. and Leach W. H, 1983a, Effects of 0.915 and 9.4 GHz CW microwaves on meiosis in male mice, BEMS 5th Annual Meeting Abstracts, p. 57.

Manikowska-Czerska E., Czerski P. and Leach W. M., 1983b, Dominant lethal testing after exposure of mice to 0.915 GHz microwave, BEMS 5th Annual Meeting Abstracts, p. 57.

Marcickiewicz J., Chazan B., Niemiec T. Sokolska G., Troszynski M.,Laczak M. and Szmigielski S., 1986, Microwave radiation enhances teratogenic effect of cytosine arabinoside in mice, Biol. Neonate, 50:75.

Marmor J. B., Hahn N., Hahn G. M., 1977, Tumor cure and cell survival after localized radiofrequency heating, Cancer Res., 37:879.

Mayers C. P. and Habershaw J. A., 1973, Depression of phagocytosis: A nonthermal effect of microwave radiation as a potential hazard to health, Int. J. Radiat. Bio., 24:449.

McRee D. I. , Thaxton J. P. and Parkhurst C. R.,1983, Reproduction in male Japanese quail exposed to MW radiation during embryogeny, Radiation Res., 96:51.

Nakao M., 1981, Effects of repeated 2450 MHz microwave radiaton on the acetylcholinesterase activity in developing rat brain, Period Biol., 81:173.

Nawrot P. S.,1981, Effects of 2.45 GHz CW microwave radiation on embryo-fetal development in mice, Teratology, 24:303.

Nawrot P. S., McRee D. I. and Galvin M. J., 1985, Teratogenic, biological and histologtical studies with mice prenatally exposed to 2.45 GHz microwave radiation, Radiat. Res., 102:35.

O'conor M. E., Bartsch S. A., Chrobak J. and Proska J. C., 1983, Behavioral and development evaluation of rats exposed in utero to microwave fields, BEMS 5th Annual Scientific Session Abstracts, p. 60.

Rama Rao G., Cain C. A., Lockwood J. and Tompkins W. A. F., 1983, Effects of microwave exposure on the hamster immune system, II. Peritoneal macrophage function, Bioelectromagnetics, 4:141.

Rama Rao G., 1984, Effects of MW exposure on the hamster immune system, III. Macrophage resistance to vesicular stomatitis virus infection, Bioelectromagnetics, 5:377.

Roberts N. J. and Steigbigel R. T., 1977, Hyperthermia and human leucocyte functions: Effects on response of lymphocytes to mitogen and antigen and bactericidal capacity of monocyte and neutrophils, Infec. Immun., 18:673.

Roszkowski W., Wrembel J. K., Roszkowski K., Janiak M. and Szmigielski S., 1980, Does whole-body hyperthermia therapy involve participation of the immune system? Int. J. Cancer, 25:289.

Rou Z. Q., Ren H. X., Zhou Z. Y. and Hang Y., 1986, Spermatogenesis supression by microwave diathermy on testes of dogs, J. Biomed. Engin., 3:108.(Chinese)

Saunders R. D.,1981, Effects of 2.45 GHz MW radiation and heat on mouse spermatogenic epithelium, Int. J. Radiat. Biol., 40:623.

Saunders R. D., Darby S. C. and Kowalczuk C. I., 1983, Dominant lethal studies in male mice after exposure to 2.45 GHz microwave radiation, Mutation Res., 117:345.

Schmidt R. E., Merritt J. H. and Hardy K. H.,1984, In utero exposure to low-level microwaves not effect rat foetal development, Int. J. Radiat. Biol., 46:383.

Shandala M. G. and Vinogradov G. I., 1982, Autoallergic effects of exposure of rats to superhigh frequency electromagnetic field and their action on the fetus and offspring, Vestn. Akad. Med. Nauk SSSR, (10):13.

Shandala M. G., 1987, Nonionizing microwave radiation as the inductor of autoallergic processes, Abstracts of BEMS 9th Annual Meeting, p. 74.

Shao B. J. and Wang Y. F., 1985, A study on effect of MW on immune system -- The effects of MW on hemagglutination titer test in mice, Chin. J. Biomed. Engin., 4:109. (Chinese)

Shao B. J., Lou Y., Wang Y. F. and Chiang H., 1986, Preliminary observation on hematological and immunological effects of MW on Wistar rats, J. Zhejiang Med. Univ., 15:12. (Chinese)

Shao B. J., Wang X. H., Lou Y. and Li N. G., 1988, A study on effects of microwave on immune system: immumological effects of prolonged low power microwave irradiation on mice and rats, Chin. J. Biomed. Engin., 7:213. (Chinese)

Smialowicz R. J., 1979, Hematologic and immunologic effects of nonionizing electromagnetic radiation,Bull. N. Y. Acad. Med., 55:1094.

Smialowicz R. J., Kinn J. B. and Elder J. A., 1979, Perinatal exposure of rats to 2450 MHz microwave radiation: Effects on lymphocytes, Radio Sci., 14(6s):147.

Smialowicz R. J., Riddle M. M., Weil C. M., Brugnolotti P. L. and Kinn J. B., 1982a, Assessment of the immune responsiveness of mice irradiated with continuous wave or pulse-modulated 425 MHz radiofrequency radiation, Bioelectromagnetics, 3:467.

Smialowicz R. J., Weil C. M., Kinn J. B. and Elder J. A., 1982b, Exposure of rats to 425 MHz (CW) radiofrequency radiation: Effects on lymphocytes, J. Microwave Power, 17:211.

Smith J. B., Knowlton R. P. and Agarwal S. S., 1978, Human lymphocyte responses are enhanced by culture at 40°C, J. Immunol., 121:691.

Szmigielski S., 1975, Effect of 10-cm (3 GHz) on granulocytes in vitro, Ann. N. Y. Acad. Sci., 247:275.

Szmigielski S., Pulverer G., Hryniewicz and Janiak M., 1977, Inhibition of tumor growth in mice by MW hyperthermia, Streptolysin S. and Colcemide, Radio Sci. 12(suppl):185.

Szmigielski S., Sokoloska G. J., Marcickiewicz J., Chazan B,Niemiec T. and Troszynski M., 1984, Microwave radiation enhances teratogenic potency of cytosine arabinoside (ara-c) in mice, XXIst General Assembly of the International Union of Radio Science, Florence, Italy.

Tofani S., Agnesod G. and Ossola P., 1986, Effects of continuous low-level exposure to radiofrequency radiation on intrauterine development in rats, Health Physics, 51:489.

Wiktor-Jedrzejczak W., Ahmed A., Czerski P., Leach W. M. and Sell K. W., 1977, Immune response of mice to 2450 MHz microwave radiation: overview of immunology and empirical studies of lymphoid splenic cells, Radio Sci., 12(suppl):209.

Wiktor-Jedrzejczak W., Ahmed A., Sell K. W., et al., 1977, MWs induce an increase in the frequency of complement receptor-bearing lymphoid spleen cells in mice, J. Immunol., 118:1499.

Wiktor-Jedrzejczak W., Ahmed A., Czerski P., et al., 1977, Increase in the frequency of Fc receptor (FcR) bearing cells in the mouse spleen following a single exposure of mice to 2450 MHz microwaves, Biomedicine, 27:250.

Wu C. and Chen X. H., 1985, Effects of dominant lethal mutatioon in mice induced by irradiation microwaves, Reproduction & Contraception, 5:41, (Chinese)

Zou R. B., Liu Y. H., Sun H. S., Tan B. Y., Fang B. R. , Lu Q. and Zhang S. C., 1980, Effects of microwave heating on rabbit spermatogenesis, Bulletin of Human Medical College, 5:131. (Chinese)

PULSED RADIOFREQUENCY FIELD EFFECTS IN BIOLOGICAL SYSTEMS

James C. Lin

Department of Bioengineering
University of Illinois
Chicago, IL 60680-4348

INTRODUCTION

The possibility that pulsed fields produce biological responses other than those elicited by continuous-wave field of the same average power has been conjectured since the early years of research into the biological effect of radiofrequency (RF) energy.   However, because of the limited availability of experimental results, few protection guides and exposure standards promulgated by various private organizations or governmental agencies attempted to specify limits to guard against potential hazards of pulsed radiofrequency fields.   Indeed, available results have led some to conclude that there is no compelling evidence that pulsed microwave, of the type produced by radar transmitters, cause biological effects not found following exposure under conditions of continuous-wave radiation at the same average power density (Postow and Swicord, 1986).   Nevertheless, the accumulation of recent experimental evidence on the biological effects of pulsed and modulated RF field suggests a need to put such interactions in a more meaningful context and a closer examination of the mechanism(s) of such interactions.

That pulse modulated radiation may penetrate more deeply and therefore, be absorbed more strongly than continuous-wave radiation having the same carrier frequency arises from the fact that pulse modulation provides a series of harmonics whose fundamental components coincides with the modulation frequency.   While higher order harmonics are strongly attenuated by biological tissues, harmonics whose frequencies are lower than the sinusoidal carrier frequency will generally penetrated deeper than continuous waves.   Moreover, the frequency spectrum of a transient field of short duration may span a wide band from zero to a few GHz.   Aside from the effects that are elicited exclusively by pulsed radiation, it is conceivable that the above mentioned difference in energy distribution may be sufficient to produce biological responses from whole organisms that are functions of modulation characteristics of the impinging RF radiation.

This paper will begin with the 1982 ANSI C95 recommendation for safety level of radiofrequency fields with respect to personnel and discuss pertinent findings of pulsed and modulated RF field interaction with single cells and whole-body structures.   The objective is to provide a succinct introduction to a variety of peak power effects attending pulsed RF radiation.   It should be noted that there is clear indication that the study of pulsed RF field effects in biological systems will continue and even accelerate.

ANSI SAFETY GUIDE

In 1974, the American National Standards Institute issued a standard concerning the safety level of electromagnetic radiation with respect to personnel. Major revisions of this standard were made in 1982 to take into account the significant expansion of scientific knowledge base. Changes include a wider frequency coverage, incorporation of dosimetry and frequency dependence resulting from whole body resonance absorption. This standards prescribes recommended radiation protection guides to prevent biological injury from exposure to RF electromagnetic radiation. Specifically, for human exposure to electromagnetic energy at radio frequencies from 300 KHz to 100 GHz, the protection guides in terms of squared electric field strengths and in term of the equivalent plane-wave, free-space power density, as a function of frequency, are given in Table 1. For both pulsed and non-pulsed fields, the permissible exposure levels are averaged over any 0.1 hour period and the time averaged values should not exceed the values given in Table 1.

The applicability of these safety quides to situations involving short pulsed of RF energy with low pulse repetition frequency is questionable (Lin 1978). The plane-wave, free-space power density allowed by the safety guide for a 0.1-microsecond to 6-minute pulse repeated at once every 0.1 hour is shown in Table 2. The dielectric breakdown field strength or power density of air is 3×10^6 V/m or 1.2×10^{10} W/m^2, respectively. It can be seen that permissible exposure levels for 0.1 microsecond pulse repeated once every 0.1 hour would exceed the breakdown field strength of air in all cases. Clearly, the safety standard needs to be refined to account for peak power and modulation to provide the protection promised by the standard.

Table 1. ANSI C95.1 - 1982 Radio Frequency Protection Guides

Frequency Range (MHz)	Electric Field Strength E^2 (V^2/m^2)	Magnetic Field Strength H^2 (A^2/m^2)	Power Density (mW/cm^2)
0.3 – 3	400,000	2.5	100
3 – 30	4,000 $(900/f^2)$	0.025 $(900/f^2)$	$900/f^2$
30 – 300	4,000	0.025	1.0
300 – 1500	4,000 $(f/300)$	0.025 $(f/300)$	$f/300$
1500 – 100,000	20,000	0.125	5.0

Note: f = frequency (MHz), 1 mW/cm^2 = 10 W/m^2

Table 2. Permissible Levels for Pulse Power According to ANSI C95.1 - 1982

Exposure Duration (Sec)	Power Density (W/m^2)			
360	10	50	100	10^3
60	60	300	600	6×10^3
1	3.6×10^3	1.8×10^4	3.6×10^4	3.6×10^5
0.1	3.6×10^4	1.8×10^5	3.6×10^5	3.6×10^6
0.01	3.6×10^5	1.8×10^6	3.6×10^6	3.6×10^7
10^{-3}	3.6×10^6	1.8×10^7	3.6×10^7	3.6×10^8
10^{-4}	3.6×10^7	1.8×10^8	3.6×10^8	3.6×10^9
10^{-5}	3.6×10^8	1.8×10^9	3.6×10^9	3.6×10^{10}
10^{-6}	3.6×10^9	1.8×10^{10}	3.6×10^{10}	3.6×10^{11}
10^{-7}	3.6×10^{10}	1.8×10^{11}	3.6×10^{11}	3.6×10^{12}

Note: Dielectric Strength of Air: 3×10^6 V/m, 1.2×10^{10} W/m^2.

PULSED RADIATION

This chapter is concerned mainly with pulse-modulated RF radiation. Figure 1 shows the waveform of rectangular pulses with a pulse width of t_o and a period of T. The pulse repetition frequency is given by 1/T. It is customary to characterize an RF pulse by its duty cycle, which is defined as the ratio of pulse width to the period, i.e. t_o/T. A duty cycle of 1.0 corresponds, therefore, to CW operation. The average power (averaged over a period) is given by the product of the peak power and the duty cycle. For short pulses with low pulse repetition frequency, the average power can therefore be very low, even though the peak power may be in the gigwatt (GW) region.

The peak power output from RF sources has grown by an order of magnitude every decade since 1940 (Florig, 1988). Current high-power laboratory sources range in pulse width from 10 nanoseconds to continuous wave; in frequency from 0.5 GHz to over 100 GHz; in pulse repetition frequency from single shot to thousands of pulses per second; and in power output from several megawatts for continuous wave to many gigawatts for single shot pulsed units (see Table 3). Pulsed power sources with these capabilities are in use today in particle acceleration, inertial-confinement fusion, electromagnetic pulse (EMP) simulation, and in experiments directed toward assessing troop and weapon vulnerability.

Electromagnetic pulses with electric field strength up to 500 kV/m or 663 MW/m^2 and with frequency spectra of 0-100 MHz are produced by nuclear

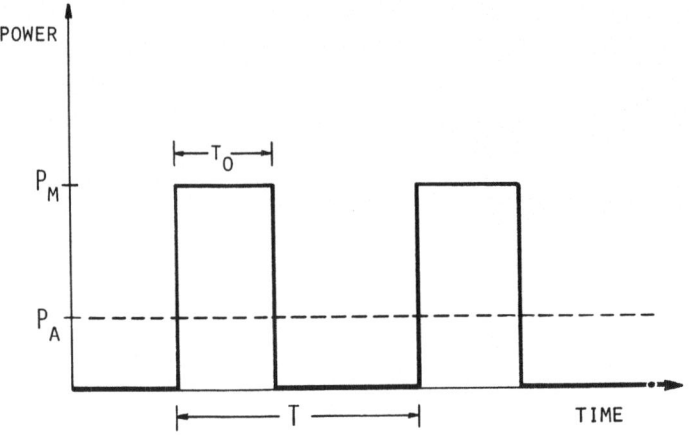

Fig. 1   Characteristics of a rectangular pulse waveform.

detonation and, of course, by EMP simulators.   Indeed, some recent
simulators have frequency contents that exceed one GHz.   A typical EMP
waveform can be characterized by a triple exponential time function (Lin,

Table 3.   The Frontier of High Peak Microwave Power Generation

Frequency (GHz)	Peak Power (MW)	Generating Devices
1	20,000	Vircator
3	10,000	Magnetron
	5,000	Gyrotron
	200	Klystron
	100	Beam-Plasma Device
10	6,000	Magnetron
30	1,000	Free Electron Laser
100	800	Free Electron Laser

From Florig, 1988

et al., 1975). Indeed, the electric field waveform shown is Fig. 2 represents an average measured EMP time function which has already been exceeded by a significant amount in time (shorter) or in strength (higher).

SINGLE PULSE EFFECTS

A number of intriguing single-pulse exposure effects have been reported in recent years. A few of these are briefly described in the following paragraphs of this section.

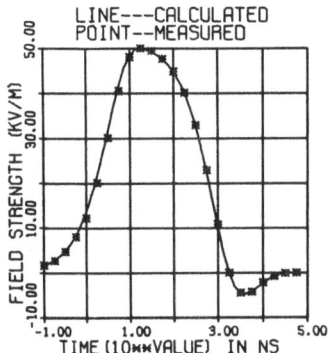

Fig. 2 A representative electric field waveform of EMP signal.

Short bursts (1 sec. or less) of high power microwave energy is used for rapid in vivo inactivation of brain enzymes prior to analysis for neurochemicals. The technique is based on the principle that many neurochemicals are relatively heat-stable substances, while the enzymes that both produce and degrade them are heat labile and denature irreversibly at temperatures around $85^{\circ}C$. In fact, at the present time the most accurate and most widely accepted measurement of many critical neurochemicals depends on the use of microwave to fix brain tissue within a fraction of a second with heat inactivation. The peak power and burst duration reported for mice and rats are given in Table 4. It can be seen that the peak power densities sufficient for sacrifice of laboratory animals in a 2450 MHz waveguide is less than 500 kW/m^2.

Exposure of heads of laboratory animals and human subjects to pulsed microwave radiation evoke auditory sensations in the exposed subject (Lin, 1978). The studies concerning microwave hearing phenomenon have emphasized demonstration of auditory responses and delineation of interactive mechanism. This is as it should be inasmuch as the effect is so very different from those associated with responses to CW radiation, so much so that it implied the possibility of direct modes of interaction that may be neurophysiologically significant. The accumulated results indicate that there is little likelihood that microwave hearing phenomenon arises from an interaction of microwave pulses directly with the cochlear nerve or neurons at higher structures along the auditory pathway, but rather the pulsed microwave energy initiates a thermoelastic wave of pressure in soft tissues that activates the inner ear receptors via bone conduction (Lin, 1980; 1981; Chou and Guy, 1982). A highly pertinent question that remains is: does microwave auditory phenomenon pose a risk to the health of an exposed individual, or under what condition does the effect become a hazard? It has been shown that the threshold of audibility of 2450 MHz microwave-induced sound in humans is about 400 mJ/m^2 per pulse for pulses shorter than 30 microsecond regardless of pulse width or peak power density (Guy, et al, 1975). There exists apparently an optimal pulse width for efficient sound pressure generation which varies according to the head size and frequency of the impinging microwaves (Lin, 1977).

Table 4. Animal Brain Fixation for Rapid Enzyme Inactivation

Exposure Duration (Sec)	Net Power (KW)	Incident Power* (W/m^2)	Brain Absorption (W/g)
Rats (Lenox, et al, 1977)			
2.80	3.5	2.1×10^5	20.5
Mice (Schneider, et al, 1982)			
1.40	2.5	1.5×10^5	145
0.50	6.3	3.7×10^5	400
0.35	6.3	3.7×10^5	575

*WR 430 Waveguide Cavity at 2450 MHz

The susceptibility of rodents to pulsed microwave-induced startle and convulsive responses have been reported nearly a decade ago (Guy and Chou, 1982). Animals exposed to a single pulse of 915 MHz microwave in the range of one microsecond to 360 milliseconds were shown to exhibit seizure reactions lasting for one minute after exposure, followed by a 5-minute unconscious state during which normal reflexes were displayed. These results indicate a threshold energy density of 28 kJ/kg in the head of a rodent for convulsion, regardless of pulse width. It should be noted that a maximum brain temperature of 46°C was recorded at the threshold exposure level. The animal began moving when brain temperature returned to within 1°C of normal. Histological examination revealed some demyelination of neurons one day after exposure and some microfocal glial nodules in the brain one month after exposure.

Although the potential biological effects of EMP pulses have been suggested for sometime (Milroy, et al., 1974, Lin, et al., 1975, 1976), its importance has been recognized only in recent years. While much remains to be learned about the biological effects of EMP, it is clear that the effects are very different from responses to CW radiation. So much so that it implies the possibility of significant neurophysiological interaction. For example, it has been shown (Bernardi, et al, 1984) using the Hodgkin and Hexley nonlinear membrane model that the current density induced in biological tissues by a Gaussian EMP pulse with a energy density equivalent to the maximum permissible under the 1982 ANSI guide would produce a large alteration in the resting potential of excitable cellular membranes. Indeed, action potentials could be generated for pulse widths of one millisecond or less. However, the physiological significance is obscure. The threshold of action potential excitation varies inversely with pulse width; i.e. the required incident electric field strength would be 400 and 2000 kV/m for 1 millisecond and 10 microsecond pulses, respectively (Bernardi and D'Inzeo, 1984).

MULTIPLE PULSE EFFECTS

The literature on biological effects of RF fields modulated with a train of brief rectangular pulses of high peak power and low repetition rate while scarce is becoming increasingly more abundant. Although such irradiation has been shown to produce responses alone or in combination with other stimulants that are dependent on the animal and tissue preparations, and on peak power and pulse width, effects have often been characterized in terms of average power or average specific absorption rate (SAR). It should be noted that for short pulses with low pulse repetition rate, the average power can be very low, even though the peak power may be in the megawatt (MW) or gigawatt (GW) region. Clearly, there is a need to specify explicitly the pertinent pulse power exposure parameters. Nevertheless, there exists a few studies which attempted to quantify the relationship between biological changes and peak power and pulse repetition rate.

A behavior study involving rhesus monkey exposed at near resonant and above resonant frequencies showed that the performance of an animal trained to press a lever for food (observing -response) was impaired at a threshold of 514 kW/m^2 of 1.3 GHz energy pulsed at 370 pps with a pulse width of 3-microsecond, and 1.06 MW/m^2 of 5.8 GHz energy pulsed at 662 pps with a pulse width of 2-microsecond (deLorge, 1984). In all cases, the front surface of the upright, seated rhesus monkey was irradiated by a horizontally propagated, vertically polarized plane wave. These exposure conditions were associated with reliable increases in colonic temperatures typically in the range of 1°C above sham exposure levels.

That microwave pulses can serve as a discriminative cue in behavioral situations is supported by the works of several investigators (Frey and Feld, 1975; Johnson, et al, 1976; Hjeresen, et al, 1978). Food-deprived laboratory rats could be trained to make a specific response to obtain food during presentation of 150 kW/m^2 of 915 MHz energy pulsed at 10 pps with a pulse width of 10-microsecond (Johnson, et al, 1976). Similarly, rats tested in a two-compartment shuttlebox, where one compartment is exposed with 330 kW/m^2 of 2880 MHz microwave pulsed at 100 pps with a pulse width of 2.3-microsecond, and the other is shielded, spend a significantly higher percentage of time in the shielded side (Hjeresen, et al, 1978). Apparently, the rats found the microwave stimulus sufficiently aversive to exhibit an active avoidance response. It is interesting to note in both situations mentioned above, the animals showed continued ability to perform correctly when presented with conventional acoustic stimuli.

Pulsed microwaves have been shown to affect the action of a variety of psychoactive drugs. For example, 45 minutes of irradiation with 10 kW/cm^2 of 2450 MHz energy (SAR, 600 W/kg) pulsed at 500 pps with a pulse width of 2-microsecond enhanced apomorphine hypothermia and stereotypic behavior, morphine-induced catalepsy and lethality, but it attenuated amphetamine-induced hyperthermia (Lai, et al, 1983). Other specific and nonspecific effects of pulsed microwave on the actions of psychoactive drug and implications of the data regarding function of the nervous system can be found in a recent review (Lai, et al, 1987).

Using isolated rat lenses, a series of studies have found that irradiation in vitro with 918 MHz pulses of 10-microsecond width and 24 kW peak power delivered at different repetition rates produced histopathological damages at the lens equator (Stewart-DeHaan, et al., 1983, 1985; Creighton, et al., 1987). Although the threshold at which damage was observed in the lenses varied depending on the type of damage, the lowest SAR at which holes within the fiber cells in the equatorial region were observed occurred at 231 W/kg after 6 minutes of exposure. Moreover, the depth of damage was about 4.7 times as great as for unmodulated sinusoidal radiation. The actual ratio of damage from pulsed to CW radiation varies depending on total dose and decreases when either total dose is increased or peak power is decreased (Trevithick, J.R., 1988, private communication). It should be noted that extrapolation from in vitro to intact lens in the whole animal is speculative and difficult to substantiate. Nevertheless, these results suggest that high power pulsed microwave radiation is capable of causing lenticular damage that is not related to average temperature elevation.

MECHANISM OF INTERACTION

The mechanism(s) responsible for pulse-modulated RF interaction with biological systems is poorly understood. Several investigators have attempted to account for the responses from physical and physiological considerations (Adey, this volume; Lai, et al., 1987). While microwave-induced increase in thermal stress is clearly a contributing factor to some of the effects outlined above, a majority of the responses can not be easily related to average temperature elevation.

Absorption of high power pulsed microwave radiation can produce thermoelastic waves in biological tissues (Lin, 1978). Whether the microwave energy is delivered as a single pulse or a train of pulses, displacement and pressure are induced in target organs and propagate with a speed comparable to an acoustic wave in tissue (Lin, et al., 1988). The calculated peak displacement and pressure in a spherical model of animal or human head whose size varying from 20 to 70 mm irradiated with a 10-microsecond pulse of 918 or 2450 MHz energy at a peak SAR of 1 W/g$_2$ are shown in Table 5. It can be seen that a peak power density of 5 MW/m^2 of short pulse width radiation produced by newly developed high-peak-power microwave sources could induce in the adult human head a pressure increase and tissue displacement of 170 N/m^2 and 10 nm, respectively. The quantity of increase in pressure and displacement could conceivably cause physical damage to cell membranes and cytoplasm. Indeed, isolated rat lens has been found to displace by 10 nm when irradiated with a 10-microsecond microwave pulse having an energy density of 300 J/m^2 (Brown and Wyeth, 1983). It has been suggested that thermoelastic expansion and the resulting pressure waves in the lens is the most likely mechanism by which high-power pulsed microwave produce histopathological damage to the ocular lens (Creighton, et al., 1987).

Table 5. Peak Pressure and Displacement in Sherical Head Models Irradiated with 10 us Rectangular Microwave Pulses at a Peak Absorption Rate of one W/g

Sphere Radius (mm)	Microwave Frequency (MHz)	Species	Pressure (N/m^2)	Displacement $(10^{-4}nm)$	Incident Power (W/m^2)
20..........2450		guinea pig	0.408	2.16	4,450
30..........2450		cat, monkey	0.369	1.51	5,890
50..........918		human infant	0.961	9.34	12,820
70..........918		human adult	0.682	3.97	21,830

As mentioned previously, the microwave pulse induced hearing in humans arises from an interaction of microwave pulses with soft tissues in the head to initiate a thermoelastic wave of pressure that activates the inner ear receptors via bone conduction. While there is very little data regarding the effect on the hearing apparatus of exposure to microwave pulses, many factors in addition to microwave frequency that possibly influence the response including pulse shape, duration, peak power and pulse repetition rate. It is clear that threshold microwave auditory response would have insignificant effect on the hearing apparatus. However, the known effects of sound exposure in addition to hearing include the nonauditory, general physiological and psychological reactions.

The nonauditory effects of sound exposure are quite subtle compared with responses of the hearing apparatus. The reactions are in many aspects similar to general stress responses that can be elicited by such stimuli as pain and motion stress. Some of the bodily functions which have been reported to be affected by excessive sound exposure include respiration, digestion, and circulation. However, the most widely reported nonauditory effect of sound exposure is annoyance. In fact, criteria for limiting community noise are often based on the presence of annoyance reactions among exposed population groups (Kryter, 1970; Krichagin, 1978; Ahrlin and Ohrstrom, 1978).

Annoyance is influenced by such factors as attitude, motivation, physical surroundings, temperature, and a host of others. It generally refers to a reaction which is present after prolonged sound exposure and has been defined as a feeling of displeasure or a general adverse attitude toward a factor in the environment which the subject knows or believes could adversely affect its health or well being (Borsky, 1972).

Although annoyance reaction to microwave pulses has not been explicitly evaluated in humans or animals, the studies described above show that laboratory rats find the microwave auditory effect sufficiently annoying or aversive so that they are motivated to actively avoid the exposure (Frey and Feld, 1975; Hjeresen, et al., 1978). In fact, it can be shown that for the microwave parameters used, i.e. 2.3-microsecond wide 2880 MHz microwave pulses at 450 kW/m^2, peak power density, microwave-induced peak pressure level inside the rat's head is about 120 db. A value that is well within the hearing range and comparable to that found to be very annoying to humans.

The effects on psychoactive drug actions observed after pulsed microwave exposure may also be caused by an annoyance reaction. The auditory system could be the afferent sensory pathway that causes changes in brain functions and alteration in psychoactive drug actions. However, the present knowledge is far from adequate to unequivocally explain the altered drug effects (Lai, et al., 1987). It should be noted that microwave radiation has been speculated as a generalized stressor (Lu, et al., 1980).

CONCLUSIONS

The question of whether high power pulsed microwave poses a risk to the health of an exposed individual or under what conditions do effects become health hazards is highly pertinent and deserves urgent attention. While a meaningful consensus on the benignity or peril of pulsed microwave exposure is yet to be achieved, it is clear that exposure of laboratory animals and human subjects to pulsed microwave radiation can evoke physiological and psychological responses in the exposed subject. Moreover, these effects can occur at incident power levels that are at or below the existing ANSI C95.1-1982 guidelines for safe human exposure (Tables 2 and 6). While there is very little likelihood that the microwave auditory effect at threshold incident power can constitute a hazard, exposures at levels that are significantly higher than threshold will undoubtedly be very harmful to cell membranes, cytoplasm and whole organisms. Inasmuch as auditory effect signifies an effect on sensory function and lens damage represents an influence on tissue pathology, and both appear to stem from pulsed microwave-induced thermoelastic expansion of tissue, it seems reasonable to

Table 6. Biological Responses to Pulsed RF Energy

Responses	Exposure Duration (sec)	Incident Power (W/m^2)	Peak SAR (W/g)
Auditory Sensation	10^{-6}	4×10^5	160
Membrane Excitation	10^{-3}	4×10^8	
	10^{-5}	8×10^9	
Unconsciousness	0.1		280
Lens Damage	360 (10 us)	5×10^6	
Drug Interaction	27×10^3 (2 us)	4.5×10^4	6×10^5
Behavior Response	3600 (2.3 us)	3.3×10^5	9.1×10^6
	30 (10 us)	1.5×10^5	1.5×10^6

regard these as lower and upper bounds for a consideration of permissible limits of pulsed microwave exposure. The thresholds for behavior modification and drug interaction are lower by about one to two order of magnitude.

For example, the threshold of audibility of microwave pulses to humans is about 400 mJ/m^2 per pulse for pulse widths shorter than 30-microsecond regardless of peak power. Moreover, deformation (10 nm) of lens has been found to occur when irradiated with a 10-microsecond pulse having an energy density of 300 J/m^2 per pulse; the calculated pressure increase was as high as 170 N/m^2. These presumably caused the observed lens histopathology. It should be noted that the threshold for excitation of excitable membranes would be several orders of magnitude greater than the above mentioned values for the pulse widths of interest. Moreover, the startle and unconciousness responses observed in rodents occurred only for long pulses and in all cases, there were substantial temperature elevation (2-8°C). Thus, a quantity based on absorbed energy derived from between the values of 400 mJ/m^2 and 300 J/m^2 per pulse for pulse widths of 1 ms or less may be sufficient for protection against inadvertant exposure to pulsed RF radiation except for the microwave auditory effect.

Obviously, the above consideration is based on theoretical treatment and limited experimental evidence. The kinds of studies that would be useful are behavioral investigations of pulsed microwave exposed animals including effects on learning and performance, and morphological examinations of the central nervous system and hearing apparatus of exposed animal subjects.

REFERENCES

Ahrlin, V., Ohrstrom, 1978, Medical Effects of Environmental Noise on Humans, J. Sound and Vibrations, 59:79-87.

Bernardi, P., Blasi, F., and D'Inzeo, G., 1984, Validita Delli Attuali Normative di Sicurezza Nei Riguardi di Esposizioni a Campi Elettromagnetici Transienti, Atti della V Riunione Nazionale di Elettromagnetismo Applicato, St. Vincent, 25-27.

Bernardi, P., D'Inzeo, G., 1984, A Nonlinear Analysis of the Effects of Transient Electromagnetic Fields on Excitable Membranes, IEEE Trans Microwave Tech Theory, 32:670-679.

Borsky, P.N., 1972, Sonic Boom Exposure Effects: Annoyance Reactions, J. Sound and Vibration, 20:527-530.

Brown, P.V.K. and Wyeth, N.C., 1983, Laser Interferometer for Measuring Microwave-induced Motion in Eye Lenses in vitro, Rev. Sci Instrum, 54:85-89.

Chou, C.K. and Guy, A.W., 1982, Auditory Reception of Radiofrequency Electromagnetic Fields, J. Acoust. Soc. Am, 71:1321-1334.

Creighton, M.O., Larsen, L.E., Stewart-DeHaan, P.J., Jacobi, J.H., Sanwal, M., Baskerville, J.C., Bassen, H.E., Brown, D.O. and Trevithick, J.R., 1987, In Vitro Studies of Microwave-induced Cataract. II Comparison of Damage Observed for Continuous Wave and Pulsed Microwaves, Exp. Eye Res., 45:357-373.

deLorge, J.O., 1984, Operant Behavior and Colonic Temperature of Macaca Mulatta Exposed to Radiofrequency Fields at and above Resonant Frequency, Bioelectromagnetics, 5:233-246.

Florig, H.K., 1988, The Future Battlefield: a Blast of Gigawatts, IEEE Spectrum, 25:50-54.

Frey, A.H., and Feld, S.R., 1975, Avoidance by Rats of Illumination with Low Power Nonionizing Electromagnetic Energy, J. Comp. Physiol. Psychol. 89:183-188.

Guy, A.E., Chou, C.K., Lin, J.C. and Christensen, D., 1975, Microwave-induced Acoustic Effects in Mammalian Auditory Systems and Physical Materials, Ann. NY Acad. Sci., 247:194-218.

Guy, A.W. and Chou, C.K., 1982, Effects of High-Intensity Microwave Pulse Exposure of Rat Brain, Radio Sci., 55:169S-178S.

Hjeresen, D.L., Doctor, S.R., and Sheldon, R.L., 1978, Shuttle Side Preference as Mediated by Pulsed Microwave and Conventional Auditory Cues, in: "Electromagnetic Fields in Biological Systems," S.S. Stuchly, Ed., IMPI, Edmonton, Canada, 194-214.

Johnson, R.B., Meyers, D.E., Guy, A.W., Lovely, R.H. and Galambos, R., 1976, Discriminative Control of Appetitive Behavior by Pulsed Microwave Radiation in Rats, in: "Biological Effects of Electromagnetic Waves," C.C. Johnson and M.L. Shore, Eds., HEW Publication (FDA) 77-8010, 238-247.

Krichagin, V.J., 1978, Health Effects of Noise Exposure, J. Sound and Vibration, 59:65-71.

Kryter, K.D., 1970, The Effects of Noise on Man, New York, Academic Press.

Lai, H., Horita, A., Chou, C.K. and Guy, A.W., 1987, A Review of Microwave Irradiation and Actions of Psychoactive Drugs, IEEE Engineering In Medicine and Biology Magazine, 6:31-36.

Lai, H., Horita, A., Chou, C.K., and Guy, A.W., 1983, Psychoactive Drug Response is Affected by Acute Low-level Microwave Radiation, Bioelectromagnetics, 4:205-214.

Lenox, R.H., Meyerhoff, J.L., Gandhi, O.P., and Wray, H.L., 1977, Regional Levels of Cyclic AMP in Rat Brain; Pitfalls of Microwave Interaction, J. Cyclic Nucleotide Res., 3:367-379.

Lin, J.C., 1975, Interaction of Electromagnetic Transient Radiation with Biological Materials, IEEE Trans. Electromag. Compat., 17:93-97.

Lin, J.C., Wu, C.L., and Lam, C.K., 1975, Transmission of Electromagnetic Pulse into the Head, Proc. IEEE, 63:1726-1727.

Lin, J.C., 1976, Electromagnetic Pulse Interaction with Mammalian Cranial Structures, IEEE Trans. Biomed. Engg., 23:61-65.

Lin, J.C., 1977, Further Studies on the Microwave Auditory Effect, IEEE Trans. Microwave Theory Tech., 25:938-943.

Lin, J.C., 1978, Microwave Auditory Effects and Applications, C.C. Thomas, Springfield, IL.

Lin, J.C., 1980, The Microwave Auditory Phenomenon, Proc. IEEE, 68:67-73.

Lin, J.C., 1981, The Microwave Hearing Effect, in: "Biological Effects of Nonionizing Radiation," K.H. Illinger, Ed., Amer. Chem. Soc., Washington, D.C.

Lin, J.C., Su, J.L., and Wang, Y.J., 1988, Microwave-Induced Thermoelastic Pressure Wave Propagation in the Cat Brain, Bioelectromagnetics, 9:141-147.

Lu, S.T., Lotz, W.G., and Michaelson, S.M., 1980, Advances in Microwave-Induced Neuroendocrine Effects: the Concept of Stress, Proc. IEEE, 68:73-77.

Milroy, W.C., O'Grady, T.C. and Prince, E.T., 1974, Electromagnetic Pulse Radiation: A Potential Health Hazard?, J. Microwave Power, 9:213-218.

Postow, E. and Swicord, M.L., 1986, Modulated Fields and Window Effects, in: "CRC Handbook of Biological Effects of Electromagnetic Fields," C. Polk and E. Postow, Eds., CRC Press, Boca Raton, FL.

Schneider, D.R., Fell, B.T. and Goldman, H., 1982, On the Use of Microwave Radiation Energy for Brain Tissue Fixation, J. Neurochem., 38:749-752.

Stewart-DeHaan, P.J., Creighton, M.O., Larsen, L.E., Jacobi, J.H., Ross, W.M., Sanwal,M., Guo, T.C., Guo, W.W., and Trevithick, J.R., 1983, In Vitro Studies of Microwave-induced Cataract: Separation of Field and Heating Effects," Exp. Eye Res., 36:75-90.

Stewart-DeHaan, P.J., Creighton, M.O., Larsen, L.E., Jacobi, J.H., Sanwal,
    M., Baskerville, J.C. and Trevithick, J.R., 1985, In Vitro Studies
    of Microwave-induced Cataract: Reciprocity between Exposure
    Duration and Dose Rate for Pulsed Microwaves, Exp. Eye Res.,
    40:1-13.

PHYSICAL MECHANISMS FOR ELECTROMAGNETIC INTERACTION WITH BIOLOGICAL SYSTEMS

Paolo Bernardi and Guglielmo D'Inzeo

Department of Electronics
University of Rome "La Sapienza"
Via Eudossiana 18, 00184 Rome, Italy

INTRODUCTION

The physical interaction mechanisms are the basic mechanisms that underlie and control the interaction of electromagnetic (EM) fields with biological systems (Fig. 1). Because of the corpuscular nature of matter, primary interaction occurs at the microscopic level through forces and couples the local EM field generates on the electrically charged particles (ions and electrons) and on the electric dipoles of molecules of the biological medium; forces and couples that, in turn, lead to dynamic, electrophysical, and electrochemical consequences.

In agreement with this definition the physical mechanisms are the common basis of various subsequent processes that can be classified in different ways according to the produced effect. Figure 2 shows such a logical scheme with separate thermal and nonthermal pathways.

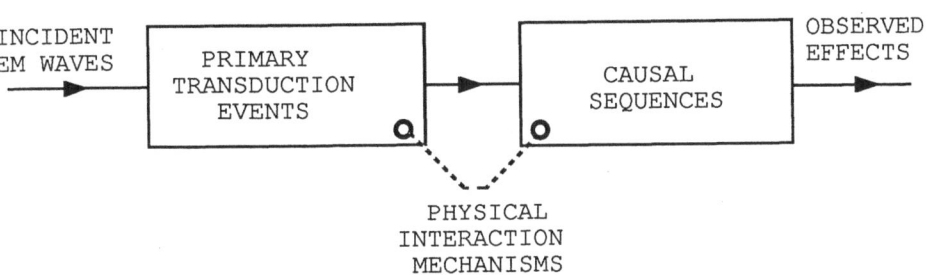

Fig.1. Electromagnetic interaction with biological systems: chain of events.

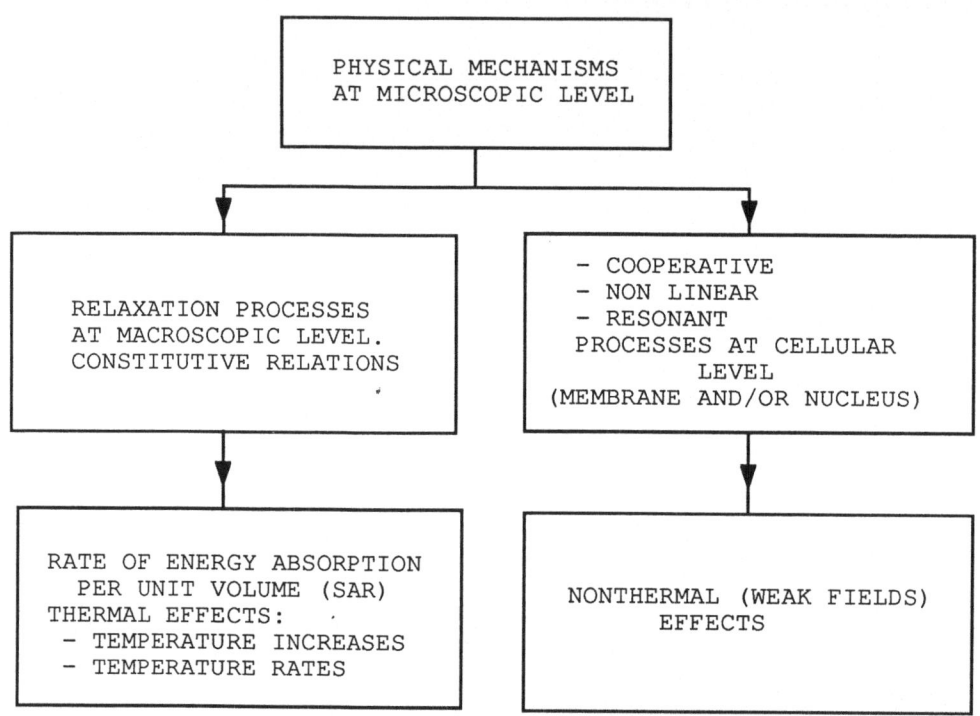

```
┌─────────────────────────────────┐
│   PHYSICAL MECHANISMS           │
│   AT MICROSCOPIC LEVEL          │
└─────────────────────────────────┘
```

Fig. 2. Link between microscopic mechanisms and biological
effects.

The two ways from physical mechanisms to biological
effects (thermal and nonthermal), even if apparently similar,
are really very different. Regarding the thermal path, the
macroscopic characterization of the dispersive properties of a
biological material allows the determination of the
constitutive relations and, therefore, the study of the
interaction using the classical electrodynamic approach
(Maxwell equations).

Concerning the nonthermal path the problems are more
complicated. For several nonthermal effects it has not been
possible to agree upon the physical mechanisms which underlie
either the primary transduction events or the consequences
leading from them to the observed effects. Furthermore, the
macroscopic characterization of a biological medium does not
contain all the information necessary for the study of the
intermediate processes of the interaction (Fig.2). These
considerations do not question the existence of weak field
effects, now widely accepted (Lerner, 1984). They only affirm
the necessity of more intense work on this argument, in order
to detect and control possible effects and to identify the
specific interaction mechanisms.

THE PHYSICAL INTERACTION MECHANISMS

From the point of view of the actions produced on charges
and dipoles, the fundamental vectors of the EM field are the
electric field **E** and the magnetic induction **B**. These fields
produce the following forces on an electric charge q:

$$F = q\,\mathbf{E} \tag{1}$$

$$F = q\,\mathbf{v}\,\mathbf{x}\,\mathbf{B} \qquad \text{(Lorentz force).} \tag{2}$$

where $\mathbf{v}$ is the charge's velocity.

In addition to the electrically charged particles (ions and electrons), a biological medium consists of molecules that, even if electrically neutral, are composed of couples of equal and opposite electric charges separated by a distance l. These charges form a dipole characterized by a vectorial quantity, the dipole moment, defined as:

$$\mathbf{P_e} = q\,l\,\mathbf{l_0} \tag{3}$$

being $\mathbf{l_0}$ the unit vector directed from the negative to the positive charge. An electric field $\mathbf{E}$ exerts a torque on a permanent dipole given by:

$$\mathbf{T_e} = \mathbf{P_e}\,\mathbf{x}\,\mathbf{E}. \tag{4}$$

This torque tends to align the dipole in the direction of the applied field. In a similar form the magnetic induction $\mathbf{B}$ generates a torque on a magnetic dipole of moment $\mathbf{P_m}$ such that

$$\mathbf{T_m} = \mathbf{P_m}\,\mathbf{x}\,\mathbf{B}. \tag{5}$$

Finally a nonuniform electric field generates a force (dielectrophoretic force) on an electric dipole given by:

$$\mathbf{F_d} = \nabla\,(\mathbf{P_e}\cdot\mathbf{E})_{\theta\,=\,\text{cost}} \tag{6}$$

where θ is the angle between the dipole moment and the direction of the electric field at the center of the dipole. This force exists only if the electric field is nonuniform.

In general, an electromagnetic field characterized by the vectors $\mathbf{E}, \mathbf{B}$ acts at microscopic level on the atoms, ions, electrons, and on the molecules that constitute the biological medium. Such actions induce movements of charges, vibrations of dipoles, energy dissipations, alterations of chemical process rate constants, that can be studied using the integro-differential equations which control the particular process considered.

RELAXATION PROCESSES AND THE MACROSCOPIC PERMITTIVITY OF BIOLOGICAL MATERIALS

The study of the thermal effects needs a macroscopic characterization of the interaction processes involved. The link between the microscopic interaction and the macroscopic model can be obtained considering the particle properties valid on the average of the particles contained in a unit volume of the medium. Such hypothesis allows us to obtain the constitutive relations for the examined medium and, as a consequence, to define the auxiliary field vectors electric displacement $\mathbf{D}$, magnetic field $\mathbf{H}$, and electric-current density $\mathbf{J}$ from the primary vectors $\mathbf{E}$ and $\mathbf{B}$ of the EM field. Knowing all the vectors of the field, the study of thermal effects induced by energy absorbed in the medium is possible using Maxwell's equations.

The electric parameters used in the macroscopic characterization of the interaction are the complex relative dielectric constant (relative permittivity):

$$\varepsilon^* = \varepsilon' - j\varepsilon'' \tag{7}$$

which in a given biological material depends on frequency and temperature, and the relative permeability generally considered equal to one. The dispersive properties $\varepsilon'(\omega)$ and $\varepsilon''(\omega)$ of different kinds of biological tissues in the range from extremely low frequency (ELF) to microwaves have been the object of several studies. A complete review of this subject can be found in Foster and Schwan (1986) and in Pethig (1979).

In this paragraph, as a simple example, we recall the methodology of passage between microscopic interaction and macroscopic characterization of a dielectric medium whose polarizability arises from electronic polarization (Collin, 1966). The simple model of an atom consisting of a nucleus with positive charge +q surrounded by a spherical electron cloud with a total charge -q (Fig. 3) is considered. The application of an electric field **E** displaces the electron cloud with respect to the nucleus of a quantity x in the direction opposite to the field. A restoring force proportional to this displacement will oppose it. In addition, dissipative processes are present and give rise to a viscous force proportional to the displacement's velocity.

If m is the mass of the electron cloud, the dynamic equation of motion induced by the applied electric field can be obtained by equating the sum of the inertial, viscous, and restoring force to the applied force, namely :

$$m\frac{d^2x}{dt^2} + m\nu\frac{dx}{dt} + kx = qE(t) \tag{8}$$

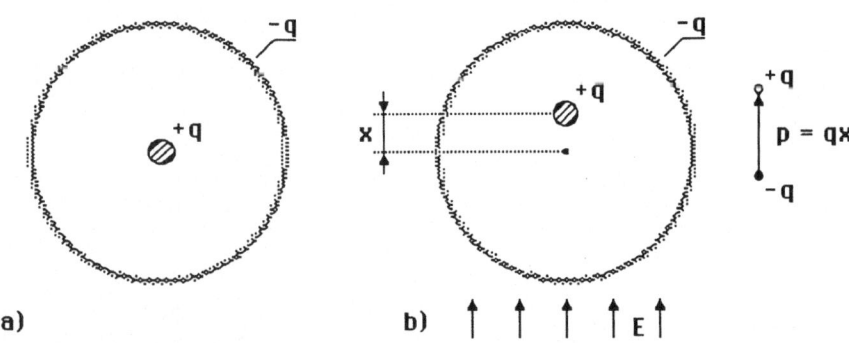

a)   b) ↑ ↑ ↑ ↑ E ↑

Fig. 3. Model of an atom with a nucleus +q surrounded by an electron cloud -q.

As a consequence of the charges' displacement a dipole moment with the same direction of the applied field is induced:

$$p_e = qx.$$

Indicating with N the number of these atoms per unit volume the macroscopic dielectric polarization P is:

$$P = N p_e = N qx.$$

Equation (8) can now be written at macroscopic level for the polarization P:

$$\frac{d^2 P}{dt^2} + v \frac{dP}{dt} + \omega_p^2 P = \frac{Nq^2}{m} E(t) \tag{9}$$

where

$$\omega_p^2 = k/m. \tag{10}$$

Solution of (9) in the frequency domain gives the following dispersive expression for the complex permittivity:

$$\varepsilon^*(\omega) = \varepsilon' - j\varepsilon'' = 1 + \frac{N q^2}{\varepsilon_0 m (\omega_p^2 - \omega^2 + j\omega v)} \tag{11}$$

In the time domain the response to an electric field step applied at t=0 (i.e. $E(t) = E u_{-1}(t)$) is

$$P(t) = \frac{Nq^2 E}{m\omega_p^2} \left(1 + \frac{p_2}{p_1 - p_2} e^{p_1 t} - \frac{p_1}{p_1 - p_2} e^{p_2 t} \right) \tag{12}$$

in which p_1 and p_2 are the roots of the characteristic equation of (9).

The previous formulation shows the dielectric response of a second order system in the frequency (Eq. 11) and time (Eq. 12) domains. In biological materials the polarization induced by an electric field step generally relax toward the steady-state as a first order process characterized by an unique relaxation time. It is interesting to show how, under certain approximations, the first order response can be obtained by Eqs.11 and 12. In fact, if $\omega_p \ll v/2$, the two roots p_1 and p_2 in Eq.12 are both real, negative, and considerably different from each other. As a consequence, the two relaxation time constants will also be considerably different. Neglecting the relaxation term associated with the higher (in absolute value) root, that corresponds to a practically instantaneous response, an expression with an unique time constant is obtained. In the frequency domain, if $\omega \ll v$, Eq.11 reduces to:

$$\varepsilon^*(\omega) = 1 + \frac{Nq^2}{\varepsilon_0 m \omega_p^2 \left(1 + j\omega \frac{v}{\omega_p^2} \right)} \tag{13}$$

This equation has the form of the dielectric response of a first order process and is typical for a material with permanent dipole moments. In this case the complex permittivity is given by:

$$\varepsilon^* = \varepsilon_\infty + \frac{\varepsilon_s - \varepsilon_\infty}{1 + j\omega\tau} \qquad (14)$$

commonly known as the Debye dispersion formula.

If the first order relaxation process does not have a unique relaxation constant but a distribution around a single mean value, different expressions of $\varepsilon^*(\omega)$ are needed. The widely used expressions include: the Cole-Cole dispersion equation which can be applied if the relaxation-time distribution is symmetric around the mean value,

$$\varepsilon^*(\omega) = \varepsilon_\infty + \frac{\varepsilon_s - \varepsilon_\infty}{1 + (j\omega\tau)^{1-\alpha}} \qquad (0 < \alpha < 1) \qquad (15)$$

and the Davidson-Cole dispersion equation which can be applied if the relaxation-time distribution is non-uniform around the mean value,

$$\varepsilon^*(\omega) = \varepsilon_\infty + \frac{\varepsilon_s - \varepsilon_\infty}{(1 + j\omega\tau)^{1-\alpha}} \qquad (0 < \alpha < 1) \qquad (16)$$

More generally the complex permittivity can be expressed by:

$$\varepsilon^*(\omega) = \varepsilon_\infty + (\varepsilon_s - \varepsilon_\infty) \int_0^\infty \frac{G(t)}{1 + j\omega t} \, dt \qquad (17)$$

where $G(t)\,dt$ is the amount of molecules having a relaxation time between t and t + dt.

Relaxation Processes in Tissues with High Water Content

The dispersion curve of the real part ε' of the complex permittivity for high water content tissues is characterized by three main relaxation processes that are well separated in frequency. The dispersion regions of these three processes are usually identified as α, β and γ regions (Fig. 4). The relaxation mechanisms responsible for the three dispersion processes are summarized in Table 1. These mechanisms have been examined extensively in the reviews previously cited and for this reason will not be discussed here. We wish to focus our attention instead on the relaxation processes of the bound water zones which produce a new dispersion region (δ-dispersion) in the frequency range between the β and γ dispersions. The study of this process begins with the experimental evidence that in the range between 100 MHz and some gigahertz the energy density dissipated by the EM field is higher in bound water than in free water. Such results have been studied closely by examining water solutions of different biomacromolecules (Grant, 1982; Grant et al.,1986).

184

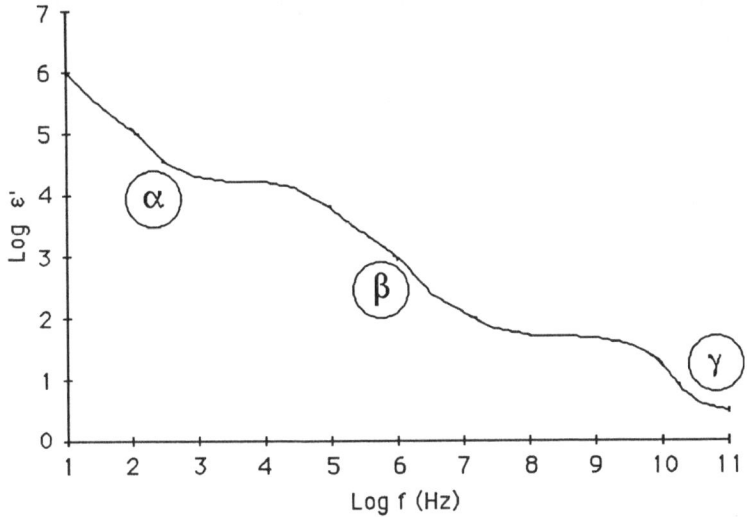

Fig.4.The dispersion curve of ε' for a high water content tissue

It is known that for numerous globular molecules in dilute aqueous solutions, bound water forms a layer 1-2 water molecules thick (~0.3 nm) around the macromolecules. This thickness is roughly constant and does not depend on the dimension of the macromolecule. As a consequence, in two different solutions with the same concentration, the amount of bound water is higher in the solution with the macromolecules of smaller dimension.

Table 1.Relaxation Mechanisms for a High Water Content Tissue

Dispersion curve	Relaxation mechanism	Center frequency(*) ($\omega\tau$= 1)
α	Mobile counterions associated with fixed charges on cell membranes (Schwarz theory)	80 Hz
β	Charging of the cell membrane through electrolytes (Maxwell-Wagner theory)	$\cong$50 kHz
γ	Dipolar reorientation of free water	25GHz

(*) Center Frequency at body temperature

It is then possible, studying solutions of different macromolecules, with different concentrations, to analyze the amount of bound water and to separate the effects of the dipolar relaxation of free water (γ region) from those of the macromolecules (β region).

The analysis of the δ-dispersion region has shown that increasing the amount of bound water, the relaxation frequency decreases with respect to that of free water, approaching the β region when the bound water is in a large amount. The interaction mechanisms underlying the δ dispersion are not completely clear. However some hypotheses have been advanced and are summarized in Table 2 together with the data taken from Grant's paper (Grant et al., 1986). It is necessary to remark that studies concerning bound water in whole biological tissue are more complicated, mainly because of the presence of membranes and the consequent large β dispersion range which extends up to several tens of MHz.

Table 2. Dielectric Behavior of the Aqueous Solutions of Three Widely Differing Macromolecules

Macromolecule conformation	δ-dispersion (T=25°C) relaxation frequency	Mechanism
Myoglobin Oblate spheroid of axial ratio 2:1 Equivalent sphere radius: 1.53nm.	15.4 ± 5.5 MHz	The bound water can be resolved into two categories: 50% of bound water has solute-solvent bonds stronger than that of normal hydrogen bonds
	4.0 ± 0.7 GHz	50% of bound water follows the Maxwell-Fricke mixture theory
LDL Spherical shape of radius 11 nm.	12.0 ± 0.1 GHz	Maxwell-Fricke mixture theory
PVP Flexible chain.	The separation of the dispersion regions resulting from PVP and bulk water is not well distinct.	The dielectric behavior of solutions of long chain molecules cannot be interpreted in terms of different forms of water present

LDL = Human serum low-density lipoprotein
PVP = Polyvinylpyrrolidone

The study of the dielectric dispersion of biological macromolecules has been recently extended up to the millimeter (mm) and sub-mm wave lengths (Genzel et al., 1983; Kremer et al., 1984). On the basis of the results obtained an effort to identify the interaction mechanisms in these frequency ranges has been done. The analysis has been performed with different spectroscopic techniques (Genzel et al., 1983): cavity-perturbation (10 GHz) and untuned-cavity (50 - 150 GHz). The range of temperatures examined was from 4 to 300 °K. In these temperature and frequency ranges the absorption coefficient $\alpha = \dfrac{\omega \varepsilon''}{nc}$ has been measured (n is the real part of the complex index of refraction and c the light velocity in vacuum). The results obtained for dried haemoglobin and dried lysozyme (Kremer, 1984) are shown in Figs 5 and 6.

Focusing our attention on the results obtained for the dried haemoglobin it can be concluded that the observed frequency and temperature dependence cannot be explained with a single Bebye relaxation term whose relaxation frequency is assumed to vary according to the theory of solids:

$$f_\tau = f_\infty e^{-U/kT} \qquad (18)$$

where U is the potential energy barrier separating the two equal energies corresponding to the two possible orientations (parallel and antiparallel) of the dipoles (symmetrical double well, Fig. 7a)

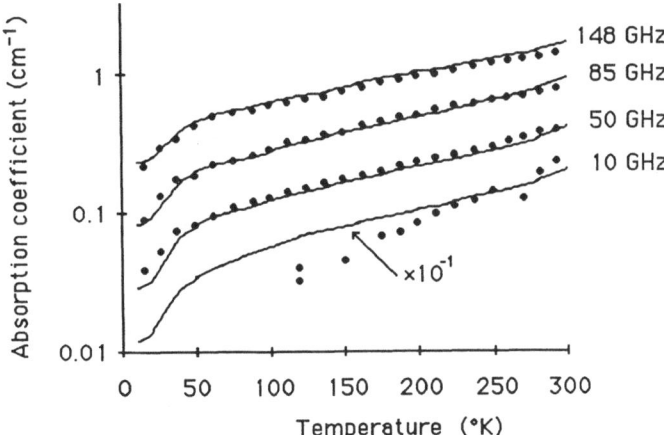

Fig.5. Absorption coefficient of dried haemoglobin from 4 to 300 °K. Experimental data (dots) and theoretical model (line).

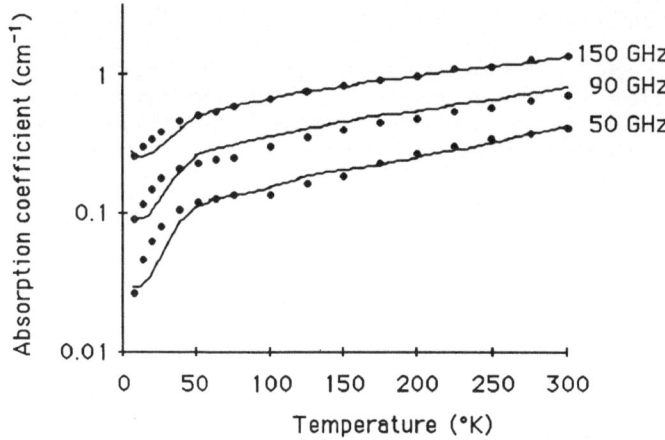

Fig.6. Absorption coefficient of dried lysozyme from 4 to 300 °K. Experimental data (dots) and theoretical model (line).

However, if the equilibrium positions of the dipoles are assumed to be unequal (asymmetric double well, Fig. 7b), the temperature-dependence of the relaxation frequency is

$$f_\tau = \frac{1}{2} f_\infty \left(e^{-\frac{U}{KT}} + e^{-\frac{U-U_2}{KT}} \right) \qquad (19)$$

Under this assumption the experimental results (dots in Figs 5 and 6) can be fitted with three Debye-type relaxation terms. The parameters for fitting the haemoglobin data (Fig. 5) are given in Table 3 (Kremer, 1984).

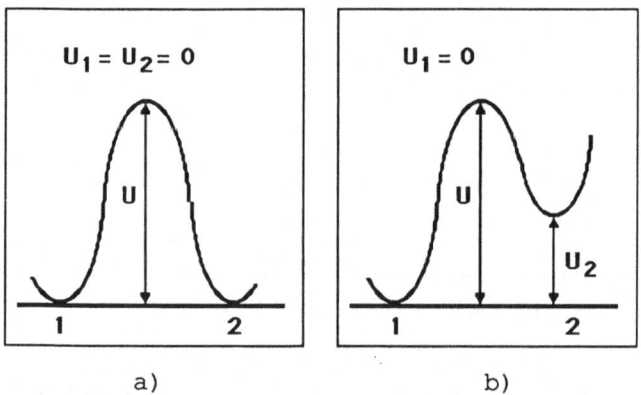

Fig.7. Potential energy barrier: a) Symmetrical double well; b) Asymmetric double well.

Table 3. Implied Mechanisms (for dried haemoglobin)

Relaxation process	High temperature relaxation-frequency limit f_∞ (GHz)	Potential energy barrier U (Kcal/mol)	μ [*] (debye)
1	370	0.29	0.21
2	350	1.11	0.59
3	300	3.60	3.70

(*) μ is the modulus of the difference between the dipole moments in the two positions.

The values in Table 3 lead us to some useful considerations about the microscopic interaction mechanisms. Considering in particular the relaxation process 3, the depth of the potential energy barrier is only slightly below the binding energies of the hydrogen bonds in the NH....OC bridges existing in haemoglobin. Moreover, the NH....OC bridge has a dipole moment of 3.7 debye that is the same value of the variation of the dipole moments in the relaxation process. This suggests (Genzel, 1983) that process 3 can be considered as a disruption of the hydrogen bond which will be changed into a weaker van der Waals bond.

LOW AVERAGE ENERGY, HIGH PEAK POWER THERMAL EFFECTS

Thermal effects produced by microwave energy absorption in biological tissues are principally of two types: high energy density effects consisting essentially in a temperature increase of the irradiated tissues and low average energy, high peak power density effects with a negligible but rapid tissue temperature rise. While for the first group we refer to the numerous literature on the subject (e.g. Baransky and Czerski, 1976; Michaelson and Lin, 1987), for the second we wish to mention one of the most interesting and widely recognized effects: the microwave-induced auditory effect(Frey and Messenger, 1973; Lin, 1977a and b, 1978). When human subjects are exposed to rectangular pulses of microwave radiation with a low average power density, a sound is perceived, apparently originated from within or behind the head. Many investigators, trying to identify the responsible mechanisms, have attempted to understand the effect with physical and physiological considerations.

The thermoelastic expansion of brain tissue has shown to be the most effective mechanism in explaining the microwave hearing phenomenon being three orders of magnitude greater than the other possible mechanisms (Lin, 1977b). A quantitative analysis of the acoustic signals generated in spherical brain models exposed to microwave pulses shows that very low (~5x10^{-6} °C) but rapid rises (~10μs) of temperature induced in the brain by microwave energy absorption create thermoelastic expansion of the brain matter that consequently launches an acoustic wave of pressure (Lin, 1977a and b). This theory also shows that there are an infinite number of resonant frequencies corresponding to the different vibration modes of the spherical brain; it further describes the acoustic waves (frequency, pressure and radial displacement) as functions of the head size and the incident wave characteristics.

Direct measurements carried out on mammalian brain (Olsen and Lin, 1983) irradiated with pulsed microwaves confirm the theoretical predictions for the first fundamental frequencies of vibration while higher order vibrations deviate from the predictions based on the homogeneous model of the head. It has been postulated by Olsen and Lin that the skull, as a thin shell of spheroidal form around the brain could contribute to the modification of the higher frequency vibrations.

The effects of pulsed field with high peak-intensity but low average power constitute the logical link between thermal and nonthermal effects. The microwave auditory effect is recognized as a low level thermal effect since the primary and most effective transduction event is the energy absorption and the consequent very rapid rise of temperature. However it may be expected that transient fields with low average energy but very high peak power interact with biological materials inducing at microscopic level local fields that give rise to a series of essentially nonthermal consequences as those discussed in the following paragraphs.

THE CELL AS THE MAIN SITE OF NONTHERMAL INTERACTION

Literature on nonthermal effects is abundant. Several results have been reported, numerous books and reviews published (Schwan and Foster, 1980; Adey and Lawrence, 1984; Chiabrera et al., 1985; Polk and Postow, 1986; Michaelson and Lin, 1987). The effects have been observed in vivo and in vitro at different levels of biological complexity. It is beyond the scope of this work to analyze the effects, but any identification of the mechanisms is based on the particular group of effects observed and on the biological sites where they are revealed. In Table 4 the different effects of the EM interaction have been classified according to the biological level where they have been observed.

During the last ten years great efforts have been applied to recognize in the observed effects a unique set of mechanisms in order to allow a unified description of the phenomena.

Table 4. Level of Nonthermal Biological Effects

Entire body	(behavioral, genetic, embryonic growth)
Physiological system	(nervous, cardiovascular, immune, endocrine)
Organs	(heart, brain, liver..)
Cells	(neurons, limphocites, vegetal..)
Molecular	(biochemical reactions, conformational changes)

The principal conclusion that can be drawn is that the cell is the primary site of interaction. Most of the hypothesized mechanisms in literature refer to alterations of the cell's behavior in terms of changes in its structure, in it's response to electrical or chemical stimuli, in the cellular capacity of growing or growth.

The effects at biological levels more complex than the cellular one can almost always be related to structural modifications or alterations of the communication processes between the cells that form the biological system or control the organ. On the other hand some molecular alterations are transferred to cellular level by the cell's capacity of integration and interpretation of biochemical signals.

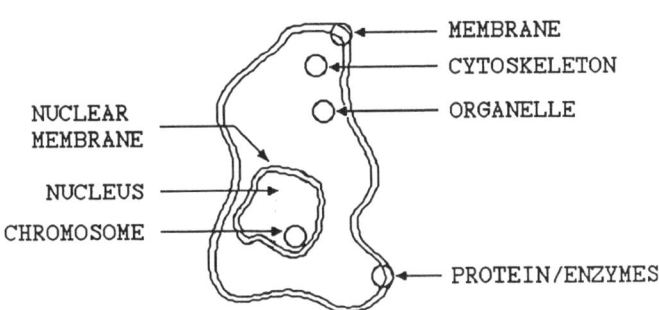

Fig.8. Main components of the cell

Some care must be used in the study of cell behavior. The cell is a complex biological system composed of different parts that can communicate with the outside and also with each other. Figure 8 shows the structure of a cell with the identification of its main components. The understanding of the cell structure is fundamental in the identification of the mechanisms of interaction. Very often, as it will be shown in the following, the field effects revealed in one site are caused by alterations induced in other sites of the cell. Several effects are based on a succession of biophysical phenomena that modify the internal activities of the cell. In the following we will analyze the main hypothesized mechanisms focusing particularly on: where the field can act, where the effects are induced, and which steps permit us to relate the field action to the effect.

THE HYPOTHESIZED MECHANISMS

Several theoretical studies have been developed in order to identify the basic mechanisms underlying the interaction of weak fields with biological matter and especially with cells. A comparison between the quantum of energy associated with the EM field, even in the millimeter range, and the energy associated with bonds between particles in biological materials shows that they are notably different. As an example, the difference between the binding energies of the van der Waals (about 0.06

eV per molecule) or the hydrogen (0.16 eV per molecule) bonds, two of the lowest energy bonds, and the quantum of energy in the millimeter range ($1.24*10^{-3}$ eV at 300 GHz) is of two orders of magnitude. This consideration proves that, if some effect does take place, it must be mediated by an energy conversion, an amplification, a resonant phenomenon or a cumulative process. Some of these mechanisms can be found in the electronics theory and practice, every time an amplification, a generation or an increase of the signal to noise ratio must be produced. From this point of view an electromagnetic field could interact with a biological system just as a signal acts on an electronic system and, through a conversion of energy, by filtering, or through a cumulative or a triggering process, the system modifies and reacts to the incoming signal.

We divide the mechanisms underlying nonthermal effects in three main categories according to the kind of process involved:

a) Resonant interactions
b) Nonlinear interactions
c) Cooperative interactions of Electric and Magnetic fields.

Several hypothesized theories suppose the presence of more than one of these different types of processes, for example, some of the resonant processes in biological systems are also accompanied by nonlinear interactions. In these cases to classify the interaction we will choose the process that is most directly influenced by the basic mechanism; i.e. the process were an action on particles (ions or molecules) can be identified. The alterations induced by this first action are considered as a perturbation of the biological system that, in turn, can integrate, transform, and transmit it's effect to other processes, including induction of the observed final effect.

Looking at these categories from an electronic point of view we observe that these interaction induced processes are frequent in the practice of an electronic engineer; only as an example we list some electronic devices and circuits that work on the basis of the interaction processes mentioned: a) oscillators and filters; b) detector diodes and transistors c) magnetrons and travelling-wave tubes. These devices are able to induce effects far more detectable than the input signal using the energy provided or stored by the electronic system itself. Not forgetting the biophysical nature of the systems and the nature of the applied actions, we believe that this approach may help to understand the intrinsic nature of the involved mechanisms.

Resonant Interactions

Some of the mechanisms described in literature refer to resonant interactions between the electromagnetic field and biological systems. In particular, the electromagnetic field can modify the spontaneous activity of a resonant phenomenon bringing it to a different oscillating situation, or the field can couple with the system at a resonant frequency and transfer energy to this oscillating process.

One of the theories that considers a nonlinear resonant interaction has been proposed by Frohlich (1968-84). Three different types of interaction were supposed (1984): a) coherent excitation of membrane characteristic vibrations in the range of 100-300 GHz; b) excitation of metastable polar states present in biological reactions and c) alterations of self-oscillating biological processes induced by the two previous kinds of processes. This theory is mainly based on the hypothesis that a biological system like a cell composed of polar molecules and proteins that auto-oscillate at a characteristic frequency as a consequence of a particular metabolic situation (e.g. during duplication) could be resonantly stimulated by an applied field. The applied EM field could behave as a selective trigger supplying the small amount of energy necessary to start-up a biological modification.

Mathematical support to this approach has been given by Kaiser in his studies on the modification of limit cycles in oscillating systems (1980-1984). Experiments supporting Frohlich's theory have been given by the works on cell absorption or growth in the millimeter wave region (Webb and Dodds, 1968; Keilmann, 1982; Grundler et al., 1977,83) (see recent review by Postow and Swicord, 1986). However we agree with these last authors' conclusion that the main problem in these experiments "is the lack of attempts or the inability to replicate or substantiate the reported results by a second investigator". Moreover, while the works from Grundler, Keilmann and coauthors (1977-1985) have not yet been confirmed by other groups, a recent paper from Furia et al. (1986) denies the possibility of "any detectable effect on either the growth rate or the viability of yeast cells exposed for 4 h to ultrastable millimeter waves between 41.650 and 41.798 GHz".

Another theory based on resonant interactions has been proposed to justify the genetic effects produced by RF and MW absorption (Postow and Swicord, 1986). In 1983 Swicord et al. reported the evidence of an increase of energy absorption at the X-band in soluted DNA with a wide distribution of molecular chain lengths. In order to justify such results they hypothesized that the irradiated DNA chain can dynamically resonate creating stationary waves along its longitudinal main axis, as a string fixed at its ends, and, as a consequence, can absorb energy at microwave frequencies. The mechanism is highly selective at a frequency value dependent on the length of the DNA chain and on the viscous bonds between the cells' components (Kohly et al.,1981; Swicord et al., 1983-84; Van Zandt et al., 1982). Some experiments supporting the validity of this theory (Edwards et al. 1984, 1985) have been recently questioned (Epstein et al., 1987; Gabriel et al., 1987).

In order to explain the alterations of the transmembrane voltage due to MW exposure (Wachtel et al., 1975; Seaman and Wachtel, 1978) a resonant mechanism has been proposed by Pickard and Rosenbaum (1978). They start analyzing the behavior of a gating molecule that is supposed to control the opening of a ionic channel. This molecule of mass m and charge q, tethered at one end, is held at its initial position by the E-field associated with the membrane resting potential. A thermal collision gives it the energy necessary to move into the region of opposing field and to reach a maximum deflection.

Subsequently the particle goes back into its initial position. The frequency of this oscillation is given by $f_0 = \dfrac{qE}{2\sqrt{mkT}}$ and, for a tetravalent particle of molecular weight around 600, this frequency is in the range of tens of GHz. An exposure at the resonant frequency f_0 could produce larger displacements of the gating molecule increasing the probability of driving the channel "on", and altering the percentage of channels that are in a conducting state. As a consequence, the global membrane permeability and the transmembrane voltage could be altered.

Nonlinear Interactions

Nonlinear interaction processes have been proposed to justify several nonthermal effects induced at cellular level by exposure to CW or modulated fields. The principal effects are: alteration of the transmembrane voltage, alteration of the membrane conductance, alteration and synchronization of the firing activity in neuronal and pacemaker cells. These effects, reported in literature since the beginnings of the 70' (Lords et al., 1973; Wachtel et al., 1975; Arber, 1976; Tinney et al.,1976), have been confirmed in the following years (Seaman and Wachtel, 1978; Arber, 1981; Barsoum and Pickard, 1981; Campbell and Brandt, 1981; Seaman et al., 82; Arber and Lin, 1983-85; Caddemi et al.,1986; Galvin and Mc Ree, 1986).

One of the first effects explained on the basis of a nonlinear approach was the variation of the firing activity in isolated neurons exposed to microwave fields. These alterations, in opposition to those induced by thermal heating, can be justified with the existence of a rectifying mechanism in or around the membrane that can create a low frequency current that, flowing through the membrane, modifies it's firing activity.

Barnes and Hu (1977) examine the effect of RF and MW fields on the ions' concentrations on the two opposite sides of a membrane. In unexposed conditions the equilibrium concentrations of the ions are maintained different by a potential barrier according to the nonlinear Nernst equation and no net current crosses the membrane. If an ac field is superimposed on the static field existing at the resting state, the equilibrium concentrations are changed and a net current crosses the membrane. This nonlinear process causes a variation of the dc membrane potential and, if stimulated by a modulated signal, will generate the sums and the differences of the signal frequency components. The difference signal assumes particular importance if it's frequency is close to the natural rate of a nerve system. Barnes and Hu determined the field values at membrane level using the nonlinear steady-state I-V characteristic for a squid axon membrane (Fig.9) and showed that an external field of about 200 V/m in the gigahertz region can induce a voltage drop across the membrane of about $9\mu V$ and a rectified current of 6×10^{-11} A/cm^2 that corresponds to an ion flow of about 4×10^8 ions/(cm^2·s) or 400 ions/s for a cell with a 10^{-6} cm^2 surface area.

Another model for frequency conversion considers the voltage-dependent membrane capacitance (Berkowitz and Barnes,

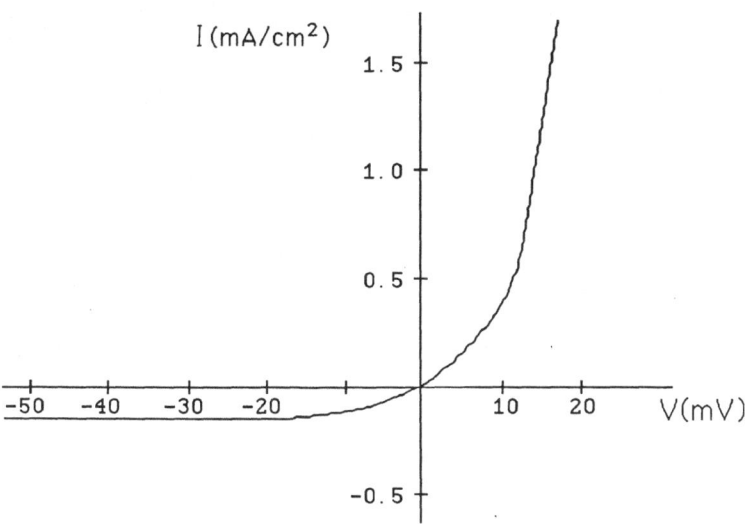

Fig.9. The steady-state current-voltage characteristic for a squid axon membrane according to Barnes and Hu (1977).

1979). If the nonlinear capacity of the membrane is assumed of the type

$$C_m = C_0 + aV + bV^2 \tag{20}$$

an applied field (RF or MW), amplitude modulated, gives rise to a signal mixing and, in particular, a current at the frequency of the modulating signal flows through the membrane. According to this model, typical pulsed fields in the vicinity of a radar antenna induce across the membrane a current density of about 5×10^{-10} A/cm^2 at a frequency of 10 Hz.

A direct rectification of RF incident fields by the cell membrane has also been hypothesized by Pickard and Rosembaum (1978) with particular attention to the cutoff frequency of the response due to the transit-times of the ions that cross the membrane. Barsoum and Pickard (1981) identified this rectification process in Characean plant cells separating the thermal response of the membrane potential from the nonthermal one. A field of about 6000 V/m produced on electrogenic cells a depolarization of 140μV at 250 kHz with a cutoff frequency of 10 MHz and an hyperpolarization of 1100 μV at the same frequency with a cutoff of 5 MHz on nonelectrogenic cells (Barsoum and Pickard, 1982). The values of these cutoff frequencies are in the same order of magnitude as those predicted by the theory, while it is more difficult to justify the opposite voltage effects in electrogenic and nonelectrogenic cells, probably due to differences in the metabolic processes.

A simple spherical cell, with a nonlinear membrane, embedded in an infinite homogeneous medium has been considered by Franceschetti and Pinto (1984) in order to evaluate the effects of RF harmonic fields. They expand the membrane current density in a Volterra series describing the most general nonlinear non-permanent relationship between the current density and the transmembrane excess potential (the change from the potential difference at the resting state). The transmembrane excess potential is computed following a quasi-static approach by solving a nonlinear boundary value problem, wherein the boundary conditions are obtained by a nonlinear membrane model. They conclude that transmembrane excess potentials of the order of 100 μV dc should be expected below 100 MHz at about 100 V/m inside the tissue.

The nonlinear process modeled by Hodgkin and Huxley (HH) equations, linking the current density to the transmembrane voltage for a squid axon (1952), has been assumed by various authors (Cain, 1980,1981; Bernardi and D'Inzeo, 1984; Casaleggio et al., 1984) as a rational tool to explain nonthermal effects at membrane level. The cited works schematize the influence of an EM field on the electrical parameters of the HH model in different ways.

Cain (1980) supposes the irradiating field to be the cause of a variation of the instantaneous voltage across the membrane. This variation shifts the mean level of the rate constants that control the sodium and potassium conductances of the ionic channels. A dc displacement of the membrane potential from its resting value is the final consequence of this process that is supposed to occur without any change of the total current through the membrane. As a biophysical support of this approach Cain (1981) suggests a mechanism of modification of the channel open and closing probabilities similar to the one discussed by Pickard and Rosenbaum (1978): the oscillating component of the membrane potential can interact with the voltage sensing charged groups of the protein macromolecules that form voltage-sensitive ion channels. This nonlinear interaction results in alterations of the channel conductance and, as a consequence, in an inhibition of the neuronal cell activity.

Casaleggio et al. (1984) and Marconi et al.(1985) used the Volterra method for the solution of the nonlinear HH equations. The first order term of the Volterra expansion is used in order to evaluate the different currents crossing a spherical model of a neuronal cell. This approach gives as a result a modification of Na$^+$ fluxes through the membrane due to low-level low-frequency fields.

The influence of transient EM fields at membrane level is evaluated with a different approach by Bernardi and D'Inzeo (84-86). They assume that the current density induced inside the tissue by an external field crosses the membrane causing an alteration of the ionic currents and, as a consequence, modify the transmembrane voltage. In this model the field is supposed to interact with the charged particles outside the membrane inducing a net current through it. This current is therefore considered as the exciting term and the transmembrane voltage as the response in the time domain solution of the HH equations.

Similar calculations have been performed for snail neurons on a HH-like model proposed by Connor and Stevens (1971) in order to compare this theoretical model with some experimental

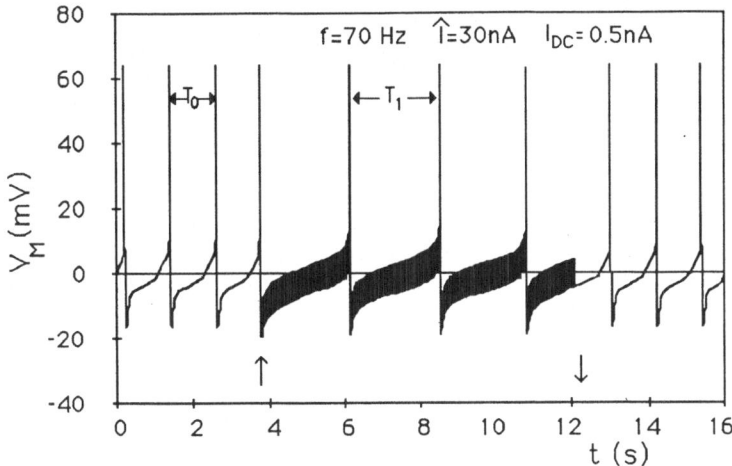

Fig.10. Calculated temporal behavior of the transmembrane voltage of a neuronal cell.

results obtained by applying ELF currents directly to the cell(Bernardi et al.,1985a).This neuronal cell is a pacemaker cell with a self sustained firing activity and the induction of currents through it can modify its firing frequency. In the model the unperturbed firing activity is simulated by applying a dc current I_{DC}. In Fig. 10 the low frequency perturbing current is applied between the arrows and the effect is a reduction of the firing activity ($T_1 > T_0$). However since the ionic currents behavior is controlled by the ac-current amplitude $\hat{I}$ and frequency (Bernardi et al., 1985b), both an increase ($F_1/F_0 > 1$) or a decrease ($F_1/F_0 < 1$) of the firing frequency can be obtained (Fig.11).

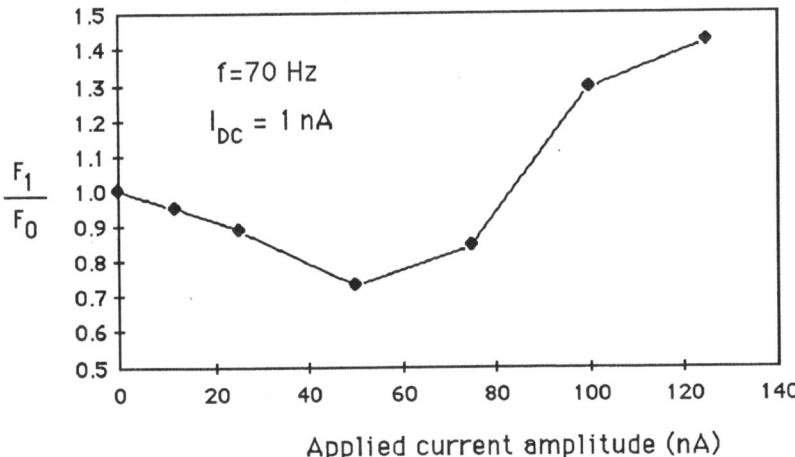

Applied current amplitude (nA)

Fig.11. Alteration of the firing frequency ($F_1 = 1/T_1$) induced by a perturbing current on a pacemaker cell simulated by the Connor and Stevens model. The data are shown as the ratio of the perturbed F_1 to the unperturbed F_0.

These experimentally confirmed results (Bernardi et al., 1986) outline the importance of the low-frequency modulating signals and might furnish a theoretical support to some recent findings by Lin et al. (1987) on the alteration of snail neuronal cells' behavior during pulsed microwave exposure and by Caddemi et al.(1986) on the behavior of chick embryo hearts which can be synchronized with the low-frequency signal modulating a microwave carrier.

Another interaction mechanism based on nonlinearities has been proposed by Adey and Lawrence (1982-84) as a possible explanation of the several results obtained by Adey's group on the EM induced flux of Calcium in vivo and in vitro in different excitable cells (see this book and the complete review made by Adey (1984)). The results on calcium ions influx or efflux from the cerebral cortex of cats in vivo (Kaczmarek and Adey, 1974) and in isolated chick cerebral hemisphere (Bawin and al. 1975) produced by modulated VHF, or similar subsequent results in the ELF-range(Bawin and Adey, 1976,77) and with ELF modulated microwaves (Adey et al, 1982), have been some of the better organized experiments that support the existence of weak field effects. These experiments have been confirmed and partially expanded by other groups (Blackman et al., 1979-1985; Athey, 1981; Merrit et al., 1982; Dutta et al., 1984) and, as will be shown in the following, have given experimental bases to new hypothesized interaction mechanisms.

Adey, in order to explain the observed effects, suggested an interaction process that involves several stages as schematized in Fig. 12. The first hypothesis is that the dipoles outside the membrane are organized in domains (patches) grouped and oriented in a common direction around the protruding part of the proteins that span the membrane (point a in Fig. 12).

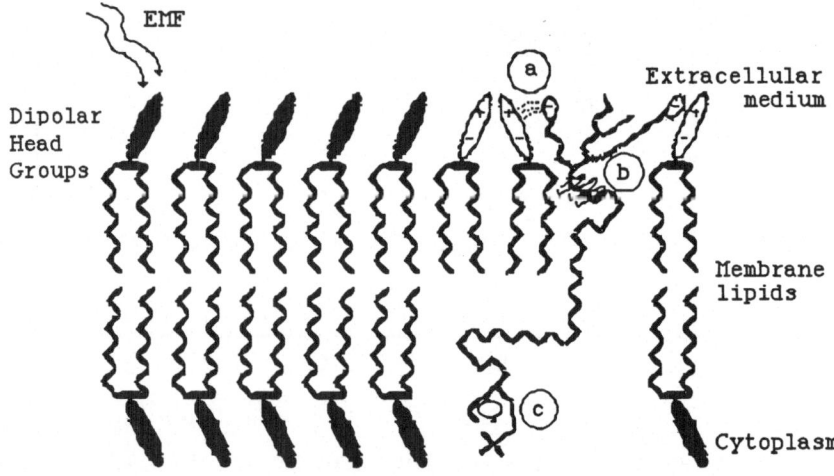

Fig.12. Model of interaction between an EM field and the cellular system. (Lawrence and Adey, 1982).

The applied electromagnetic field acting on the cell membrane interacts with the protruding dipolar heads causing a domino cooperative effect that modifies the domains and the Ca^+ bindings with release of extracellular calcium. This first step can justify the results obtained on the ionic fluxes but not the several modifications of the cell's global activity. So, considering that this domino effect can cede energy to other parts of the system, Adey supposed a transfer of energy to the glyco-proteins that connect the outside of the membrane to the intracellular space(point b).

The transfer of energy develops along the chain of molecules with a nonlinear mechanism: the excitation of solitons in DNA, as proposed by Davydov (1979). The third stage (point c) consists in a modification of some of the intracellular chemical reactions (e.g. enzymatic reactions) which can greatly modify the cell behavior. This last step can also be described with a nonlinear process (Albanese and Bell, 1984). These authors examined an enzyme catalyzed chemical reaction ($X \rightarrow C$) occurring in a fixed volume. Indicating with $k(T)$ the rate at which the reaction takes place and with $X(T,t)$ the concentration of the reactant, the time rate of change of product formation is:

$$dC/dt = k(T) \ X(T,t) \qquad\qquad (21)$$

Equation 21 together with the nonlinear differential equations that control the enzymatic process simulate the reaction dynamics. The heating produced by a low-frequency modulated microwave signal can indirectly modify the time-rate of the enzymatic reaction by altering the thermodynamics of the reaction.

The computed results for a microwave field at 400 MHz reveal a slight perturbation of the temperature (ΔT less than $0.01°K$) caused by the irradiating field, but, on the contrary, a significant alteration of the reaction rate constants.

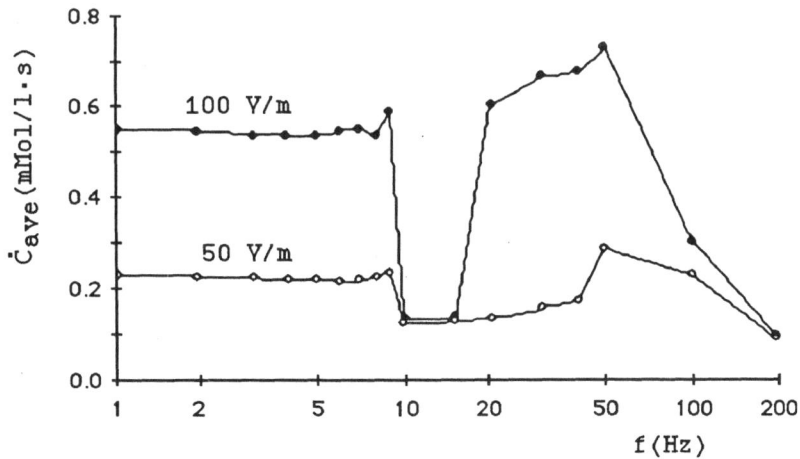

Fig.13. Average of the rate of change of product formation in an enzymatic reaction induced by a modulated microwave field versus the frequency of the modulating signal.

The analysis shows a strong dependence of the reaction rate on the amplitude of the modulated field and on the frequency of the modulating signal. In particular, in Fig.13, the arithmetic average $\overset{\circ}{C}_{ave}$ of 200 computed values of $\overset{\circ}{C} = k(t)X(T,t)$ equispaced in time over one period of the modulating frequency is represented as a function of the modulating frequency for field strengths of 50 V/m and 100 V/m.

The cited processes can justify the existence of the amplitude and frequency windows in the Calcium ion flux identified by Adey and Blackman in their experiments, but not directly support the dependence of these windows on the steady magnetic field, recently outlined by Blackman et al.(1985). Noting that their previous results (Blackman et al.,1980) only partially confirmed Adey's effects, Blackman and coauthors (1985) demonstrated that the frequency windows of the EM field strongly depend on the value of the static magnetic field present in the experiment. The conclusion is that two experiments using the same signal (amplitude and frequency) may give different results if the environmental static magnetic field differs. This conclusion not only can justify some disparities in the experimental results, but outlines the importance of the **E** and **B** cooperative interaction as another mechanism present in the observed alterations of exposed biological systems.

Cooperative E and B Interactions

A mechanism that can justify Blackman's results has been proposed by Chiabrera et al. (1985) by analyzing the EM field local effects on proteins and ions that interact with each other on the surface of the cell.

In a previous work the effects induced by a static E-field on the membrane proteinic receptors of the cell has been examined (Mc Laughlin and Poo, 1981). The field induces a grouping of the proteins on one side of the cell with consequent enhancement of the probability of Ca^{2+} channels formation. This mechanism could be explained with the action of a microelectrophoretic force that moves all the proteins towards the same side along the membrane. Chiabrera's group considered this mechanism in studying the mitogenic capacity of low-level low-frequency fields in the presence or absence of PHA (Phytohemagglutinin) (Chiabrera et al., 84). The ac field modifies the lectine-PHA receptor's mean life-time or/and causes a Ca^{2+} influx. In the first case the mitogenic capacity decreases, in the second it increases.

This theory has been extended by considering the action of electric and magnetic forces on different ionic messengers and binding sites. In absence of any applied field a first messenger Φ_e external to the cell (ion, lectin, hormone) acts on a binding site (mouth of a passive channel μ, head of a proteinic receptor E, membrane pump-release e), to induce changes in the concentration of a second internal messenger Φ_i (ion, enzymatic product) (Fig 14). The binding process is clearly a statistical event: the binding site moves bidimensionally on the cell surface, while the external messenger (ligand) moves in a three-dimensional path in themicroenviroment outside the cell.

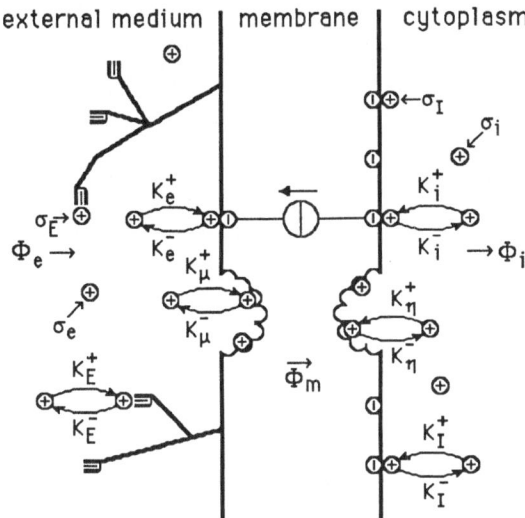

Fig.14. Sketch of the cell surface model for the evaluation of EM-field effects on Ca^{2+} flux (from Chiabrera et al., 1985b, simplified). σ_e represent positive counterions, σ_i free Ca^{2+} ions, Φ_m Ca^{2+} ion transport mechanism.

The association and dissociation rate constants of this process control the cell behavior. The equation modelling the binding process is :

$$dp_b/dt = K^+ \, p_f(1-p_b) - K^- p_b \qquad (22)$$

where p_f and p_b are the probability for free and bound ligands respectively, K^+ and K^- the association and dissociation rate constants which are related to the mean time necessary to reach the critical binding distance. An applied EM field modifies the velocity and the path of the external messenger moving near the binding site and will therefore change the association and dissociation rate constants.

The Lagrangian representation of the particle's motion is:

$$d\mathbf{v}/dt + \beta\mathbf{v} + \gamma_B\mathbf{B_t} \; \mathbf{x} \; \mathbf{v} = \mathbf{n_r} + \gamma_B\mathbf{E_{end}} + \gamma_B\mathbf{E} \qquad (23)$$

where $\mathbf{v}$ is the ligand velocity, β is the ratio of the local viscous coefficient to the ligand mass, γ_B is the ratio of the ligand charge to the mass, $\mathbf{E_{end}}$ is the local endogenous field, $\mathbf{B_t}$ and $\mathbf{E}$ are the local induced fields and $\mathbf{n_r}$ is a noise vector. In Eq. (23) effects of the applied EM field are appreciable only if the interaction takes place in water free zones so that the field can effectively act on the ligand through the Lorentz and electric forces. Equation (23) is generally difficult to solve analytically and must be solved

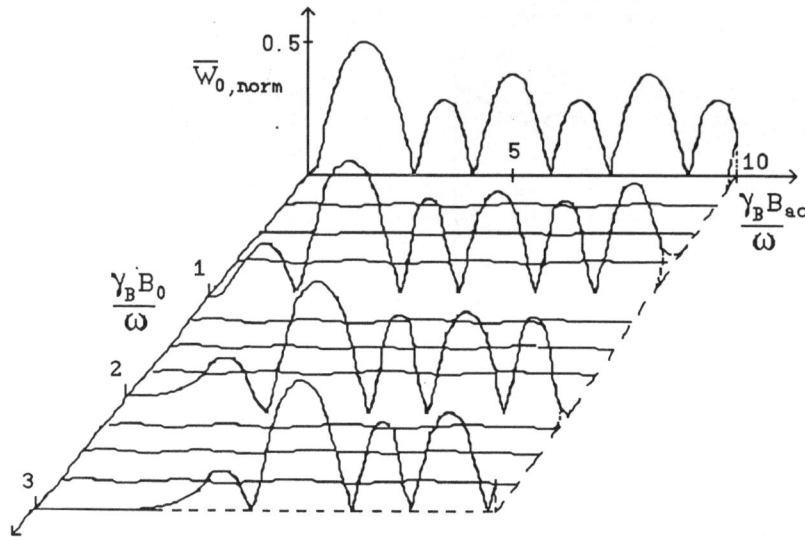

Fig. 15. Normalized time average of the velocity change W_0 as a function of the static and ac magnetic field amplitudes. The velocity change is normalized according to the exposure condition (Chiabrera and Bianco, 1987)

numerically, but in some particular cases results have been obtained in a closed form both for an harmonic electromagnetic field and for an ELF field superimposed on a static magnetic field (Chiabrera el al., 1985b; Chiabrera and Bianco, 1987). In Fig.15 are shown the effects on the velocity of a messenger due to an EM field of the type $\mathbf{E} = E_1 \sin \omega t\ \mathbf{x_o}$, $\mathbf{B} = (B_0 + B_{ac} \sin \omega t)\mathbf{x_o}$.

The results show the existence of either sensitive or insensitive zones depending on the mass and charge of the messenger, the frequency and the amplitude of the applied ac field, and the amplitude of the static magnetic field. In particular, alteration are obtained for integer values of the ratio between the cyclotron resonant frequency $\omega_0 = \gamma_B B_0$ and the angular frequency ω of the ac component. With this theory the results obtained by Blackman can be justified accordingly as follows (Chiabrera et al., 1985). The EM field modifies the association and dissociation rate constants of the binding processes that control the ionic fluxes through the cell membrane: a) channel-mouth escaping system (Ca^{2+} in), and b) pump-release system (Ca^{2+} out). The static magnetic field and the frequency of the applied field can modify in different ways the two processes producing globally either an influx or an efflux of calcium.

Preliminary results confirming this theory have been presented (Chiabrera et al., 1986; Pilla et al., 87). They show the possibility of modifying the swimming direction of a unicellular structure (Paramecium) whose velocity is determined by the flux of calcium through its membrane; the chosen environmental exposure (ac and dc fields) and the frequency allow the modification of the paramecium behavior in a predictable manner by a factor ten.

Also to justify the results obtained by Blackman et al. (1985) a mechanism has been proposed by Liboff(1985) that hypothesizes a frequency selective effect based on the cyclotron resonance frequency (CRF) of the ions. Considering an ion moving with a velocity $\mathbf{v}$ in a static $\mathbf{B_0}$ field, a Lorentz force will induce it to cover an helicoidal path with the axis parallel to $\mathbf{B_0}$. The motion can be divided in a drift displacement at a constant velocity along the $\mathbf{B_0}$ direction and in a uniform circular movement in the plane normal to $\mathbf{B_0}$ at the cyclotron resonance frequency. Table 5 shows the CRF of some ions (unhydrated) in presence of a static field of the same order of magnitude as the geomagnetic field (50 μT). As it can be observed these frequencies are in the ELF range where several effects have been revealed, directly or using a modulated RF carrier (Bawin et al., 1975; Adey et al., 1982; Blackman et al., 1985). A superimposed oscillating magnetic or electric field produces a force on the moving ion with components both in the direction of the helicoidal axis and on the plane normal to it. If components on the transverse plane have a frequency near the cyclotron resonance or its odd multiples, they can act synchronously on the particle and compensate the damping of energy in the ions movement. In this manner the path and the resonance frequency remain unaltered.

Some doubts have been advanced on the validity of this model especially on the possibility of existence of free paths for the ions in a biological environment and how these ions could effectively cross the membrane. McLeod and Liboff (1986) suppose that the ionic movements are guided by helicoidal pathways present in the proteins' structure of the channel walls, or that the quasi cylindrical structure of the channels (Hucho, 1986) could screen the ions from the thermal noise

Table 5. CRF for Some Unhydrated Ions at B_0 =50 μT.

Ionic Species	q/m (C/Kg) $*10^8$	CRF (Hz)
H^+	0.958	762
Li^+	0.138	110
Mg^{2+}	0.079	63.2
Ca^{2+}	0.048	38.3
Na^+	0.042	33.4
K^+	0.025	19.6

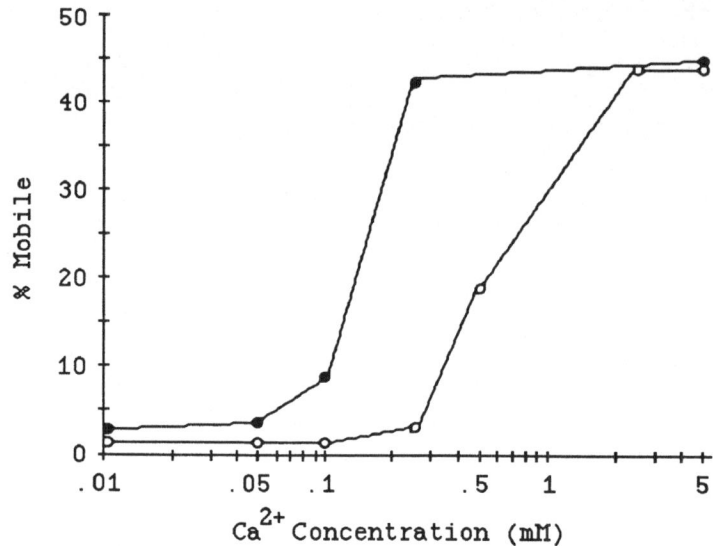

Fig.16. Mobility (in percent of cells that moved more than
10 μm) versus Ca^{2+} concentration. Open dots = control
(no field), black dots = experiments tuned for Ca^{2+}
resonance condition (CRF =16 Hz at B_0 =20.9 μT, B_{ac} =
20.9 μT). The effect is significant at 0.1, 0.25 and 0.5
mM, at lower values the effect on calcium was too low,
at higher values the phenomenon saturates.

present on both sides of the membrane. In this way the ions are
easily accelerated by the field and can quickly cross the
channel increasing the activation of biological phenomena
driven by ionic fluxes. Liboff (1985) justifies Blackman's
results by indicating that the responsible ion for the
cyclotron resonant effect is the K+ ion. In 1986 Thomas et al.
gave a support to the CRF theory analyzing the behavior of rats
under different exposure conditions: sham (no field), static
magnetic field, oscillating field, both static and oscillating
fields. The frequency of the oscillating component in these
last exposure conditions was chosen equal to the CRF of Li^+
ions. The rats were tested by evaluating their ability to
obtain food; the results showed a modified behavior only in the
fourth experiment. The authors relate this result "to the
resonant efflux of lithium ions from cells in rat brains".

A confirmation of the Liboff and McLeod theory has been
recently given by Smith et al.(1987) with experiments on
diatoms (Amphora coffeaeformis) whose mobility is controlled by
the calcium that crosses the membrane. The CRF condition has
been produced in samples that contain diatoms at a fixed
concentration of Ca^{2+}, the comparison between exposed samples
and sham gives a tenfold increase in the mobility of the
exposed diatoms at the CRF condition (Fig.16). A resonance
curve similar to the one calculated with the theoretical model
has been obtained experimentally (Fig. 17). An increase of the
diatoms mobility has also been observed at a frequency three
times the CRF condition.

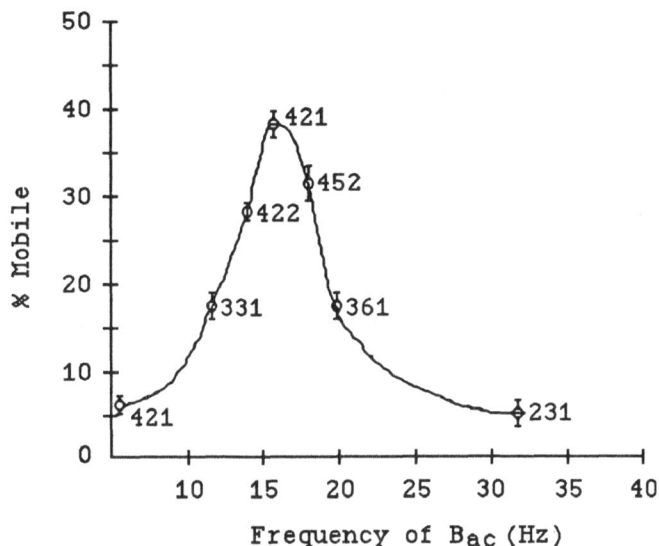

Fig.17. Mobility versus frequency of B_{ac}. The exposure conditions are the same as Fig.16. The numbers are the cell counts.

CONCLUSION

In this work a review of the principal physical interaction mechanisms that could underlie the most recognized and meaningful thermal and nonthermal biological effects is presented. In order to identify the physical mechanism at microscopic level or, in any case, at the lowest biological level possible we have selected among the thermal effects only those induced by direct dissipation of energy density, while among the nonthermal ones only those revealed at cellular level have been considered. Also, we have omitted all those physical mechanisms, typical of the interaction between electromagnetic fields and matter, that could be present in biological materials, but have not been linked as yet to any specific, experimentally supported, biological effect.

The review shows that it has not always been possible, particularly for the nonthermal results, to relate a given effect to an unequivocal responsible mechanism; more than one, probably closely linked together, could be present instead. Moreover it is not always easy to clearly identify the microscopic site where the interaction takes place and what physical parameter is primarily influenced by the field. Therefore considerable efforts must still be exerted to establish a clear link between the nonthermal effects and their responsible mechanisms. These efforts require accurate experimental work over the whole frequency range from ELF to millimeter waves with well defined protocols in both the exposure conditions and biological samples, thus allowing replication of experiments in different laboratories.

REFERENCES

Adey, W.R., and Bawin, S.M., 1977, Brain interactions with weak electric and magnetic fields, Neurosci. Res. Prog. Bull., 15:7.

Adey, W. R., 1980, Frequency and power windowing in tissue interactions with weak electromagnetic fields, Proc. IEEE, 68: 119.

Adey, W.R., 1981, Tissue interactions with nonionizing electromagnetic fields, Physiol. Rev., 61: 435.

Adey, W.R., and Bawin, S.M., 1982, Binding and release of brain calcium by low level electromagnetic fields: a review, Radio. Sci., 17(5S): 149.

Adey, W.R., Bawin, S.M., and Lawrence, A.F., 1982, Effects of weak amplitude-modulated microwave fields on calcium efflux from awake cat cerebral cortex, Bioelectromag.,3: 295.

Adey, W.R., 1984, Nonlinear, nonequilibrium aspects of electromagnetic field interactions at cell membranes, in: "Nonlinear electrodynamics in biological systems," W.R. Adey, and A.F. Lawrence, eds., Plenum Press, New York.

Adey,W.R., and Lawrence, A.F., 1984, "Nonlinear electrodynamics in biological systems," Plenum Press, New York.

Albanese, R.A., and Bell, E.L., 1984, Radiofrequency radiation and chemical reaction dynamics, in: "Nonlinear electrodynamics in biological systems," W.R. Adey, and A.F. Lawrence, eds., Plenum Press, New York.

Arber, S.L., 1976, Effect of microwaves on resting potential of giant neurons of mollusk, Helix pomatia, Elektron, Obrab, Mater, 6:78.

Arber, S.L.,1981, The effect of microwave radiation on passive membrane properties of snail neurons, J. Microwave Power, 16:15.

Arber, S.L., and Lin, J.C., 1983, Microwave enhancement of membrane conductance in snail neurons: role of temperature, Physiol. Chem. Phys., 15:259.

Arber, S.L., and Lin, J.C., 1984, Microwave enhancement of membrane conductance in snail neurons: effects of EDTA, Caffeine and tetracaine, Physiol. Chem. Phys., 16:469.

Arber, S.L., and Lin, J.C., 1985a, Microwave- induced changes in nerve cells: effects of modulation and temperature, Bioelectromag., 6:257.

Arber, S.L., and Lin, J.C., 1985b, Extracellular calcium and microwave enhancement of membrane conductance in snail neuron, Radiat. Environ.,24: 149.

Athey, T.W., 1981, Comparison of RF-induced calcium efflux from chick brain at different frequencies: do the scaled power density windows align?, Bioelectromag., 2:407.

Baransky, S., and Czerski, P., 1976, "Biological effects of microwaves," Dowden, Hutchinson & Ross, Inc., Stroudsburg, Pennsylvania.

Barnes, F.S, and Hu, C.J., 1977, Model for some nonthermal effects of radio and microwave fields on biological membranes, IEEE Trans. MTT, 25:742

Barsoum, Y.A., and Pickard, W.F., 1981, Radiofrequency bioeffects at the membrane level: separation of thermal and athermal contributions in the characeae, J.Membrane Biol., 61:39.

Barsoum, Y.A., and Pickard, W.F., 1982, The vacuolar potential of characean cells exposed to electromagnetic radiation in the range 200-8200 MHz, Bioelectromag., 3: 393.

Bawin, S.M., Gavalas-Medici, R.J., and Adey, W.R., 1973, Effects of modulated very high frequency fields on specific brain rhythms in cats, Brain Res. 58:365.

Bawin, S.M., Kaczmarek, K.L., and Adey, W.R., 1975, Effects of modulated VHF fields on the central nervous system, Ann. N.Y. Acad. Sci., 247: 74.

Bawin, S.M., and Adey, W.R., 1976, Sensitivity of calcium binding in cerebral tissue to weak environmental electric fields oscillating at low frequencies, Proc. Natl. Acad. Sci. U.S.A., 73: 1999.

Bawin, S.M., and Adey,W.R., 1977, Calcium binding in cerebral tissues, in: "Biological effects and measurement of Radiofrequency Microwaves," D.G. Hazzard, ed., HEW Publ., FDA.

Bawin, S.M., Sheppard, A.R., and Adey, W.R., 1978, Possible mechanisms of weak electromagnetic field coupling in brain tissue, Bioelectrochem. Bioenerg., 5:67.

Berkowitz, G.C., and Barnes, F.S., 1979, The effects of nonlinear membrane capacity on the interaction of microwave and radio frequencies with biological materials, IEEE Trans. MTT, 27:204.

Bernardi, P., and D'Inzeo, G., 1984, A nonlinear analysis of the effects of transient electromagnetic fields on excitable membranes,IEEE Trans. MTT, 32:670.

Bernardi,P., D'Inzeo, G., and Eusebi, F., 1985a, Response of a neuronal membrane to applied sinusoidal currents, Cell Biophysics, 7:185.

Bernardi,P., D'Inzeo, G.,and Pisa, S., 1985b, Alteration of the firing frequency in neuronal membranes stimulated with sinusoidal currents, Proc. of the 7° IEEE/EMBS Conference, Chicago, 1: 74.

Bernardi,P., D'Inzeo, G., and Pisa, S., 1986, Effects of modulated microwave and RF fields on the membrane of neuronal cells, Proc. of the 16th European Microwave Conference, 1:581.

Blackman, C.F., Elder, J.A., Weil, C.M., Benane, S.G., Eichinger, D.C., and House, D.E., 1979, Induction of calcium ion efflux from brain tissue by radiofrequency radiation: effects of modulation, frequency and field strength, Radio Sci., 14(6S): 93.

Blackman, C.F., Benane, S.G., Elder, J.A., House, D.E., Lampe, J.A., and Faulk, J.M., 1980, Induction of calcium ion efflux from brain tissue by radiofrequency radiation: effect of sample number and modulation frequency on the power density window, Bioelectromag., 1: 35.

Blackman, C.F., Benane, S.G., Rabinowitz, J.R., House, D.E., and Joines, W.T., 1985, A role for the magnetic field in the radiation-induced efflux of calcium ions from brain tissue in vitro, Bioelectromag.,6: 327.

Caddemi, A., Tamburello, C., Zanforlin, L., and Torregrossa, V., 1986, Microwave effects on isolated chick embryo hearts, Bioelectromag., 7:359.

Cain, C.A., 1980, A theoretical basis for microwave and RF field effects on excitable cellular membranes, IEEE Trans. MTT, 28:142.

Cain, C.A., 1981, Biological effects of oscillating electric fields: role of voltage sensitive ion channels, Bioelectromag., 2:23.

Campbell, N.L., and Brandt, C.L., 1981, Response of aplysia californica neurons to 2.45GHz, 3th Annual Meet. of Bioelectromag. Soc., (Abstr.), Washington.

Casaleggio,A., Marconi, L., Morgavi, G., Ridella, S;, and Rolando, C., 1984, Evaluation of ionic fluxes in a cell with non linear membrane stimulated by an electric field, J. of Bioelectricity, 3 (1 & 2):305.

Chiabrera, A., Grattarola, M., and Viviani, R., 1984, Interaction between electromagnetic fields and cells: Microelectrophoretic effect on ligands and surface receptors, Bioelectromag., 5:173.

Chiabrera, A., Nicolini, C., and Schwan, H.P., 1985a, "Interaction between electromagnetic fields and cells," Plenum Press, New York.

Chiabrera, A., Bianco, B., Caratozzolo, F., Giannetti, G., Grattarola, M., and Viviani, R., 1985b, Electric and magnetic field effects on ligand binding to the cell membrane., in:"Interaction between electromagnetic fields and cells", A. Chiabrera, C. Nicolini, H.P. and Schwan, eds., Plenum Press, New York.

Chiabrera, A., Gyebyi, K., Kaufman, J., Ryaby, J., Smith Sonneborn, J., and Pilla, A.A., 1986, Lorentz magnetic force effect on biological systems: application to paramecium, Abstr., 6th Annual Meet. of BRAGS, Utrecht, The Netherlands.

Chiabrera, A., and Bianco, B., 1987, The role of the magnetic field in the em interaction with ligand binding, in " Mechanistic approaches to interaction of electromagnetic fields with living systems", M. Blank, and E. Findl, eds., Plenum Press, New York.

Collin, R.E., 1966, "Foundations for Microwave Engineering", Mc Graw-Hill, New York.

Connor, J.A., and Stevens, C.F., 1971, Inward and delayed outward membrane currents in isolated neural somata under voltage clamp, J. Physiol., 213: 1

Davydov, A.S., 1979, Solitons in molecular systems, Phys. Scripta, 20: 387.

Dutta, S.K., Subramoniam, A., Ghosh, B., and Parshad, R., 1984, Microwave radiation-induced calcium ion efflux from human neuroblastoma cells in culture, Bioelectromag., 5: 71.

Edwards, G.S., Davis, C.C., Saffer, J.D., and Swicord, M.L., 1984, Resonant microwave absorption of selected DNA molecules, Phys. Rev. Lett., 53:1284

Edwards, C.S., Davis, C.C., Saffer, J.D., and Swicord, M.L., 1985, Microwave-field-driven acoustic modes in DNA, Biophys. J., 47:799.

Epstein, B.R., Gealt, M.A. Foster, K.R., 1987, The use of coaxial probes for precise dielectric measurements: a reevaluation, Proc. of the IEEE-MTT Symp., 255.

Foster, K. R., and Schwan, H. P., 1986, Dielectric properties of tissues, in:"CRC Handbook of Biological Effects of Electromagnetic Fields", C. Polk, and E. Postow, eds., CRC Press, Boca Raton.

Franceschetti, G., and Pinto, I., 1984, Cell membrane nonlinear response to applied electromagnetic field, IEEE Trans. MTT, 32:653.

Frey, A.H., and Messenger, R., 1973, Human perception of illumination with pulsed ultra-high frequency electromagnetic energy, Science, 181:356.

Frohlich, H.,1968, Long-range coherence and energy storage in biological systems, Int. J. Quantum Chem., 2:641.

Frohlich, H.,1978, Coherent electric vibrations in biological systems and the cancer problem, IEEE Trans. MTT, 26:216.

Frohlich, H.,1980, The biological effects of microwaves and related questions, Adv. in Electron. Electron Phys, 53:85.

Frohlich, H.,1984, General theory of coherent excitations in biological systems, in :"Nonlinear electrodynamics in biological systems," W.R. Adey, and A.F. Lawrence, eds., Plenum Press, New York.

Furia, L., Hill, D.W., and Gandhi, O.P., 1986, Effects of millimeter-wave irradiation on growth of saccharomyces cerevisiae, IEEE Trans. BME 33:993.

Gabriel, C., Grant, E.H., Tata, R., Brown, P.R., Gestblom, B.,and Norenland, E., 1987, Microwave absorption in aqueous solutions of DNA, Nature, 328: 145.

Galvin, M.J., and McRee, D.I., 1986, Cardiovascular, hematologic and biochemical effects of acute ventral exposure of conscious rats to 2450-MHz (CW) microwave radiation, Bioelectromag., 7:223.

Genzel, L., Kremer, F., Poglitsch, A., and Bechtold, G., 1983, Relaxation processes on a picosecond time scale in hemoglobin and poly (L-Alanine) observed by millimeter-wave spectroscopy, Biopolymers, 22:1715.

Grant, E.H., 1982, The dielectric method of investigating bound water in biological material: An appraisal of the technique, Bioelectromag., 3:17.

Grant, E.H., McClean, V.E.R., Nightingale, N.R.V., Sheppard, R.J., and Chapman, M.J., 1986, Dielectric behavior of water in biological solutions: studies on myoglobin, human low-density lipoprotein, and polyvinylpyrrolidone, Bioelectromag., 7:151.

Grundler, W., Keilman F., and Frohlich, H., 1977, Resonant growth rate response of yeast cells irradiated by weak microwaves, Phys. Lett., 62A:463.

Grundler, W., and Keilmann, F., 1978, Nonthermal effects of millimeter microwaves on yeast growth, Z.Naturforsh, 33:15.

Grundler, W., Keilmann, F., and Strube, D., 1982, Resonant-like dependence of yeast growth rate on microwave frequencies, Br. J. Cancer, 45 (5): 206.

Grundler, W., Keilmann, F., Putterlik, V., Santo, L., Strube, D., and Zimmermann, I., 1983, Non-thermal resonant effects of 42-GHz microwaves on the growth of yeast cultures, in: "Coherent excitations in biological systems", H. Frohlich, and F. Kremer, eds., Springer-Verlag, Berlin.

Hodgkin, A.L., and Huxley, A.F., 1952, A quantitative description of membrane current and its application to conduction and excitation in nerve, J. Physiol., 117:500.

Hucho, F., The nicotinic acetylcholine receptor and its ion channel, J. Biochem., 158:211.

Kaczmarek, L.K., and Adey, W.R., 1974, Weak electric gradients change ionic and transmitter fluxes in cortex, Brain. Res., 66:537.

Kaiser, F., 1980, Nonlinear oscillations in physical and biological systems, in : "Nonlinear electromagnetics," P.L.E. Uslenghi, ed., Academic Press, New York.

Kaiser, F., 1982, Theory of resonant effects of RF and MW energy, in: "Biological effects and dosimetry of nonionizing radiation," S.M. Michaelson, ed., Plenum Press, New York.

Kaiser, F., 1984, Entrainment- quasiperiodicity - chaos - collapse: bifurcation routes of externally driven self sustained oscillating systems, in :"Nonlinear electrodynamics in biological systems," W.R Adey, and A.F. Lawrence, eds., Plenum Press, New York.

Keilmann , F., 1982, Experimental rf and mw resonant nonthermal effects, in: "Biological effects and dosimetry of nonionizing radiation," S.M. Michaelson, M. Grandolfo, and A. Rindi , eds, Plenum Press, New York.

Kohli, M., Mei, W.N., Prohofsky, E.W., and Van Zandt, L.L., 1981, Calculated microwave absorption of double-helical B-conformation Poly (dG) Poly (dC), Biopolymers, 20:853.

Kremer, F., Koschnitzke, C., Santo, L., Quick, P., and Poglitsch, A., 1983, The non-thermal effect of millimeter wave radiation on the puffing of giant chromosomes, in: "Coherent Excitations in Biological Systems", H. Frohlich, and F. Kremer, eds., Springer-Verlag, Berlin.

Kremer, F., Poglitsh, A., and Genzel, L., 1984, Picosecond relaxations in proteins and biopolymers observed by MM-wave spectroscopy, in :"Nonlinear electrodynamics in biological systems," W.R. Adey, and A.F. Lawrence, eds., Plenum Press, New York.

Lawrence, A.F., and Adey, W.R., 1982, Non-linear wave mechanisms in interaction between excitable tissue and electromagnetic fields, Neurol. Res., 4, 115.

Lerner, E.J., 1984, Biological effects of electromagnetic fields, IEEE Spectrum, 21:57.

Liboff, A.R., 1985, Cyclotron resonance in membrane transport, in: "Interaction between electromagnetic fields and cells", A. Chiabrera, C. Nicolini, and H.P. Schwan, eds., Plenum Press, New York.

Lin, J.C., 1978, "Microwave Auditory Effects and Applications", C.C. Thomas, Springfield.

Lin, J.C., 1977a, On Microwave-Induced Hearing Sensation, IEEE Trans. MTT, 25:605.

Lin, J.C., 1977b, Further Studies on the Microwave Auditory Effect, IEEE Trans. MTT, 25:938.

Lin, J.C., O'Neill, W.D., Field, A., and Ginsburg, K.,1987, Pulsed high power microwave effects on spontaneous firing activities of snail neurons, 9th Annual Meet. of Bioelectromag. Soc., (Abstr.) G-3, Portland, USA.

Lords, J.L., Durney, C.H., Borg, A.M.,and Tinney, C.E., 1973, Rate effects in isolated hearts induced by microwave irradiation, IEEE Trans. MTT, 21:834.

Marconi, L., Morgavi, G., Ridella, S., and Rolando, C., 1985, Nonlinear ionic fluxes in EMF exposed cells, in:"Interaction between electromagnetic fields and cells," A. Chiabrera, C. Nicolini, and H.P. Schwan, eds., Plenum Press, New York.

Mc Laughlin, S., and Poo, M.M., 1981, The role of electro-osmosis in the electric field -induced movement of charged macromolecules on the surface of cells., Biophys. J., 34:85.

McLeod, B.R., and Liboff , A.R., 1986 , Dynamic characteristics of membrane ions in multifield configurations of low frequency electromagnetic radiation, Bioelectromag.,7:177.

Merritt, J.G., Shelton, W.S., and Chamnes, A.F., 1982, Attempt to alter $^{45}Ca^{2+}$ binding to brain tissue with pulse-modulated microwave energy, Bioelectromag., 3:475.

Michaelson S.M., and Lin, J.C., 1987, " Biological effects and health implications of radiofrequency radiation," Plenum Press, New York.

Olsen, R.G., and Lin, J.C., 1983, Microwave-induced pressure waves in mammalian brains, IEEE Trans BME, 30:289.

Pethig, R., 1979, "Dielectric and Electronic Properties of Biological Materials," John Wiley & Sons, Chichester.

Pickard, W.F., and Rosenbaum, F.J., 1978, Biological effects of microwaves at the membrane level: two possible athermal electrophysiological mechanisms and a proposed experimental test, Math. Biosci.,39:235.

Pilla, A.A., Chiabrera, A., Kaufman, J.J., and Ryaby J.T., 1987, A unified electrochemical approach to electrical and magnetic modulation of biological processes: application to the paramecium ciliary movement, 9th Annual Meet of Bioelectromag. Soc., (Abstr.) F-6, Portland, USA.

Polk, C., and Postow, E., 1986, "Handbook of biological effects of electromagnetic fields," CRC Press, Boca Raton, Florida.

Postow, E.,Swicord, M.L.,1986, Modulated fields and "window" effects, in: "Handbook of biological effects of electromagnetic fields," C. Polk, and E. Postow, eds, CRC Press, Boca Raton, Florida.

Schwan, H.P.,and Foster, K.R., 1980, RF-field interactions with biological systems: Electrical properties and biophysical mechanisms,Proc. IEEE, 68:104.

Seaman, R., and Wachtel, H., 1978, Slow and rapid response to CW and pulsed microwave radiation by individual Aplysia pacemakers, J. Microwave Power, 13: 77.

Seaman, R.L., Ajer, R.K., and DeHaan, R.L., 1982, Changes in cardiac-cell membrane noise during microwave exposure, Proc. of the IEEE-MTT Symp., 1:436.

Smith, S.D., McLeod, B.R., Liboff, A.R., and Cooksey, K., 1987, Calcium cyclotron resonance and diatom mobility, Bioelectromag., 8:218.

Swicord, M.L., Edwards, G.S., Sagripanti, J.L., and Davis, C.C., 1983, Chain-length-dependent microwave absorption of DNA, Biopolymers, 22:2513.

Swicord, M.L., and Davis, C.C.,1984, Microwave absorption of DNA between 8 and 12 GHz, Biopolymers , 24: 21.

Thomas, J.R., Schrot, J., and Liboff, A.R., 1987, Low-intensity magnetic fields alter operant behavior in rats, Bioelectromag., 7:349.

Tinney, C.E., Lords, J.L.,and Durney, C.H., 1976, Rate effects in isolated turtle hearts induced by microwave irradiation, IEEE Trans. MTT, 24:18.

Van Zandt, L.L., Kholi, M., Prohofsky, E.W.,1982, Absorption of microwave radiation by the double elix in aquo, Biopolymers, 21:1465.

Wachtel, H., Seaman, R., and Joines, W., 1975, Effects of low-intensity microwaves on isolated neurons, Ann. N.Y. Acad. Sci., 247: 46.

Webb,S.J., and Dodds, D.E., 1968, Inhibition of bacterial cell growth by 136-GHz microwaves, Nature, 218: 374.

PART III

SAFETY GUIDES AND RATIONALES

PROTECTION GUIDES FOR RF RADIATIONS:

RECENT DEVELOPMENTS IN THE U.S.A.

Don Justesen

Medical Research Laboratories (151)
USVA Medical Center
Kansas City, Missouri 64128 U.S.A.

INTRODUCTION

The history of U.S. standards for controlling human exposure to radio-frequency electromagnetic (RF) fields is largely an account of efforts by Committee C95.1 and Subcommittee C95-IV of the American National Standards Institute (ANSI). Other agencies and institutions have been involved in standard-setting, but the ANSI efforts have left an indelible imprint on all such activities (see, e.g., Report 86[1] of the National Council on Radiation Protection and Measurements, and the review[2] by Elder and Cahill).

To be useful, a protective standard must appeal to and reflect reliance on a body of biophysical data. In the ANSI tradition, the harvesters and interpreters of the RF data are the voluntary members of ANSI Subcommittee C95-IV. Actually, there have been several Subcommittees IV, the members of which have been drawn from disciplines that range from physics through the life and medical sciences. The products of subcommittee deliberation have eventuated in formal, albeit "voluntary," standards after approval by the parent, C95 committee.

THE PAST AND THE PRESENT

Two distinct epochs characterize ANSI's RF standards. Before 1982, heavy emphasis was placed on thermophysiological data and the corollary, *a priori* assumption that thermal insult--excessive elevation of body temperatures--is the principal hazard incumbent with overexposure to RF radiations. Also, limits on exposure were couched solely in terms of field strength or power density of incident radiations. With promulgation of ANSI Standard C95.1-1982[3], the postulation of hazards underwent a significant change. *A posteriori* in character, hazardous exposure was redefined by outcome--by the experimental demonstration of ill effects in laboratory animals. A thermal mechanism of insult is not ex-cluded by the 1982 standard, but it is not assumed. Conjoined with an empirical philosophy of insult is the adoption of dosimetry. Limits on exposure are couch-ed, as before, in terms of field strength and power density, but both measures are grounded in dosimetry: Energy dose and dose rate as labelled, respectively, *Specific Absorption* (SA, in S.I. units of J/kg) and *Specific Absorption Rate* (SAR, in S.I. units of W/kg). In prescribing dosimetrically based limits on exposure, ANSI inaugurated a frequency-dependent standard, which forced recognition of data on whole-body resonance from the pioneering studies, analytical and experimental, of Prof. Om Gandhi and colleagues at the University of Utah.

215

Independent behavioral experiments on rats and primates had revealed that thresholds of *behavioral incapacitation* occurred during exposures of a few tens of minutes when microwave or shortwave irradiation resulted in whole-body-averaged SARs near 4 W/kg. Although reversible when durations of exposure are short--less than one hour--behavioral incapacitation was adopted by ANSI as its most reliable end point for indexing harmful levels of RF radiation. To insure a margin of safety, a tenfold reduction from the SAR threshold of harm to an SAR of 0.4 W/kg was adopted for limiting exposure of human populations.

Capture of RF energy is maximal at whole-body resonance, which will occur in human beings at frequencies that range from approximately 30 MHz (tall adult) to 300 MHz (short infant). Any given individual has two modes of whole-body resonance, one mode based on an ungrounded body, and one on grounding, which effectively doubles the body's wavelength and halves its resonant frequency. To limit the SAR to 0.4 W/kg within the range of resonant frequencies, the power density of incident radiation must not exceed 10 W/m^2. At frequencies below 30 MHz and above 300 MHz, power-density limits are relaxed.

PROGRESS (?) ON A NEW ANSI STANDARD

Although ANSI policy mandates a decision on its standards at five-year intervals--reaffirm, revise, or reject--the promulgation of a standard in 1987 did not materialize. And it is unlikely that one will materialize in calendar year 1988. Several factors are responsible for the delay, each of which deserves examination.

Unwieldy Evaluative Procedures

In evaluating the biomedical literature on RF radiations in the period before adoption of the 1982 standard, members of Subcommittee C95-IV utilized a narrow-but-focused sample of pertinent reports. The criteria for acceptance of a report were a) concurrence by physical, engineering, medical, and life scientists that the report's data are based on technically and scientifically sound procedures; b) the report must contain experimental findings of an SAR threshold of putative harm in a mammalian species; and c) other reports must provide independent confirmation of the threshold of harm in the same or another species. Because of the relatively small number of reports that met criteria--only a few dozen among a data base of thousands--Subcommittee C95-IV came under intense criticism after publication of the 1982 standard.

To counter such criticism in a revised standard, a decision was made to perform a large-scale evaluation of data, virtually all that had been reported in the archival literature between 1982 and 1985. Two committees were formed to provide a first-pass round of evaluations, Physical Principles Validation and Biological Principles Validation, which were chaired, respectively, by Prof. Arthur W. Guy of the University of Washington (Seattle) and by me. As is their wont, the engineers and physicists of Prof. Guy's committee responded admirably to their charge: to determine which of 300-plus archival reports contain reliable data on SAR thresholds. I suggested that my committee await the findings of Prof. Guy's committee--why burden biological scientists with reports that can't pass dosimetric muster?--but was overruled. A massive mailing to chairmen of select working groups ensued. Unfortunately, my concern that the burden would eventuate in prolonged delays in completion of evaluations was confirmed.

Biological experiments, in virtue of methodological nuance and subject-matter complexity, can sorely tax the time of scientists that are attempting to read reports and reach interpretive consensus. If my reading of the box score is accurate, somewhere between one-third and one-half the reports failed to get a passing grade from Prof. Guy's committee. Demoralizing, then, were after-the-fact discoveries that reports, on each of which one had spent one or two hours, had been declared null and void by the physical scientists. Hindsight also dictates another argument against large-scale evaluations: Why include reports, however valid and virtuous, that are based on high levels of energy dosing? An experiment in which rats were decimated by radiation at 40 W/kg has doubtful implications for a standard that has limited the SAR at 0.4 W/kg.

As a footnote with implications for timeliness of future ANSI revisions, I suggest anew that dosimetric evaluations precede biological review. Screening of reports to winnow those irrelevant to exposure standards would also reduce the workload and accelerate the evaluative pace.

The original intent, once the Physical and Biological Principles Committees completed their work, was to forward recommendations to a Risk Evaluation Committee--or, if statistical problems were evident, to a special committee that would address questions of data analysis. At the time this paragraph was written (June of 1988), the Risk Evaluation Committee had not issued a report of its activities, which may indicate that its task is yet to be completed.

Indemnification

Citizens of the United States justly enjoy a reputation for being litigious. Although the threat of legal action against any given ANSI committee or subcommittee might be low in probability, the existence of hundreds of standards that bear on the public's health constitutes a valid economic concern. Consequently, the administrators of ANSI have backed away from bonding or insuring the volunteers that generate and sanction its standards. The possibility that one might stand naked in the adversarial courtyard has understandably chilled the enthusiasm that members of Subcommittee C95-IV have brought to their standard-setting endeavours. Even if all the evaluative committees had completed grading of all papers in the bioelectromagnetics literature, and even if the Risk Evaluation Committee had issued final guidance, a new ANSI standard would not be forthcoming until indemnification of individual participants had been achieved. To this end, Prof. Arthur Guy and the Secretary of Subcommittee C95-IV, Dr. John Osepchuk of the Raytheon Company, have devoted much effort. They have been aided by the Committee on Man and Radiation (COMAR) and by the Technical Activities Council (TAC) of the Institute of Electrical and Electronic Engineers (IEEE). A former co-sponsor of ANSI Committee C95 with the Department of the Navy, the IEEE has been invited to become the Committee's sole sponsor. In this capacity, the IEEE would provide legal protection of Committee and Subcommittee members.

On the Shedding of a Tier

ANSI Standard C95.1-1982 contains what many public-health professionals consider a paradox. The standard applies to civil as well as working populations, but no distinction is made with respect to recommended limits on exposure to RF radiations. Traditionally, public-health specialists inveigh more restrictive limits for members of civil populations than they do for workers in so-called controlled environments. Workers are doubtless healthier than older, more sickly members of the general population, and they may also be more aware of hazards in the working environment and means to mitigate them. These and other arguments have been waged in efforts to generate a two-tiered standard, one that would have less stringent limits for working populations. The counter argument is that an ANSI SAR limit, if "safe" for workers, could hardly be "safer" if made more stringent for the general population. Sentiments for and against a second tier are about evenly split among members of Subcommittee C95-IV. Because ANSI policy requires a substantial consensus on limits and guidance among its volunteering specialists, the division of opinion on a one- vs. two-tiered policy constitutes a troublesome obstacle in the elaboration of a revised standard.

What is the Hazard Threshold?

Two of the major entities with missions that entail development of RF standards--the NCRP and the U.S. Environmental Protection Agency (EPA)--published lengthy reviews of the bioelectromagnetics literature[1,2] after promulgation of ANSI C95.1-1982. In both reviews, the critical end point of behavioral incapacitation, which defines ANSI's 4-W/kg threshold of harm, was reaffirmed. Recently, however, in reporting an updated evaluation of the literature, the EPA has sounded a more conservative note:

> [I]t is concluded that exposure to RF radiation causes biological effects at SARs above and below 1 W/kg; some of the effects [that] occur at about 1 W/kg may be significant under certain environmental conditions. The biological significance of the effects [that] occur below 1 W/kg, including those. . .at specific temperatures different from the physiological temperature range, specific frequencies or at specific amplitude-modulation conditions is not established.[4]

The reference to "effects" that are "significant" circa 1 W/kg are based on a thermal-insult model in which in-house data from murine experiments were extrapolated to higher species, including man. This largely analytical model, which does not account for differing thermophysiological capabilities among the species, has not been validated in independent experimentation. Although Subcommittee C95-IV has yet to implement a lower SAR threshold of harm in consequence of the EPA's advice, any such move would subvert the 1982 criteria, which mandated use of independently verified experimental data. Caveat!

There are data that have created justifiable concern that the 4-W/kg threshold of harm may need to be revised downward. Reference is, first, to studies of mice in Poland by Szmigielski and colleagues[5] in which three different types of experimental malignancies were *promoted*--but not induced--by long-term, but intermittent, microwave irradiation at SARs near 2 W/kg. Second, I refer to the celebrated, but-yet-to-be-archival report of experiments on rats by Prof. Arthur W. Guy and colleagues at the University of Washington[6]. These experiments, which subjected animals to near life-long exposures to microwaves, revealed that SARs circa 0.4 W/kg were associated with a reliable increase of malignancies above control incidence. Remarkably, despite a nearly fourfold greater incidence of malignant tumors in exposed animals as compared with controls, average life spans did not differ. These data and those from the study performed in Poland have not been independently confirmed. The putative threshold of hazardous irradiation would doubtless be driven lower if either set of malignancy data met the critical scientific test of confirmation. It is ironic, to say the least, that levels of funding of research on microwaves and other high-frequency RF radiations have decreased so dramatically that attempts to confirm the malignancy studies may never be undertaken. The dilemma for participants in the standard-setting process is obvious: One should not use unconfirmed findings; but one is justifiably uneasy when well-executed experiments generate a portent of malignant disease.

Burns and Shocks

Profs. Om Gandhi and Arthur Guy have both confirmed in the laboratory what long ago a health physicist with the E.P.A., Richard Tell, learned by direct experience when climbing highly powered, low-frequency antennas. The lesson is that electric shocks and RF burns can be induced in the human body at relatively low power densities. The reader familiar with the "U" shaped power-density and field-strength curves that characterize ANSI Standard C95.1-1982 should be prepared to see some downward revision of limits, especially those at the lower end of the spectrum.

CONCLUDING SUMMARY

Six years have passed since publication of the current ANSI standard for protection of human populations from excessive RF radiations. Much work on a revised standard has been performed by evaluative committees and working groups, which have attempted, with mixed results, to perform a highly ambitious review of the bioelectromagnetics literature. There are other barriers that have delayed completion of a new standard, not the least of which has been disagreement by evaluators on whether to establish different limits on exposure of general and working populations. Perhaps the most foreboding barrier to completion of a new standard--one that reliably reflects the reality of field--body interactions and that validly protects the human population--is the decline of support for scientific inquiry. Trite but true: A protective standard is no better than its data base.

Acknowledgements

My thanks to Prof. James C. Lin of the University of Illinois for his diligence and efforts in convening a special program for URSI--and for his patience whilst I coaxed my printer to turn out copy. Thanks, too, to Prof. Arthur W. Guy, who read and repaired several glitches in an early draft of this manuscript. I am solely responsible for any errors that remain.

REFERENCES

1. NCRP Report No. 86, "Biological effects and exposure criteria for radio-frequency electromagnetic fields." National Council on Radiation Protection and Measurements: Bethesda MD (1986).
2. J. A. Elder and D.F. Cahill (Eds.), "Biological effects of radiofrequency radiation." U.S. Environmental Protection Agency: Research Triangle Park NC (1984).
3. ANSI C95.1--1982, "American National Standard: Safety levels with respect to human exposure to radio-frequency electromagnetic fields, 300 kHz to 100 GHz." The Institute of Electrical and Electronics Engineers, Inc.: New York (1982).
4. J. A. Elder, "A reassessment of the biological effects of radiofrequency radiation: Non-cancer effects." U.S. Environmental Protection Agency: Research Triangle Park NC (1987).
5. S. Szmigielski, A. Szudzinski, A. Pietrazek, M. Bielec, M. Janiak, and J. K. Wremble, Accelerated development of spontaneous tumors and benzopyrene-induced skin cancer in mice exposed to 2450-MHz microwave irradiation. _Bioelectromagnetics_ 3:179 (1982).
6. A. W. Guy, C.-K. Chou, L. L. Kuntz, J. Crowley, J. Krupp, "Effects of long-term low-level radiofrequency radiation exposure on rats." USAF School of Aerospace Medicine: Brooks Air Force Base TX (1985).

EASTERN EUROPEAN RF PROTECTION GUIDES AND RATIONALES

Stanislaw Szmigielski

Department of Biological Effects of Non-ionizing
Radiations, Center for Radiobiology and Radiation Safety
128 Szaserow, 00-909 Warsaw, Poland

INTRODUCTION

Electromagnetic fields (EMFs) are the relatively new, but rapidly
intensifying ecologic factors. For millions of years the biological
life on the Earth has been developing and undergoing evolution under
influence of complex natural EMFs generated by the external sun radiation and natural electric fields in the atmosphere. The exposure was
relatively weak, as the natural EMF intensities does not exceed the
order of $10^{-8} - 10^{-9}$ W/m^2. The situation has changed however dramatically some 50 years ago with introduction of various devices generating microwaves (MWs) and radiofrequencies (RFs) first for communication and navigation and later for multiple industrial and household
purposes. The steady increasing occupational groups and whole populations of people living in certain areas (e.g., close to power lines
and stations, TV/Radio broadcasting antennas, air force bases,etc.)
are continously being exposed to EMF intensities that by few magnitude orders exceed the natural fields of the Earth. At present only
the second-third generation of human beings is exposed to artificially
generated EMFs and the long-term effects of these exposures are still
difficult for forecasting. Despite numerous experimental investigations and epidemiologic studies, it is still not possible to prove
the existence and character of any specific molecular, cellular or
system-related damage that may be evoked by long-term exposure in
low-level EMFs. Excluding the well defined thermal effects and reactions to whole body or local hyperthermia evoked by exposure in
high-level EMFs at certain frequencies, it became clear from the very

beginning of bioelectromagnetic studies that biological effects related to single or long-term exposures in non-thermal EMFs are inconsistent, transient and difficult both for confirmation and interpratation.
Review of the available American, Russian and European literature (for
most recent comprehensive summaries, see Dawidow et al.,1984; Elder
and Cahill,1984; Romero-Sierra,1984) revealed that for biological effects related to EMFs it is not possible to determine the primary target organ or system that may be considered as "biological dosimeter"
in terms of health hazards. On the other side, under certain well controlled conditions of exposure in experimental EMFs, a variety of behavioural, neurological, endocrine, hemato-immunological as well as reproductive abnormalities were demonstrated (Elder and Cahill,1984).
However, it is still not possible to establish valid thresholds of
fields power density or energy absorption for the above phenomena.

Medical examinations of personnel exposed occupationally to EMFs
and epidemiological analysis of large groups of people working in this
environment also did not show conclusive health hazards related to
EMFs, although a variety of non-specific symptoms, including liability
of vegetative nervous system and increased frequency of neuroses have
been reported (Baranski and Czerski,1976; Minin,1974).

Lack of generally accepted and repeatable biological effects and
health hazards related to long-term exposures in low-level EMFs considerably impedes establishing of satisfactory safety levels of nonionizing radiations and maximal permissible intensities of EMFs that
may be considered safe for occupational exposure. The first basic rules for EMF safety levels have been established in the early fifties
and from the very beginning a distinct philosophies have been accepted in USA and West European countries versus USSR and Commecon Eastern European countries. The subject was from time to time reviewed
(INIRC-IPRA,1984; Romero-Sierra,1984) and both in the West and in the
East there are recent tendencies for amendments of the operating standards with a search for more uniform safety rules (see INIRC-IPRA,
1984). The present situation is additionally complicated by the fact
that virtually nothing is known for sure about possible long-term and
delayed effects of exposures at different levels of EMFs intensity
(e.g. life span, carcinogenesis, influence on future generations),as
well as about biological effects of complex EMFs and combined action
of EMFs and other harmful environmental or occupational risk factors.
E.g., the experimental data indicating that MW radiation, not being
carcinogenic per se, may enhance a potency of certain carcinogens

(Balcer-Kubiczek and Harrison,1985; Szmigielski et al.,1982) and possesses a tumor-promoting activity (Szmigielski et al.,1987) need further investigations and considerations in terms of safety levels. The recently reported better or worse documented and acceptable epidemiologic data on increased risk of neoplastic diseases in human beings exposed to MW/RF radiation (Szmigielski et al.,1987) or to extremly low frequencies (ELFs) (Mc Dowell,1983; Milham,1982) should be also stressed as one of the possible clues influencing future safety levels of EMFs.

The aim of the present paper is to describe and critically discuss rationales and tendencies for amendments of EMF safety rules being accepted at present in Poland and in other European Commecon (Council for Mutual Economic Aid) countries. The material presented here is based on rules and acts operating at present in Poland, proposals for amendments of EMF safety rules, personal experience of the present author in this field and on discussions of experts from the Commission of Biological Effects of Non-ionizing Radiations, Committee of Medical Physics of the Polish Academy of Sciences (chaired by the present author), Interdepartmental Commission for Amendment of Safety Levels for EMFs at the Central Institute for Labor Protection in Warsaw, Poland and the Commecon Countries Expert Working Group for Unifying of Safety Rules for Physical Factors.

RATIONALES FOR ESTABLISHING EMF SAFETY RULES

Data being acceptable for considerations concerning safety levels of EMFs may arise from:
- theoretical considerations of power absorption and possible interactions with living materia;
- experimental investigations of cells or whole organisms exposed in artificial EMFs under various conditions;
- periodic medical examinations of personnel exposed occupationally to EMFs with a search for relations of the observed effects to intensity and period of exposure;
- epidemiologic analysis of health risk assessment in large groups of human beings either exposed occupationally or living in areas with above the average environmental intensities of EMFs.

Unfortunately, none of these sources of informations provides till now data that may be directly used as sure indicators of safety levels and thus, in any case it is a matter of arbitrary decision at what level of EMF intensity should be the bar settled. E.g., the situation occuring in animals irradiated with MW/RFs is additionally complicated by the concomitant stress reaction (Fig.1). Thus, the possible behavioural effects may be due to discomfort caused by non-specific stress reaction (there exists relations that small rodents, being the main subject of experimental MW/RF exposures can in an unknown way "percieve" low-level of these radiations and seem to be aware of being irradiated in weak fields) and in turn influence the neurohormo-nal pathways. It may be easy to relate the behavioural effect caused by non-specific stress reaction to action of low-level EMFs.

Discussing rationales for EMF safety levels a possibility of certain interactions connected with pulse modulation of the carrier wave should be also considered. In older Russian and East European literature (for review, see Baranski and Czerski,1976; Minin,1974) there exist equivocial opinions that pulse modulated MW radiation exertrs stronger neurologic, behavioural and hemato-immunologic ef-fects, compared to continous wave of the same frequencies. Actually, there are no convincing evidences that millisecond pulses of MWs at mean power densities not leading to detectable thermal effects may influence function of living organisms to a higher degree than continous wave. However, certain cellular disturbances attributed to specific interactions of RF/MWs modulated sinusoidally at the 1-100 Hz frequencies have been recently postulated. Adey and his group (Adey,1981; Byus et al.,1984,1986) on base of 20-year exper-ience have gained numerous evidences that the amplitude modulation characteristics appear to be a prime determinant of the nature of interaction at the cellular level. The authors have stressed the "windowing" of many of the observed interactions in both the frequency and amplitude (of pulses) domains. The effects are strongly dependent on the modulation frequency. E.g. for the 450 MHz carrier wave the strongest effects (inhibition of certain enzymes activity in cultured human lymphocytes) were observed at sinusoidal modulation of 16 Hz with diminishing responses at 40 and 60 Hz and no response at 80 or 100 Hz (Byus et al.,1984). These and other not fully understandable and unresolved events occuring in cells exposed to low-level MW/RF modulated fields stresses the complexity of interactions and indi-cates caution in accepting the safety levels.

There exists at present a general agreement among Polish experts that in case of MW radiation, either continous or pulse modulated no significant shifts in function of the organism of experimental ani-

mals occur at field power densities below 10 W/m^2 (1 mW/cm^2), even in
mice exposed to the resonant frequency of 2450 MHz. In earlier Russian
and East European literature (summarized by Baranski and Czerski,1976;
Minin,1974) there are numerous reports on a variety of behavioural,
neurologic, biochemical and hemato-immunological effect occuring in
animals exposed for a long time (few weeks - few months) to very weak
(microWatts/cm^2) MW fields. However, in the most recent Russian monograph (Dawidow et al.,1984) immunologic alterations occuring after
exposure of mice and rats in very weak MW fields are no more presented
and discussed. Instead of this the authors claim that inconsistent and
transient immunologic alterations may be observed in animals exposed
for a longer time to power densities exceeding 5 W/m^2 (0.5 mW/cm^2),
while clearly demonstrable immunologic effects are observed only at
thermogenic power densities. Still more the above authors state, although without evidencing this view, that ... "high adaptability of
the immune system causes that alterations observed after exposure
of animals in weak MW/RF fields are meaningless from the point of
view of safety standards"... (Dawidow,Tichonczuk and Antipow,1984,
page 55).

It was already mentioned that considering the results of experimental investigations in animals exposed to EMFs one has to take into
advance the concomitant stress reaction (Fig.1). Thus, the effects
being observed after exposure of animals in MW/RF chambers may result
from the stress reaction and/or adaptation mechanisms (Fig.1) and
cannot be directly accepted as indicators for EMF safety levels.

The other important premise for safety criteria that originates
from experimental investigations of animals exposed to EMFs is a lack
of linear correlation between the observed biological effects and intensity of EMFs, at least at the RF/MW frequencies. E.g., reaction of
the immune system to exposure in MW/RF fields (for the most recent
review, see Szmigielski et al., 1987), including natural antibacterial,
antiviral and antineoplastic resistance,appears to be biphasic (Fig.2)
with a phase of immunostimulation, followed by transient suppression
of certain immune reactions. Russian authors (Sawin et al.,1983) in
the recent guidelines for amendments of RF/MW safety criteria also
stress the non-linear correlation between observed biological effects
and field power density. These authors claim that non-linear depen-
dences occur at power densities 4 - 10 mW/cm^2, with closer linearity
below 4 mW/cm^2.

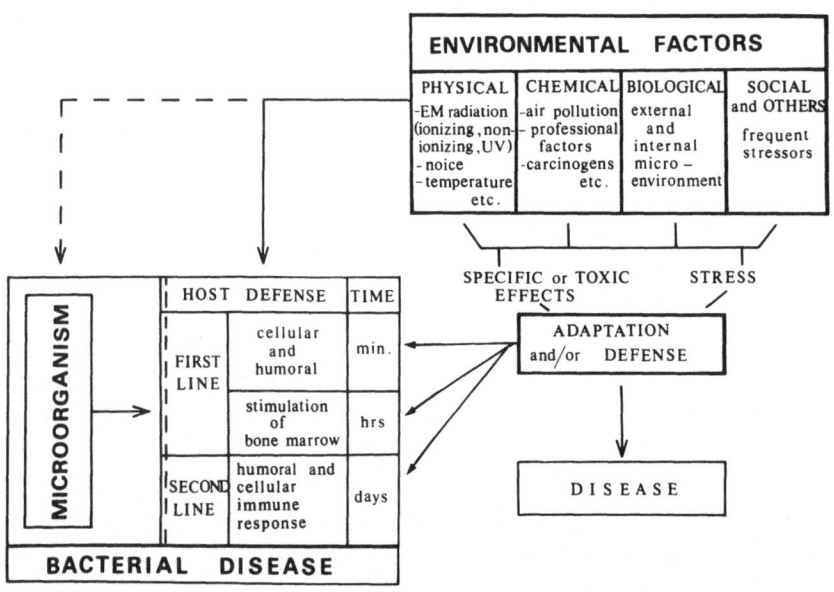

Figure 1

POSSIBLE INFLUENCE OF ENVIRONMENTAL FACTORS,INCLUDING
ELECTROMAGNETIC FIELDS ON HOST'S DEFENSE MECHANISMS
AND DEVELOPMENT OF BACTERIAL DISEASES.

Note that environmental factors may evoke both specific
(toxic) effects and the concomitant non-specific stress
reaction. Both the specific and the stress reactions
trigger adaptation and/or defense mechanisms that cannot
be regarded as definite health hazards. However, the above
adaptation mechanisms may temporary lower the host's defense mechanisms (transient suppression of certain immune
functions) that in turn results in higher risk of bacterial
infections. There exists also a possibility of direct interaction of environmental factors with virulence of micro-
organisms.
Note that the disease related to environmental factors
develops only after overpassing of the adaptation/defense
mechanisms of the host.

Medical examinations of personnel exposed occupationally to EMFs, at least those performed till now in Poland, are still not fully conclusive in terms of safety criteria, although all people starting to work in the EMF environment undergo selective general, ophtalmologic and neurologic medical examinations that are later (during employment) repeated every 1 - 3 years. The syndrom of "microwave disease" reported in earlier Russian and Polish literature (Baranski and Czerski,1976) is no more recognized in Poland, although the recent

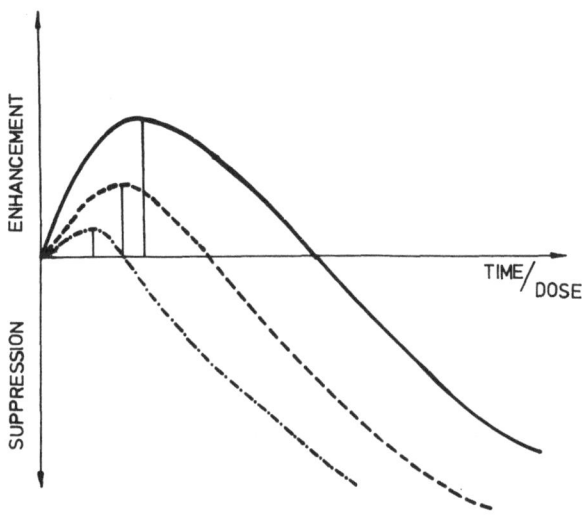

Figure 2

BIPHASIC REACTION OF THE IMMUNE SYSTEM OF MAMMALS TO
LONG-TERM EXPOSURE IN RF/MW FIELDS AS AN EXAMPLE OF
NON-LINEAR BIOLOGICAL EFFECT TO EMFs.

Note stimulation of immune functions in the early phase of
exposure with enhancement upon intensity of RF/MW fields
and the following phase of progressing (but transient and
reversible) suppression of immune functions. For details,
see Szmigielski et al., 1987.

medical statistics in this country still list more frequent vegetative neuroses and a variety of non-specific symptoms (liability of blood pressure, fatigue, headaches,etc.) in personnel exposed occupationally to MW/RF radiations. However, these symptoms do not show close relation with period of employment or intensity of exposure.

Considering the health hazards related to occupational exposure to EMFs Polish medical experts stress two facts important from the point of view of safety criteria:

1. Presence of single subjects being hypersensitive ("allergic"?) to very weak EMFs in a wide range of frequencies;

2. More frequent occurance of non-specific symptoms during the first one-two years of employment in the EMF environment with disappearance of the symptoms during further years of employment (adaptation period ?).

These facts cannot be still convincingly proved statistically, but there exists a general agreement that both of them should be taken into advance in selection of personnel for employment in the EMF environment. Subjects considered as hypersensitive to EMFs suffer from a variety of non-specific, but severe symptoms (fatigue, migraines, speech problems, depression, rarely convulsions) when exposed to weak (microWatt - milliWatt) MW/RF fields. Some of these subjects suffer also from similar symptoms when find themselfs in the vicnicity of power lines, high-power TV/Radio transmitters or even near VDTs. In our material the hypersensitivity to EMFs, being still a non-understandable phenomenon, occurs rarely and counts for one-few cases per 10 000 human beings. Recently, similar observations on hypersensitivity to EMFs were reported by Choy et al (1987).

In summary, there still exists quite a long list of unresolved and/or uncertain issues in assessment of biological effects and possible health hazards related to both environmental and occupational exposure to EMFs (Table I). The present knowledge in this field, although based on relatively broad experimental and medical data, does not provide sufficient backgrounds for establishing satisfactory and generally acceptable safety levels, untill at least the basic unresolved problems (Table I) will be better understandable. Thus, all the safety levels for EMFs have to be based on arbitrary decisions with a sufficient "safety coefficients" to avoid appearance of unsuspected and/or still unknown health effects. Certainly, the arbitrary established safety levels should be a subject for amendments with progress in bioelectromagnetics.

In practice, there are two main problems in safety criteria for EMFs that in our opinion should be considered separately and with distinct philosophies:
- environmental uncontrolled exposure;
- occupational exposure.

Looking at the safety standards proposed or operating in various countries we have a feeling that these two kinds of exposure are not suffi-

Table I.

BASIC UNRESOLVED AND UNCERTAIN PROBLEMS CONCERNING BIOLOGICAL
EFFECTS AND POSSIBLE HEALTH HAZARDS RELATED TO EXPOSURES
IN ELECTROMAGNETIC FIELDS (EMFs).

Unresolved and/or uncertain issues	Claims for significance of the issue for assessment of health hazards
THRESHOLDS FOR SPECIFIC BIOLOGICAL EFFECTS OF EMFs AT VARIOUS FREQUENCIES	Non-linear biological effects (Sawin et al.,1983); biphasic reaction of the immune system (Szmigielski et al.,1987)
BIOLOGICAL EFFECTS OF ELECTRIC (E) AND MAGNETIC (H) FIELDS AT ELF AND LF	Preliminary informations on biological activity of very weak magnetic fields
BIOLOGICAL EFFECTS OF COMPLEX AND MODULATED EMFs	Cellular effect of sinusoidally-modulated RF and MWs with "windowing" of the phenomena in the 1-100 Hz modulation (Adey,1981)
INTERACTION OF EMFs WITH OTHER ENVIRONMENTAL RISK FACTORS	Tumor-promoting activity of MWs, enhancement of carcinogenesis by MWs (Szmigielski et al.,1982; Balcer-Kubiczek & Harrison,1985)
DIFFERENTIATION OF EFFECTS RELATED TO SPECIFIC INTER-ACTION OF EMFs WITH BIOLOGICAL MEDIA AND TO THE CONCOMITANT STRESS REACTION	Similar effect observed in rodents exposed to weak MW fields and to positive control of non-specific stress situations; "perception" of EMFs by rodents and other animals ?
LONG-TERM and/or DELAYED EFFECTS OF EXPOSURE TO EMFs	Increased number of neoplasms in rats exposed for a life-time in MW fields (Guy et al.,1985; Increased morbidity of neoplasms in personnel exposed occupatio-nally to RF/MWs (Szmigielski et al.,1987)
EFFECTS OF EMF EXPOSURES ON DEVELOPMENT AND COURSE OF DISEASES CAUSED BY OTHER FACTORS AND INTERACTION OF EMFs WITH PHARMACOLOGICAL DRUGS	

ciently differentiated and in some cases there is a tendency to establish one level for both.

As it was already mentioned, for a long time only natural EMFs accounted for exposure of living organisms to non-ionizing radiations. With appearance of artificial EMFs the environmental exposure has increased by few orders of magnitude and due to rapid growth of both number and power of devices generating EMFs directly to the atmosphere as well as to reflexions from certain terrain objects (e.g. buildings, hills, metallic constructions,etc.) there exist at present places where intensity of EMFs of various frequencies has reached a value of few-several microWatts per cm^2. Safety levels for uncontrolled environmental exposure to MWs are still a matter of controversis, as in this case a very complex fields, difficult for both measurement, prediction and interpretation occur. Environmental exposure covers in practice all frequencies of EMFs (Table II), including both ELF, LF, RF and MWs and a variety of modulations, although in most cases a certain and measurable frequency predominates, depending upon vicnicity of EMF sources. In contrary to the really occuring conditions there is a tendency toward establishing safety levels for single ranges of EMFs, at least for ELF, RF and MWs (Table II). As the present knowledge on biological effects of complex EMFs and on interaction of weak EMFs with other environmental risk factors (e.g. chemical pollutants, carcinogens) is very scarce and fragmentaric, the only logical, although tentative, solution is application of a"safety coefficient" of about 10 - 100 from what may be concluded from known biological effects with an unproven hope that this will protect the whole mankind, including high risk groups (e.g. children, pregnant women,etc.) against unsuspected and/or yet unknown long-term effects. In another words, it is postulated to establish the environmental safety levels for EMFs at the "unwarranted" low levels, untill more will be known on interactions of EMFs with other risk factors and on bioeffects of complex EMFs. Certainly, the best safety level would be "zero" with limiting of the environmental exposure to natural EMFs, but this is no more possible in any place of the world. Thus, a logical compromise is needed, taking into advance the economic costs of the introduced safety levels. E.g., lowering of the RF/MW environmental safety levels to 10 mW/cm^2 (1 μW/cm^2) would close large areas in the vicnicity of radar stations, air force bases and TV/Radio antennas for house building, abiding of people, agriculture,etc., that would be economically groundless. On the other side, establishing of the RF/MW environmental safety levels at 10 W/m^2 (1 mW/cm^2), that solves

Table **II.**

SPECTRUM OF ELECTROMAGNETIC FIELDS (EMFs) FROM THE POINT OF VIEW OF SAFETY CRITERIA.

EM radiation frequency (Hz)	ranges covered by the present Polish safety regulations	ranges proposed for uniform safety regulations for Commecon countries	
		frequency (Hz)	No. of range
EXTREMLY LOW FREQUENCIES (ELF) 1		below	
10		30 Hz	1
50/60 Hz power frequency ---10^2	power frequency		
		300 Hz	2
-------------------- 10^3- 1 kHz		3 kHz	3
LOW FREQUENCIES 10^4	1 kHz - 100 KHz (pending)		
(LF) 10^5		30 kHz	4
-------------0.3 MHz 10^6- 1 MHz	0.1 - 10 MHz	300 kHz	5
RADIOFREQUENCIES 10^7		3 MHz	6
(RF) 10^8	10 - 300 MHz	30 MHz	7
--------- 300 MHz 10^9- 1 GHz		300 MHz	8
MICROWAVES $\underline{dcm}$ 10^{10} cm	300 MHz - - 300 GHz	3 GHz	9
(MW) ---- 10^{11} mm	with separate levels for stationary and rotating fields	30 GHz	10
----------- 300 GHz 10^{12}		300 GHz	11

most of the above problems, would be considered too high in view of possible health hazards. So, the environmental MW/RF safety level has to be arbitrary set somewhere in the range of 1 μW/cm^2 - 1 mW/cm^2, taking into advance a complexicity and variability of pollution by EMFs. Still more, there is a need for establishing safety levels for at least few frequencies that evoke bioeffects at different levels of intensity, and for the frequencies below RFs (below 0.1-0.3 MHz) a separate levels for E and H fields intensities should be considered. The last remarks are more important for occupational safety standards, but to some extent should be also taken into advance for environmental exposures.

At the environmental EMF exposure the physical principles of far field (plane wave) are operating and thus, it is postulated for the future to establish one integrated safety level for the whole spectrum of EMFs (except of ELFs) of 1 kHz - 300 GHz with one measurement of E field (V/m), H fields (A/m) or field power density (W/m^2) covering the whole range and additional level for power frequency (kV/m). The present tendencies in East European countries are however far from this postulate. E.g. the recent proposal for Commecon countries of EMF environmental safety levels (Table IV) sets the levels for 6 sub-ranges, covering only the RF/MW frequencies. In our opinion this is an echo of occupational safety standards that should be precised for certain frequencies occuring in the place of work and the authors of this proposal did not take into advance the complexicity of EMFs in the environment.

GUIDELINES FOR OCCUPATIONAL SAFETY CRITERIA

In case of occupational exposure the situation appears to be different from that discussed for environmental EMF pollution. The main differences can be summarized in the following points:

1. There exists a possibility for reasonably precise characterization of the EMFs occuring at the place of occupational exposure;

2. EMFs occuring in occupational exposure are normally much less complex than those in the environment and in most cases there is a very strong predominance of one frequency, although with a variety of modulations and a possibility of interferences;

3. There exists a chance for reasonably easy and cheap lowering of the occupational exposure intensities by ergonomy of the working place, screens, individual protecting devices (suits,glasses,etc.);

4. Period of occupational exposure can be controlled and limited in time;

5. High risk groups of people (children, pregnant women, senior citizens, subjects with health disorders) are excluded from occupational exposures;

6. With proper organization of the industrial/military medical services it is relatively easy (but also reasonably expensive) to select the personnel for occupational exposure and to control the health status periodically during employment.

It is the present author's feeling that in case of occupational exposure the main importance should be attached to medical/epidemiological data (including possible delayed health hazards !!) , while theoretical considerations and experimental observations should have only the adjuvant application. In another words, the final effect should be decisive for evaluation whether or not the really occuring levels of occupational exposure are safe. This assumes however the precise evaluation of individual exposures during relatively long periods of time (e.g. records of exposure, measurements of power density and characterization of EMFs in the place of work), careful and specialistic selective and periodic medical examinations, as well as further registration and medical examinations of subjects that from various reasons quit their employment in the EMF environment. Application of modern epidemiological techniques to evaluate the data and use of valid control groups are further crucial points for obtaining conclusive results. The best solution seems to be to control the whole population (e.g. the whole population of career military personnel) living under similar social conditions, differentiate and well define (individual integrated exposure to EMFs) subpopulation(s) exposed and non-exposed to EMFs and to perform prospective epidemiologic study during 10-20 years. A variety of other environmental and occupational factors (e.g. consumption of tabacco and alcohol, chemical pollutants, occupational and social stressors) should also be included in the analysis. This is however the ideal (but expensive) solution and in the present literature we were not able to find a single study meeting this requirements. We are even not aware of any such study being in progress.

In this situation, as in the case of environmental exposure, the occupational safety levels have to be established arbitrally, basing on scarce and fragmentaric medical data with over-estimating of results of experimental investigations on animals. As there exists a general agreement that for RF/MWs no significant harmful effects occur at resonant frequencies below 10 W/m^2 (1 mW/cm^2), e.g., in mice exposed to 2450 MHz MW fields (except of transient adaptation phenomena, that cannot be regarded as health hazards), it seems logical to consider this level as suitable for occupational exposure without time limits (8-12 hr daily). With limiting time of exposure and with use of individual protection (suits, glasses) it seems to be safe to pass the MW fields of about 10 times higher intensity (10 mW/cm^2) that still does not result under experimental conditions in hyperthermia of the whole body.

In the recent Russian guidelines for occupational exposure standards to MWs (Sawin et al., 1983; Dawydow et al., 1984) there is stated that the 1 W/m^2 for 2 hr daily exposure (a safety coefficient of 10 from the 10 W/m^2 intensity) and the energetic load resulting from field power density of 2 W/m^2 per hour should be advised as occupational safety levels. Both the 10 W/m^2 (1 mW/cm^2) limit as power density not leading to harmful effects and the "safety coefficient" of 10, being a certain echo of our habit of using the decadic system, have been set arbitrally, but at the present state of art nothing better can be adviced. Certainly, as in the case of environmental safety levels, the occupational exposure criteria are subject for amendments in the future.

EMF SAFETY CRITERIA IN POLAND AND COMMECON COUNTRIES

The environmental and occupational safety levels of EMFs operating currently in Poland and proposed recently for Commecon countries are presented in Tables III through VII. Since 1972, when the currently operating act on safety criteria for MWs was edited there appeared the concept of exposure zones (safety, intermediate, hazardous and dangerous zones for occupational exposure; I^o and II° zones for environmental exposure) with upper and lower limits for each zone and defined limitations of exposure in each zone. Establishment of the above exposure zones facilitates marking of regions with unlimited and limited dwelling of human subjects both in the working place and in the environment.

The safety levels operating at present in Poland have been elaborated relatively long ago - 1972 for occupational exposure in MW fields 1977 for occupational exposure in RF fields and 1980 for environmental exposure to ELFs (power frequency), RF and MWs. Since that only slight changes in interpretation of the zones and executory regulations were introduced, but the levels were not changed. Thus, it is generally accepted among experts in this country that the operating safety regulations do not meet criteria of the present knowledge and there exists a need for amendments. However, as at present there is a tendency for unification of EMF safety levels for European Commecon countries and the International Working Group of Experts is engaged in this issue it was decided not to change the Polish regulations in this matter until the uniform levels will be definitively accepted.

Safety Levels for Environmental Exposure

For environmental exposure only power frequency (50 Hz in Poland)
RFs (ranges of frequency 0.1-10 MHz and 10-300 MHz) and MWs (300 MHz -
300 GHz with separate levels for stationary and rotating fields) were
taken into advance in the Polish act edited in 1980. For each of the
above frequencies the minimal and maximal environmental safety levels
were arbitrary established and two exposure zones (I^o and II°) with
field intensities between the above minimal and maximal levels have
been delimited (Table III). It is assumed that below the minimal levels (1 kV/m for ELFs, 5 V/m for 0.1-10 MHz, 2 V/m for 10-300 MHz,
0.025 W/m^2 - 2.5 μW/cm^2 for stationary 300 MHz - 300 GHz and 0.25 W/m^2
- 25 μW/cm^2 for rotating 300 MHz-300 GHz MW fields) there is no need
for controlling the exposure and it is believed that EMFs below these
intensities are safe and do not cause any effects, including the delayed health disorders. On the other side, maximal EMF intensities
allowed for environmental exposure (10 kV/m for ELFs, 20 V/m for
0.1- 10 MHz, 7 V/m for 10 - 300 MHz, 0.1 W/m^2 - 10 μW/cm^2 for stationary 300 MHz - 300 GHz and 1 W/m^2 - 100 μW/cm^2 for rotating 300 MHz -
300 GHz MW fields) are also considered being safe, but not advisable
for permanent, uncontrolled exposure (e.g., location of buildings, hospitals, schools, etc.). At field intensities between minimal and maximal stated by the act there are in force certain limitations (see
definition of the II° exposure zone); it is also assumed that exposures at the above EMF intensities will be in practice limited in time
(to few hours daily), although still not controlled.

The safety levels of environmental exposure to RF/MW radiation
proposed recently for Commecon countires by Russian experts are presented in Table IV. This proposal does not follow the rule of two exposure zones and sets only one environmental level (maximal), although
the 30 kHz - 300 GHz range of frequencies is divided into 6 subranges
(Table IV), while in the Polish restrictions there are only 3 subranges (0.1-10 MHz, 10-300 MHz and 300 MHz-300 GHz, Table III). Thus, the
above Russian proposal (Table IV) is still a subject of criticism by
Polish experts, as it would be desirable from practical point of view
to establish only one integrated environmental safety level for the
whole range of RF/MWs (0.3 MHz - 300 GHz) due to complexity of EMFs
in the environment. As the single safety level for MWs and RFs is
still a controversial issue, we propose at present 5 V/m for 0.3 -
300 MHz and 0.5 W/m^2 (50 μW/cm^2) for 300 MHz-300 GHz (without

Table III.

CURRENTLY OPERATING AND PROPOSED ENVIRONMENTAL SAFETY LEVELS
IN POLAND.

Frequency range	Operating environmental safety levels			Proposed environmental safety levels		
	L e v e l		Exposure zone	L e v e l		Exposure zone
	Mini-mal	Maxi-mal	II°	Mini-mal	Maxi-mal	II°
Power frequency (50 Hz)	1 kV/m 10 kV/m		1-10 kV/m	no changes, untill more medical informations will be collected and problem of ELF relation to cancer (leukemia) will be solved		
1 - 100 kHz	not established			50 V/m 50-100 V/m 0.5 A/m 0.5-1 A/m 100 V/m 1 A/m		
0.1 - 10 MHz	5 V/m 20 V/m		5-20 V/m	5 V/m 5 - 15 V/m 15 V/m		
10 - 300 MHz	2 V/m 7 V/m		2- 7 V/m	for the whole range 0.1 - 300 MHz		
300 MHz - 300 GHz stationary fields rotating fields	0.025 W/m^2 (2.5 μW/cm^2) 0.1 W/m^2 (10 μW/cm^2) 0.25 W/m^2 (25 μW/cm^2) 1 W/m^2 (100 μW/cm^2)		0.025-0.1 W/m^2 0.25 - 1 W/m^2	0.5 W/m^2 (50 μW/cm^2) 2 W/m^2 (200 μW/cm^2) for all MW fields		0.5-2 W/m^2

differentiation of stationary and rotating MW fields) with the II°
exposure zone at 5 - 15 V/m for RFs and 0.5 - 2 W/m^2 (50 - 200 μW/cm^2)
for MWs. These proposals, summarized in Table IV significantly increase
environmental safety levels for MWs and thus the proposed criteria
are still a matter of discussion.

Table IV.

PROPOSAL (1987) OF UNIFORM ENVIRONMENTAL EMF SAFETY LEVELS
FOR EAST EUROPEAN COUNTRIES.

Frequency (Hz)	Range No.	Maximal intensity permissible for uncontrolled environmental exposure
1		
10 below	1	not defined
10^2 --- 30 Hz	2	not defined
10^3 --- 300 Hz - 1kHz	3	not defined
10^4 --- 3 kHz	4	not defined
10^5 --- 30 kHz	5	25 V/m
10^6 ---300 kHz - 1MHz	6	15 V/m
10^7 --- 3 MHz	7	10 V/m
10^8 --- 30 MHz	8	8 V/m
10^9 -- 300 MHz - 1GHz	9	0.1 W/m^2 (10 μW/cm^2)
10^{10}--- 3 GHz	10	0.1 W/m^2 (10 μW/cm^2)
10^{11} -- 30 GHz	11	not defined
10^{12}-- 300 GHz		

The 1980 act, operating at present in Poland, defines the I$^\circ$ exposure zone (above maximal environmental safety level) as prohibited for uncontrolled exposure (dwelling of human beings is forbidden, except of those selected medically for work with EMF sources). In the II$^\circ$ exposure zone (between minimal and maximal environmental EMF intensity levels) a temporary (not stated in time period) dwelling of human subjects is allowed, connected with management, touristic or recreative activities, however in this zone it is not allowed to localize living quarters and buildings qualified for special protection (hospitals, schools, kindergartens, infants' day nurseries, boarding school, sanitarium). It is adviced that in the II$^\circ$ exposure zones parks and recreative areas, as well as agricultural farms should be organized.

Safety Levels for Occupational Exposure

The safety criteria for occupational exposure operating at present in Poland (Table V) cover only the RFs (0.1 - 300 MHz) and MWs (300 MHz - 300 GHz) and were elaborated and edited in 1977 and 1972, respectively. Maximal intensities of RF/MWs allowed for occupational exposure were established and 4 exposure zones have been differentiated:

- safety zone - uncontrolled exposure of personnel, intensities below maximal environmental safety level;
- intermediate zone - the personnel (selected for employment at EMF-generating devices on base of selective medical examinations and fit for work in the EMF environment) may dwell in this zone without time limits during working day (8-12 hr, depending on type of work or service);
- hazardous zone - the personnel (selected, as above) may dwell in this zone only for a limited time (Table V) and there is no need for individual protection (suits, glasses);
- dangerous zone - the trained and selected personnel may dwell only temporary in this zone with individual protecting devices (suits, glasses) limiting the exposure levels to at least those occuring in the hazardous zone.

On base of field measurements on working places in the vicnicity of EMF-generating devices the above 4 zones have to be boarded and signed.

Safety restrictions operating in Poland state that in employment of personnel in the EMF environment four basic principles have to be obeyed:

1. Selective general, ophtalmologic and neurologic medical examination before employment at EMF-generating devices with precise criteria for fittness to occupational exposure;

2. Periodic (every 1-3 years, depending upon type of work or service) medical examinations with individual records(each employee has his own booklet with records of exposure and medical examinations);

3. Elimination of personnel showing certain health disorders (listed in the instruction for RF/MW safety, including e.g. changes in eye lens, liability of blood pressure, vegetative neuroses, EEG changes);

4. Time limits for dwelling in the hazardous zones and obligatory use of protective suits and glasses in the dangerous zone.

Table V.

OCCUPATIONAL EMF SAFETY LEVELS CURRENTLY OPERATING IN POLAND.

Frequency range	Date of issue of the act	Safety zone*	Inter- mediate zone**	Hazar- dous zone***	Time limits for hazar- dous zone	Dange- rous zone****
Power frequency (50 Hz)	pending (1985-1987)	1 kV/m	1-10 kV/m	10-25 kV/m	not stated	above 25 kV/m
1 - 10 kHz	proposed (1987)	100 V/m 1 A/m	100-200 V/m 1-2 A/m	above 200 V/m above 2 A/m	2 hr daily for 1-100 kHz	not defined
10 - 100 kHz	proposed (1987)	50 V/m 0.5 A/m	50-100 V/m 0.5-1 A/m	above 100 V/m above 1 A/m		
100 kHz - 10 MHz	1977	20 V/m 2 A/m	20-70 V/m 2-10 A/m	70-1000 V/m 10-250 A/m	$\dfrac{560}{E(V/m)}$ hours	above 1000 V/m above 250 A/m
10 - 300 MHz	1977	7 V/m	7-20 V/m	20-300 V/m	$\dfrac{3200}{E^2(V/m)}$ hours	above 300 V/m
300 MHz - 300 GHz (MICROWAVES) stationary fields	1972	0.1 W/m²	0.1-2 W/m²	2-100 W/m²	$\dfrac{32}{P^2(W/m^2)}$ hours	above 100 W/m²
non-stationary fields (rotating, modulated, pulsed, etc.)		1 W/m²	1-10 W/m²	10-100 W/m²	$\dfrac{800}{P^2(W/m^2)}$	above 100 W/m²

* maximum level for uncontrolled exposure; ** no time limits for medically selected personnel working with EMF-generating devices; *** time limit for medically selected personnel, no need for protective suits; **** incidentally, for trained and selected personnel, protective suits obligatory.

Table VI.

1984 PROPOSAL OF THE INSTITUTE OF LABOR MEDICINE IN LODZ,POLAND
OF OCCUPATIONAL SAFETY LEVELS FOR ELECTROMAGNETIC FIELDS.

Frequency range	Maximum environmental exposure intensity*	Basic occupational exposure limit**	Additional occupational exposure limit*** (1 hr daily)
50 - 100 Hz	500 V/m 1.5 A/m	5 kV/m 15 A/m	15 kV/m 40 A/m
100 Hz - 10 kHz****	500 V/m : 0.01f 1.5 A/m : 0.01f	5 kV/m:0.01f 15 A/m:0.01f	15 kV/m:0.01f 40 A/m :0.01f
10 kHz - 10 MHz	5 V/m 0.015 A/m	50 V/m 0.15 A/m	150 V/m 0.4 A/m
10 MHz - 300 MHz	2.5 V/m 0.007 A/m	25 V/m 0.07 A/m	75 V/m 0.2 A/m
300 MHz - 300 GHz (MICROWAVES) stationary fields non-stationary fields	 0.1 W/m^2 (10μW/cm^2) 0.25 W/m^2 (25μW/cm^2)	 2.5 W/m^2 (250μW/cm^2) 5 W/m^2 (0.5 mW/cm^2)	 25 W/m^2 (2.5 mW/cm^2) 50 W/m^2 (5 mW/cm^2)

* below this intensity uncontrolled exposure of the whole population
is permissible;

** below this intensity uncontrolled exposure of trained and medically selected personnel is permissible;

*** additional occupational exposure is allowed for 1 hr daily
at EMF intensities between the basic limit (**) and that stated
as maximum without protective suits;

**** for EMF frequencies 100 Hz - 10 kHz the maximum environmental
and occupational exposure limits are calculated by dividing the
given value by 0.01 of frequency, expressed in Hz. E.g., for

1 kHz - 1000 Hz the maximum environmental exposure intensity
counts   500 V/m : 0.01 x 1000 Hz - 500 V/m : 10 - 50 V/m

The medical records of personnel exposed occupationally to RF/MW
radiations in Poland, at least those noted for the military carrer
personnel did not deliver till now alarming symptoms that may be related to the exposures and number of personnel eliminated from employment with use and/or repair of RF/MW-generating devices increases
only insignificantly the morbidity observed in personnel not exposed
to RF/MWs. Thus, the present occupational safety regulations are considered as satisfactory, although there is a general agreement that
the intensities of magnetic fields (H, expressed in A/m) for 100 kHz
- 10 MHz (see Table V) were set to high. A more recent proposal of
amendments in occupational safety levels (Table VI) established much
lower intensities of magnetic fields at the above frequencies.

Some confusion has been however thrown recently into the operating restrictions by the retrospective epidemiological analysis of
cancer morbidity among carrer military personnel in Poland during
a decade of 1971-1980 (Szmigielski et al.,1987). In general,this study has documented that the subpopulation of the personnel exposed
occupationally to MW/RFs suffered about 3 time more frequently from
various neoplastic diseases with preference for certain hemopoietic
and lymphatic neoplastic syndromes, alimentary tract and skin neo-
plasms, compared with non-exposed personnel. The risk of neoplastic
diseases has shown for the tested population a close relation with
period of exposure in MW/RF fields and the more detailed analysis
of data (Szmigielski et al., 1987) suggests that in this case the
MW/RF occupational exposure accelerated clinical appearance of neo-
plasms that without the exposure would develop few years later (postulated tumor-promoting, but not carcinogenic potency of MW/RF radiations). As this is the single and first epidemiologic study on cancer
morbidity in personnel exposed occupationally to MW/RFs, the problem
needs both confirmation and further investigations, untill it may influence the occupational safety criteria. The prospective epidemiologic studies, planned for 1985-1990 and being now in progress, should
deliver new and more convincive informations on cancer risk related
to MW/RF occupational exposure. Thus, it is the present author's
feeling that the occupational MW/RF safety criteria and levels
should not be changed untill the problem of cancer risk will became
more clear.

In 1984 the group of research workers from the Institute of Labour Medicine in Lodz,Poland, headed by Professor H.Mikolajczyk,M.D.,
proposed new occupational exposure criteria for the whole EMF spectrum

(Table VI). This proposal is at present a subject for discussion.
The same group of experts is working on occupational exposure criteria
for EMF of 1 - 100 kHz frequencies, the range not covered by the current Polish regulations. On base of field measurements in factories
using 1-100 kHz generators and in the vicnicity of VDTs, experimental
investigations in rats and medical examinations of about 200 people
working with use of industrial generators of 10-50 kHz generators
Dr H.Mikolajczyk and his coworkers from the Institute of Labour Medicine in Lodz, Poland have proposed in 1987 the safety levels for
occupational exposure at 1 - 100 kHz (Table V), being at present a
subject of discussion by Polish experts. The authors of the proposal
divided the range 1-100 kHz in two subranges 1-10 kHz and 10-100 kHz
with separate safety levels for electric (E) and magnetic (H) fields
(Table V). The proposed intensities of E fields (100-200 V/m for
1-10 kHz and 50-100 V/m for 10-100 kHz) did not stir up meaningful
reservations from Polish experts, however the intensities of H fields
(1-2 A/m for 1-10 kHz and 0.5-1 A/m for 10-100 kHz) were regarded as
too low in view of the present knowledge of bioeffects related to
the above frequencies. The subject was left for further considerations.

It may be worth noting here a very recent Russian criteria for
occupational exposure criteria for radiolocatory devices (radars),
summarized in Table VII. As the radar devices work in few well defined frequencies, it seems desirable from the practical point of
view to adjust occupational safety criteria to the particular frequency and oscillation of the antenna. The Russian criteria (Table VII)
are one of the first trials in this direction.

In summary, it may be concluded that the occupational safety
criteria in Poland are relatively old and should be reconsidered in
view of more modern tendencies. In contrary to environmental safety
levels that should be as simple as possible and limited to one level
for the whole range of 0.1 MHz - 300 GHz, the occupational safety
criteria should be adjusted to EMFs really occuring at the particular
place of work and there is a need for elaborating separate occupational exposure criteria for power frequency and ELFs (1-1000 Hz),
1-100 kHz, 100 kHz-10 MHz, 10 MHz-300 MHz, 300 MHz-30 GHz (decimeter
and centimeter MWs) and 30-300 GHz (millimeter MWs). For occupational exposure to MWs there is an additional need for establishing separate levels for stationary and rotating (in practice radar) fields.

Table VII.

RUSSIAN PROPOSAL (1986) OF OCCUPATIONAL SAFETY LEVELS FOR RADIO-LOCATION DEVICES (RADARS).

Type of the radar system	Wave length (cm)	Conditions of operation		Relation of time permissible for exposure to total daily working time	Maximal allowed exposure intensity (W/m^2)
		Frequency of antenna oscillations (Hz)	Time of irradiation with unchanged intensity		
Meteorologic RLD*	0.80 +0.12	below 0.1	below 0.03 of the brooming period	0.5	140
		0	below 12 hr daily	1.0	10
Meteorologic and other with similar conditions of operation	3.0 ± 0.6	below 0.1	below 0.o4	0.5	60
		0	below 12 hr daily	1.0	10
	10.0 ± 1.5	0	below 12 hr daily	0.5	20
	17.0 ± 2.2	0	below 12 hr daily	0.5	24
				1.0	12
Typical RLDs in aviation and other with similar conditions of operation	10.0 ± 2.0	below 0.25	below 0.05 of the brooming period	1.0	15
	23.0 ± 3.2	below 0.25	below 0.02	1.0	20
	35.0 ± 4.0	below 0.25	below 0.02	1.0	25

* RLD - radiolocation devices (radars).

REFERENCES

Adey, W.R.,1981, Ionic nonequilibrium phenomena in tissue inter-
    actions with electromagnetic fields,in : "Biological Effects of
    Non-ionizing Radiation, K.H.Illinger,ed.,ACS Symposium Series,
    157: 271.
Balcer-Kubiczek,E.K.,and Harrison,G.,1985, Evidence for microwave
    carcinogenesis in vitro, Carcinogenesis, 6: 859.
Baranski,S., and Czerski,P.,1976, "Biological Effects of Microwaves",
    Dowden,Hutchinson and Ross, Stroudsburg,Pa.
Buys,O.V., Kartun,K., Pieper,S., and Adey,W.R., 1986, Microwaves act
    at cell membranes alone or in synergy with cancer-promoting
    phorbol esters to enhance ornithine decarboxylase activity,
    Bioelectromagnetics, 7:
Buys,O.V., Lundak,R.L., Fletcher,R.M., and Adey,W.R., 1984, Altera-
    tions in protein kinase activity following exposure of cultured
    human lymphocytes to modulated microwave fields, Bioelectro-
    magnetics, 5: 341.
Choy,R., Monro,J., and Smith,C., 1987, Electrical sensitivity in
    allergy patients, Clinical Ecology, 4: 93.
Dawidow,B.I., Tichonczuk,W.S., and Antipow,W.W.,1984 "Biological
    Action, Safety and Protection Against Electromagnetic Radia-
    tions" (in Russian), Elektroatomizdat, Moscow.
Elder,J.E., and Cahill,D.F.,eds.,1984, "Biological Effects of Radio-
    frequency Radiation",EPA Report 600/8-83-026F,US Environmental
    Protection Agency,Research Triangle Park,N.C.
Guy,A.W., Chou,C.K., Kunz,L.L., Crawley,J. and Krupp,J.,1985, "Effects
    of Low-level Radiofrequency Radiation Exposure on Rats,Vol.9,
    Summary", Report No. USA FSAM-TR-85-64, USAF School of Aerospace
    Med., Brooks Air Force Base.
INIRC-IPRA, 1984, International Non-ionizing Radiation Committee of
    the International Radiation Protection Association, Interim
    Guidelines on Exposure Limits to Radiofrequency Electromagnetic
    Fields in the Frequency Range of 300 kHz to 300 GHz, Health
    Physics, 46 : 4.
Mc Dowall, M.E., 1983, Leukemia mortality in electrical workers in
    England and Wales, Lancet, 1: 246.
Milham,S.,Jr., 1982, Mortality from leukemia in workers exposed to
    electrical and magnetic fields, New England J.Med., 307: 249.
Minin,B.A., 1974,"Ultra-high Frequencies and Human Safety"(in Russian),
    Elektroatomizdat,Moscow.
Romero-Sierra,C., 1984,Bioeffects of electromagnetic waves, in "Review
    of Radio Science 1981-1983", Bowhill,S.A.,ed.,Intern.Union of
    Radio Sciences (URSI), Brussel,Belgium,p.K-1.
Sawin,B.M.,Nikonowa,K.W., Lobanowa,E.A., Sadczikowa,M.N., and
    Lobed,E.K.,1983, Nowoje w normirowanij elektromagnitnych izlu-
    czenii mikrowolnowo diapazona (in Russian, Novelties in safety
    standards of EM radiations of the microwave range), Gigiena
    Truda, 3: 1.
Szmigielski,S., Bielec,M., Lipski,S.,and Sokolska,G., 1987, Immuno-
    logic and cancer-related aspects of exposure to low-level micro-
    wave and radiofrequency fields, in : "Fundamentals of Modern
    Bioelectricity", Marino,A.A.,ed., M.Dekker Co., N.York.
Szmigielski,S., Szudzinski,A., Pietraszek,A., Bielec,M., Janiak,M.,
    and Wrembel,J.K., 1982, Accelerated development of spontaneous
    and benzopyrene-induced skin cancer in mice exposed to 2450 MHz
    microwave radiation, Bioelectromagnetics,3: 179.

WESTERN EUROPEAN POPULATION AND

OCCUPATIONAL RF PROTECTION GUIDES

Kjell Hansson Mild

National Institute of Occupational Health
Box 6104, S-900 06 UMEÅ, Sweden

INTRODUCTION

Very few Western European countries - if any - have a legislation regarding both occupational and population exposure to radiofrequency electromagnetic fields. However, most countries have some form of regulation of the occupational exposure for frequencies above about 10 MHz, either as limits, standards or as guidelines. There is also an accordance in the values and generally it can be said that most countries tend to follow the values given by ANSI C95-1982 with only minor divergences.

The setting of standards for the working environment is always a political decision. Usually the body that issues the standard has a board which consists of representatives for all parties on the labour market, for instance the trade unions and the trade associations. The proposal for a new standard is thus discussed on a risk-benefit basis, where the economical impact is put against the risk of a possible health hazard. The practical aspects of the enforcement of the standard are also taken into account. Here things like available instruments for measurements of RF-field strengths and frequency, calibration, accuracy etc. also must be considered in the standard setting.

Thus, with this in mind it is understandable that it is not always possible to achieve a standard that for all possible exposure situation would give for instance SAR-values below a certain level, but those worst case conditions when the SAR values could be exceeded may have been considered to occur so seldom that the hazard involved is estimated to be minimal and acceptable.

In the last few years several countries have presented proposal for new standards and guidelines extending down to the ELF-region, and values are given for both electric and magnetic fields separately and not necessarily in far field proportionality. Interesting proposal has come from Western Germany and Great Britain, and these will be presented below. In Sweden new limits were recently adopted for frequencies down to 3 MHz, and these values will be presented. The recently accepted Finnish emission standard for high frequency apparatus will also be discussed.

The general population exposure is mainly caused by various products such as microwave ovens, burglar alarm, walkie-talkies etc. and in many countries this is dealt with by use of emission standards.

SWEDEN

In Sweden the limits are set by the National Board of Occupational Safety and Health, and for the radiofrequency radiation in workplaces a standard has been in effect since 1977. The permissible values are given as 6-minutes time-average values and ceiling values - rms values for 1 sec. For frequencies from 10 MHz to 300 MHz the long term values is 50 W/m^2 and from 300 MHz to 300 GHz 10 W/m^2 applies. The ceiling value is 250 W/m^2 for the whole frequency range.

Recently a new set of values were issued by the Board. The new values which will be valid from January 1, 1988, cover the frequency range down to 3 MHz. The limits are based on thermal considerations and the documentation behind the IRPA interim guidelines (IRPA, 1984) and the ANSI C95.1-1982 have been used as starting points for the standard setting. In contrast to the previous Swedish regulation giving the values in terms of power density the electric and the magnetic field strength are now the fundamental quantities being restricted. The recent finding of whole body resonance at about 30-40 MHz for persons in contact with high frequency electric ground has been taken into account as a lowering of the values in these cases (Grønhaug and Busmundrud, 1982; Hill,1984). In Tables 1 and 2 the new values are summarized.

An exclusion paragraph has also been included to take care of the mobil radio equipment operating on power less than 7 W. In the frequency range up to 1 GHz the radiation emitted from an extended antenna is excluded from the limits given above.

The manufacturer of various high frequency apparatus are also requested under the new law to provide safety instructions for the operating of the equipment in such a way that the exposure is less than the given limits.

The reason for not using the frequency dependent limits as is done by IRPA and ANSI for the high frequency band is that it would be very cumbersome for the labour inspectorate to beside field measurements also include frequency measurements of for instance glue dryers and RF sealers. These machines are often not frequency stable during the operation cycle, and thus a value would have to be picked for each machine and different limits given. Furthermore, it would be very difficult for the men operating the different machine to make the distinction between these different limits. In view of this it was thought better to have limits constant over a wide frequency range than a slope in frequency. The constraints imposed on for instance 13 MHz machines by this is more than compensated for by the gain in simplicity at 27 MHz.

The guidelines for the general population exposure in Sweden is issued by the National Institute of Radiation Protection. In a recommendation from 1978 it is said that the power density should be kept below 10 W/m^2 , but if possible the level of 1 W/m^2 should be applied for long term exposure. The Institute have declared that very shortly new guidelines will be issued.

FINLAND

In June 1985 the government in Finland issued a law regarding stray fields from high frequency (10 MHz-100 MHz) apparatus. The basic exposure limits are 60 V/m and 0.2 A/m for a typical working day. These values are rms field strength determined in a period of 6 minutes or the mean from

TABLE 1

Swedish occupational exposure limits as given by the National Board of Occu pational Safety and Health ordinance 1987:2, valid from Jan 1, 1988. The field strengths values are for positions where the operator can be, but should be measured as undisturbed fields. The values are time-averaged over any 6 minute period during the working day.

Frequency	E (V/m)	H (A/m)
3 - 30 MHz	140	0.40
30 - 300 MHz	60	0.16
0.3 - 300 GHz	60	--

TABLE 2

Ceiling values never to be exceeded . For pulsed fields these are the time-average during 1 second.

Frequency	E (V/m)	H (A/m)
3 - 300 MHz	300	0.8
0.3 - 300 GHz	300	--

TABLE 3

When the worker can be closer than 10 cm to what can be considered as high frequency ground the values in Tables 1 and 2 should be replaced by the values below for frequencies up to 60 MHz.

Frequency	E (V/m)	H (A/m)
3 - 30 MHz	47	0.13
30 - 60 MHz	20	0.05
Ceiling values:		
3 - 60 MHz	100	0.27

five cycles. Ceiling values never to be exceeded are also given; 300 V/m and 0.8 A/m, respectively. Measurements should be taken immediately after installation of the equipment and the protocol should be sent to the National Board of Occupational Safety and Health. New measurements are to be taken at least every third year or whenever a change has been made on the machine that might influence the stray fields. The accuracy in the measurements should be within 1 dB.

For exposure times shorter than 1 h of the working day the highest permissible values are obtain by the formula

$$t = k/s$$

where $k = 36000$ Ws/m^2, t the total time in seconds, and s is the equivalent power density.

These regulations are valid for all new equipment installed after January 1, 1986, and from January 1, 1989, they are also valid for older equipment.

According Dr K. Jokela at the Finnish Centre for Radiation and Nuclear Safety no other official exposure standard or even proposal exist for RF-radiation. A preliminary paper have been presented regarding occupational exposure and general public. The paper follows in principle the IRPA guidelines.

UNITED KINGDOM

In 1982 The National Radiological Protection Board (NRPB) published a proposal regarding permissible limits for exposure to RF fields (see further NRPB,1982, and Allen and Harlen, 1983). The principal basis for the Board's proposal was that exposure at levels likely to cause harmful effects should be prevented. For occupational exposures levels at which perceptible but harmless effects occur are permissible but should be avoided if possible. For the general population also perceptible but harmless effects should be avoided. For microwave and radiofrequency radiation the Board accepts whole body SAR less than 0.4 W/kg.

Now NRPB has prepared a consultative document based on the earlier proposals and comments received on them (NRPB, 1986). The draft consists of a set of basic limitations on electric currents and current densities in the body which apply to frequencies roughly below 500 kHz, and above this frequency a set of restrictions on the rate of power dissipation in the body.

In the Tables 4 and 5 the guidelines for whole body exposure are given. It can be noted that the proposal includes even static fields. Also to be noted is that the ratio of E/H, i.e. space impedance, varies from 377 at microwave frequencies to 20 at 50 kHz.

Guidelines for public exposure are also proposed, and at frequencies below 1 MHz the limits are set to avoid electric shocks from large ungrounded metal objects in electric fields. The guidelines also take into account the theoretically different absorbtion characteristics of small children compared to adult from 100 MHz to 1.5 GHz. The values shown in Tables 5 and 6 are about a factor 2-3 lower than for occupational exposure, and for further details see NRPB (1986).

TABLE 4

Guideline levels for whole body occupational exposure to frequencies below
30 MHz (rms values for alternating fields) for an avereage total period
not exceeding 2 hours per day.

Frequency	Electric fields V/m	Magnetic fields A/m
Static	40,000	16,000 (20 mT)
< 10 Hz	30,000	7,500
10 Hz - 50 Hz	30,000	$7.5 \times 10^{**4}/f$
50 Hz - 750 Hz	$1.5 \times 10^{**6}/f$	$7.5 \times 10^{**4}/f$
750 Hz - 50 kHz	2,000	100
50 kHz - 300 kHz	2,000	$5.0 \times 10^{**6}/f$
300 kHz - 10 MHz	$6.0 \times 10^{**8}/f$	$5.0 \times 10^{**6}/f$
10 MHz - 30 MHz	60	$5.0 \times 10^{**6}/f$

f is the frequency in hertz (Hz).

TABLE 5

Guideline levels for whole body occupational exposure to frequencies in
the range 30 MHz to 300 GHz (rms values) for an average total period not
exceeding 2 hours per day.

Frequency	Power density W/m^2	Electric field V/m	Magnetic field A/m
30 MHz-100 MHz	10	60	0.16
100 MHz-500 MHz	f/10	$6.0 \sqrt{f}$	$0.016 \sqrt{f}$
500 MHz-300 GHz	50	135	0.36

f is frequency in megahertz (MHz).

TABLE 6

Guidelines for areas of public access for frequencies below 30 MHz (rms values for alternating fields) for an average total period not exceeding 5 hours per day.

Frequency	Electric fields V/m	Magnetic fields A/m
Static	16,000	6,700
< 10 Hz	12,000	3,000
10 Hz - 50 Hz	12,000	$3.0 \times 10^4/f$
50 Hz - 750 Hz	$6.0 \times 10^5/f$	$3.0 \times 10^4/f$
750 Hz - 50 Hz	800	40
50 kHz - 365 kHz	800	$2.0 \times 10^6/f$
365 kHz - 475 kHz	800	5.5
475 kHz - 580 kHz	$3.8 \times 10^8/f$	5.5
580 kHz - 10 MHz	$3.8 \times 10^8/f$	$3.2 \times 10^6/f$
10 MHz - 30 MHz	38	$3.2 \times 10^6/f$

f is frequency in hertz (Hz).

TABLE 7

Guidelines levels for areas of public access for frequencies in the range 30 MHz to 300 GHz (rms values) for an average total period not exceeding 5 hours per day.

Frequency	Power density W/m^2	Electric field V/m	Magnetic field A/m
30 MHz - 300 MHz	4	38	0.10
300 MHz - 1.5 GHz	f/75	$2.2\sqrt{f}$	$0.006\sqrt{f}$
1.5 GHz . 300 GHz	20	85	0.23

f is frequency in megahertz (MHz).

In August 1986 a proposal to West German VDE DIN norm was published. The proposal covers the frequency range from 0 Hz to 3000 GHz. The limits are set in accordance with today's knowledge to avoid hazards from exposure to EM fields for healthy persons. The standard applies to the general public as well as to the industrial labor force. For frequencies up to 30 kHz the limits are based on the calculation of current density in human as performed by dr J. Bernhardt (1986). The main basis is to avoid stimulation of receptors and individual cells, influences of the heart activity etc. The magnetophospene phenomena is not addressed. It is said that the intention of the standard is to keep the current density below 1 mA/m^2.

For frequencies above 30 kHz the limits are set on a purely thermal basis, and such that a suffucent margin of safety is kept to SAR-values of 4 W/kg where uinwanted effects occur according to animal experiments. In brief the proposal is presented in the Figures 1, 2 , 3 and 4. For further details see VDE (1986) and Rozzell (1985).

To be noted is that special values applies to pacemaker patients in the frequency range up to 30 MHz, and for higher frequencies limits are under preparation.

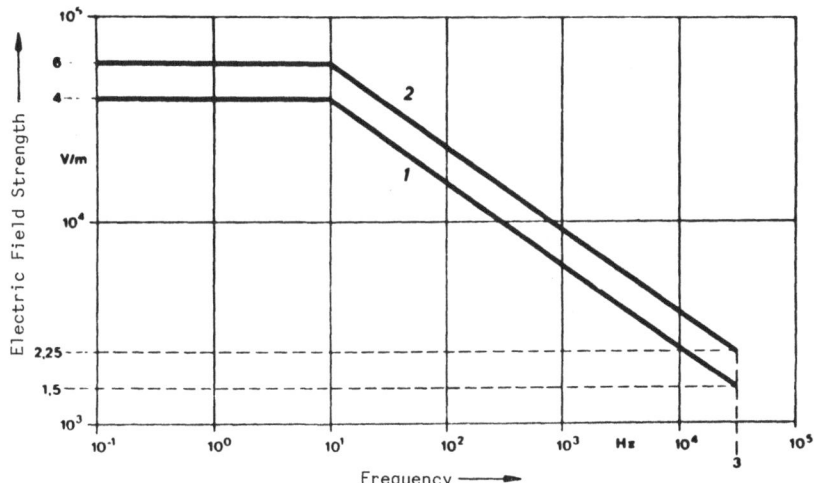

Fig. 1. Limits for the electric field strength for 0 Hz < f < 30 kHz for whole body exposure. Curve 1 shows the allowed rms-values and curve 2 the ceiling values.

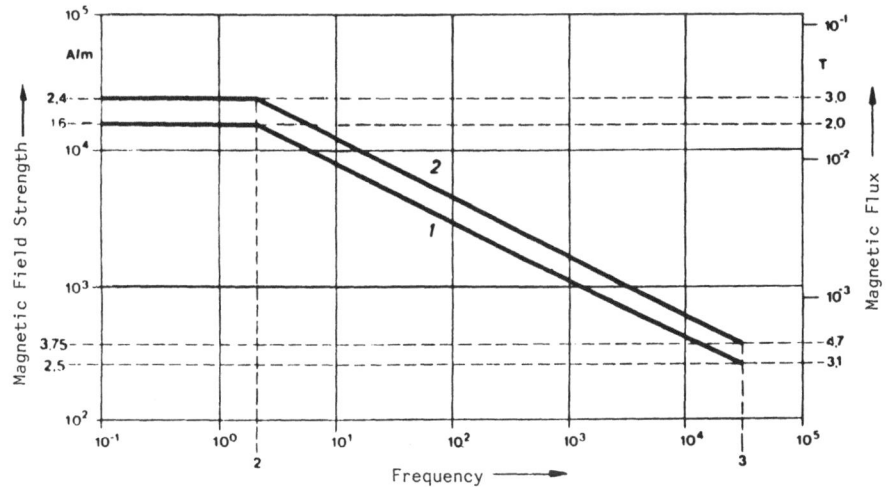

Fig. 2. Limits for the magnetic field strength for 0 Hz < f < 30 kHz for whole body exposure. Curve 1 shows the allowed rms-values and curve 2 the ceiling values.

251

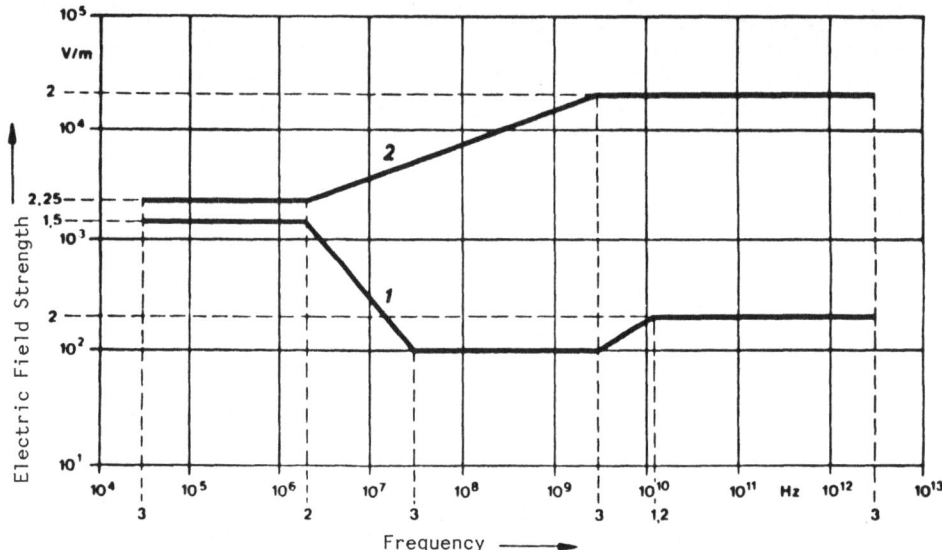

Fig. 3. Limits for the electric field strength for 30 kHz < f < 3000 GHz for whole body exposure. Curve 1 shows the allowed time-averaged values (6 min. and more) and curve 2 the ceiling values.

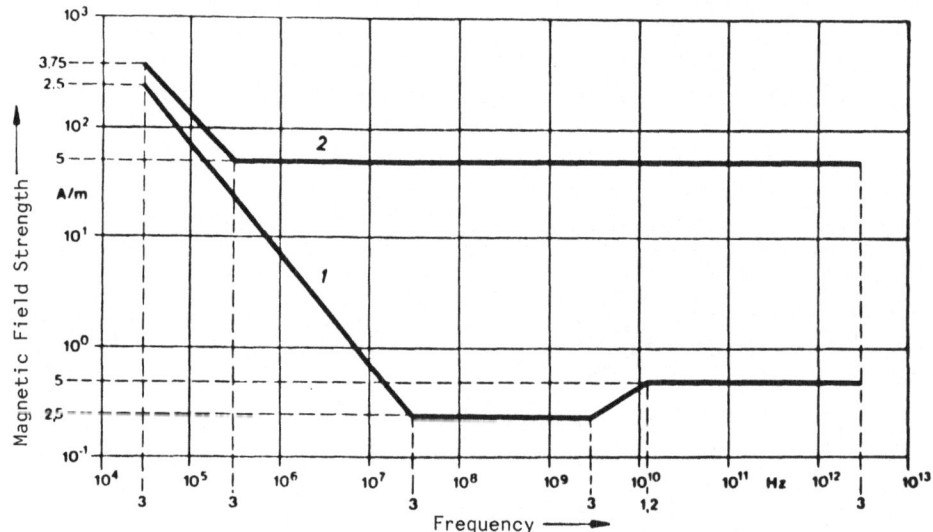

Fig. 4. Limits for the magnetic field strength for 30 kHz < f < 3000 GHz for whole body exposure. Curve 1 shows the allowed time-averaged values (6 min. and more) and curve 2 the ceiling values.

BELGIUM

The only regulation regarding RF radiation in Belgium is given as a general rule regarding protection of man and environment against harmful effects and nuisance caused by non-ionizing radiation, infra and ultra sound (July 12, 1985). This law is only the general frame work that gives the possibility to issue standards for RF fields. However, no such values have been given so far.

FRANCE

In France no rules are given for RF fields and occupational and public exposure. The Institut National de Recherche et de Securité have published a series of paper regarding the hazards from RF exposure and how to reduce stray fields (INRS, 1978, 1982, 1983, 1985). Their recommendations regarding limits follows the AGGIH values closely.

NORWAY

The State Institute of Radiation Hygiene is the body that issues the standards in Norway, and so far they only have a proposal from 1982 (Saxebøl, 1982) to administrative norms. The norm concerns whole body radiation for up to 8 h per day. The values for workplaces are given in Table 6 below. For the general population a safety factor of ten is used for the equivalent power density values.

TABLE 8

Frequency	$(E)^2$ V^2/m^2	$(H)^2$ A^2/m^2	S W/m^2
1 - 6 MHz	$10^{**}5$	0.625	250
6 - 30 MHz	$36 \times 10^{**}5/f^2$	$22.5/f^2$	$9000/f^2$
30 MHz - 1 GHz	4000	0.025	10
1 - 300 GHz	20000	0.125	50

f = frequency in MHz

The values are mean values over 1 h, and the ceiling value never to be exceed for 6 MHz - 300 GHz is given as the equivalent of 250 W/m^2.

The recommendation for general public exposure is to keep that no higher than a factor of ten below the above values for the equivalent power density.

These administrative norms are based upon technical, economical and medical assessments. The norm is not legally enforceable.

REFERENCES

Allen, S.G. and F. Harlen, 1983, Sources of Exposure to Radiofrequency and Microwave Radiations in the UK. NRPB-R144.

ANSI, 1982, ANSI C95.1-1982, American national standard- safety levels with respect to human exposure to radio frequency electromagnetic fields, 300 kHz to 100 GHz.

Bernhardt, J.H., 1985, Evaluation of human exposure to low frequency fields. In AGARD lecture series no 138, The Impact of proposed radio-frequency radiation standards on military operations, available from NATO advisory group for aerospace research and development, 7 Rue Ancelle 92200 Neuilly Sur Seine, France.

Grønhaug, K.L. and Busmundrud, O., 1982, Antenna effect of the human body to EMP. Norweigian Defence Research Establishment, Kjeller, Norway. FFIF/453/153 (In Norwegian).

Hankin, N.H., 1986, The Radiofrequency Radiation Environment: Environmental Exposure Levels and RF Radiation Emitting Sources. EPA 520/1-85-014.

Hill, D.A., 1984, The effect of frequency and grounding on whole-body absorbtion of humans in E-polarized radiofrequency fields. Bioelectromagnetics, 5, pp. 131-146.

INRS, 1978, Le rayonnement électromagnétique "radiofréquences" Applications et risques. ND 1127-92-78 (39-14) CDU 621.37: Cahiers de notes documentaires, No 92, 3e Trimestre 1978.

INRS, 1982, Risques liés aux rayonnements électromagnétiques non ionisants. Cahiers de notes documentaires, No 107, 2e Trimestre 1982.

INRS, 1983, Valeurs limites d'exposition aux agents physiques. Cahiers de notes documentaires, No 110, 1er Trimestre 1983.

INRS, 1985, Protection contre les rayonnements électromagnétiques non ionisants. ND 1552-121-85 CDU 538.56. Cahiers de notes documentaires No 121, 4e Trimestre 1985.

IRPA, 1984, Interim guidelines on limits of exposure to radiofrequency electromagnetic fields in the frequency range from 100 kHz to 300 GHz. Health Physics, 46, pp. 975-984.

NRPB, 1982, Proposal for the health protection of workers and members of the public against the dangers of extra low frequency, radiofrequency and microwave radiations: A consultative document. ISBN 085951 185 5. Chilton, Didcot, Oxon, UK.

NRPB, 1986, Advice on the protection of workers and members of the public from the possible hazards of electric and magnetic fields with frequencies below 300 GHz. A consultative document. ISBN 0 85951 267 3. Chilton, Didcot, Oxon, UK.

Saxebøl, G., 1982, Administrative normer for radiofrekvent stråling for yrkeseksponerte? SIS Rapport 1982:2, State Institute of Radiation Hygiene.

Rozzell, T.C., 1985, West German EMF exposure standard. BEMS Newsletter, no 55, dec 1984/jan 1985.

VDE, 1986, Hazards by electromagnetic fields; Protection of persons in the frequency range from 0 Hz to 3000 GHz. DIN VDE 0848, teil 2.

CANADIAN AND OTHER NATIONAL RF PROTECTION GUIDES

Maria A. Stuchly

Bureau of Radiation and Medical Devices
Health and Welfare Canada
Ottawa, Ontario K1A OL2, Canada

INTRODUCTION

In 1979 the Federal Department of Health and Welfare in Canada issued a safety code which recommended the limits of exposure to radiofrequency (RF) radiation (Health and Welfare Canada, 1979). The limits apply to occupational and general public exposures at frequencies between 10 MHz and 300 GHz. The safety procedures and installation guidelines outlined in that document are for the instruction and guidance of persons employed in federal public service departments and agencies, and those coming under the jurisdiction of the Canada Labour Code. Since the time these recommendations were made, a considerable amount of new information on interactions of RF fields with living systems has become available, and recommendations on exposure levels have been made in several countries and also by an international organization. These developments, as well as the need for a standard at lower radiofrequencies, i.e. below 10 MHz, have stimulated the proposed revision.

In development of health protection standards, such as RF exposure limits, various general rules and criteria can be selected. Preferably, there should be a sufficient data base of adverse effects on human beings and their mechanisms, which permits a quantitative analysis of health risks related to any proposed protection limit. Additionally, it may also be desirable to consider a cost/benefit analysis. This is not the situation in the field of RF hazards. Scientific data on adverse effects on humans including epidemiological investigations are so sparse and incomplete that they are insufficient for even qualitative hazard assessment. Therefore, RF exposure standards are nearly exclusively based on experimental evidence with animals and rather complex physical and physiological extrapolations. Furthermore, interaction mechanisms are not well understood for many of the effects observed. Even for a well established thermal interaction mechanism there is a considerable uncertainty involved in extrapolations of the data from animal to human systems. Under these circumstances, the rationale for the exposure limits is to a large degree based on scientific evidence but at least partly on subjective value judgment of validity of the results reported as well as their health implications. No attempt has been made to perform a risk/benefit or cost/benefit analysis. In general, a conservative approach was taken as the recommended exposure limits have been derived from "the worst-case" exposure conditions.

The proposed standard applies to frequencies from 10 kHz to 300 GHz, to occupational exposures and exposures of the general public. Exposure limits are specified in terms of the field strengths which are derived from specific absorption rates (SAR) (National Council on Radiation Protection, 1981). Limits are also prescribed on the total current passing through a person making contact with any conducting object located anywhere within areas accessible to people.

The objectives of this contribution are:

(1) to present a current proposal of a Canadian Federal RF exposure standard;
(2) to describe the rationale as based on the present state of knowledge;
(3) to outline similarities and differences between this proposal and other national protection guides.

SCIENTIFIC DATA

The main scientific findings that impacted on the proposal are: (1) high local SARs in the ankles for persons in close proximity to electrical ground, (2) thresholds for perception, electric shock and burns due to RF fields, (3) weaker coupling of the magnetic field than the electric field in the lower frequency range (below 10 MHz), and (4) data on the energy deposition from portable transmitters.

Current densities in an electrically-grounded man at radio frequencies below 50 MHz have recently been considered (Gandhi et al, 1985, 1986; Guy and Chou, 1982; Hill and Walsh, 1985, Tell et al, 1987). For a human being exposed to a low frequency electric field parallel to the body main axis, the magnitude of the current is:

$$I = AfE \tag{1}$$

where A is a factor proportional to the square of the subject's height, and also weakly dependent on frequency, f is the frequency, and E is the strength of the electric field. For a man 1.75 m tall A is about 275 $\mu A/(Vm^{-1} MHz)$ for frequencies from 60 to 700 kHz, and increasing to 300 – 350 $\mu A/(Vm^{-1} MHz)$ for frequencies up to 10 MHz (Hill and Walsh, 1985; Tell et al, 1987).

For an exposed person standing on a ground plane, the current defined by eq.1 flows through the legs, resulting in the greatest current density in the smallest cross-section, i.e. the ankle. The mean current density in the ankle (assuming uniform distribution) with the feet grounded was estimated as (Guy and Chou, 1982; Guy, 1985):

$$J_A = 3.44 \ 10^{-8} \ f \ E \tag{2}$$

In order to relate the current density to the SAR it is necessary to know the tissue conductivity (or resistivity) at various locations.

Measurements of total induced current were extended up to 50 MHz (Gandhi et al, 1985, 1986) and calculations of current density were made assuming that the ankle cross-section consists of various tissues having different conductivities. Effective cross sectional areas for current flow were calculated using anatomical data. According to these assumptions the effective cross-section area of the ankle was reduced to 9.5 cm^2. The calculated maximum SAR in the ankles for man and the estimated values for children are given in Table 1.

Table 1. Maximum ankle section Specific Absorption Rates, SARs, for a grounded person exposed to 61.4 V m^{-1} (Gandhi et al, 1986).

	Height (m)	Frequency (MHz)	SAR (W kg^{-1}) No Shoes
Adult	1.75	40	243
10-year-old child	1.38	50.7	371
5-year-old child	1.12	62.5	534

The induced current in a man standing on a ground plane depends also on the footwear and position of the body. Footwear decreases the current and the resulting whole-body-average and local SARs. Separation from ground by raising the person exposed reduces induced current and resulting SAR in the ankles (Gandhi et al, 1986). The reduction depends on separation distance and frequency.

High current densities in the extremities were shown to lead to high rates of surface temperature increase (Chen and Ganhi, 1987). It was shown that for SAR = 46 Wkg^{-1} in the high-water-content tissues the temperature of the ankle surface increased to about 36°C in 20 min from 32.5°C at the beginning of the exposure. For SAR = 23 Wkg^{-1} there was practically no temperature increase. Similarly for the wrist for SAR = 25 Wkg^{-1} only a temperature increase of about 0.5°C was reported.

Electromagnetic fields at frequencies below a few MHz can also pose an indirect hazard resulting from induction of RF charge on ungrounded or poorly-grounded metallic objects such as cars, trucks, cranes, wires, fences, etc. When a person comes in contact with such objects discharge current to ground flows through the body. Similarly, when an ungrounded person touches (with a hand or a finger) a grounded object, current flows to the object with likely high current density at the contact location.

Other hazards that have to be considered when a grounded person touches an isolated charged metallic object in an electric field are shock and burn effects. When considering shock two extreme cases may be considered, perception (a tingling or prickling sensation below about 100 kHz and a sensation of warmth above) and a possibly painful severe shock defined in terms of "let-go" current. There is a considerable amount of data on thresholds for both of these phenomena (Chatterjee et al, 1986; Dalziel and Mansfield, 1950; Guy and Chou, 1982). The current thresholds for both phenomena increase with frequency up to about 100 kHz (Guy and Chou, 1982).

Recently, body impedance and threshold currents for perception and pain were measured for over 350 persons (male and female) at frequencies from 10 kHz to 3 MHz (Chatterjee et al, 1986). Various types of contact with metallic electrodes were used to simulate the contact a person can make with a car, van, etc. For all types of contact investigated the threshold current values increased with frequency up to about 100 kHz, and remained nearly constant for higher frequencies. The measurements for adults were also used to predict the thresholds for a ten-year-old child. These results are used in setting up current limits in the proposed standard.

RF burns occur when the current entering the body through contact by a small cross-section of the body such as the fingers with an isolated object in the electric field exceeds 200 mA. Burn hazard has to be considered in setting up protection limits below about 100 MHz.

Current work in experimental dosimetry for the near-field provides considerable information directly applicable to deriving exposure limits. The radiators and exposure conditions investigated can be considered as representative of the "worst-case" conditions of exposure from portable transmitters or leaky transmitter cabinets (Stuchly et al, 1985). At the three frequencies investigated, namely, 160, 350 and 915 MHz, whole-body average SARs are within a range 4.6 - 17.8 mWkg^{-1} per 1 W of the radiator output power (Kraszewski et al, 1985; Stuchly et al, 1985), for realistic distances (0.1 wavelength) between the radiators and the body surface. Practically all the energy, however, is deposited within about 20% of the body volume close to the antenna. Knowledge of these SARs can be utilized in specifying, for instance, the maximum output power of portable transmitters that can be allowed under a selected limit of the SAR.

A considerable volume of data has also been collected on the SAR distribution for near-field exposures (Kraszewski et al, 1985; Stuchly et al, 1985, 1986). One of the most important findings is that the SAR distributions are highly nonuniform, with typical ratios between spatial peak and whole-body average SARs of the order of 150:1 to 200:1 (Stuchly et al, 1985). At all frequencies investigated, the maximum SAR is at the body surface, with lower magnitude "hot spots" located inside the body. These findings lead to a question regarding setting of standards for near-field exposures. Should a limit on the peak SAR be independent of where the peak SAR is produced? High SARs at the body surface are likely to elicit thermoregulatory responses in a manner similar to infrared radiation. Even cursory consideration of physiology would suggest that high SARs in such tissues as brain or other vital organs are likely to be more critical in producing biological effects which may be potentially hazardous.

Biological effects were recently reviewed by the Environmental Protection Agency (1986) which stated that "the data currently available on the relationship of SAR to biological effects show evidence for biological effects at an SAR of about 1 W/kg".

Another comprehensive assessment of scientific reports on RF biological effects concludes that there appears to be a threshold for adverse biological effects in animals at the SAR of about 2 Wkg^{-1} (Lary et al, 1985). Among the effects cited are (only lower thresholds cited here): death in rabbits at 4.05 Wkg^{-1} (2 h), death in dogs at 5.9 Wkg^{-1} (2-3 h), a rectal temperature increase of 3.5°C in monkeys at 3.45 Wkg^{-1} (1.5 h), staggering and other effects in chicks at 1.87 - 2.97 Wkg^{-1}. If the effects on chicks are not taken into account, and there are many reasons to question the validity of this study in the context of human exposure, the threshold for adverse effects appears to be 3-4 Wkg^{-1} rather than 2 Wkg^{-1}. However, at lower levels other effects occur, which although not clearly detrimental, may be considered undesirable (Lary et al, 1985).

Athermal effects associated with RF fields deeply amplitude modulated at extremely low frequencies (1 - 300 Hz) are heightened in the review published by the National Council on Radiation Protection and Measurements (1986). Also, they recommended lowering occupational limits to those applying to the general public for RF fields amplitude modulated at 50% or more at frequencies between 3 and 100 Hz. In view of low field intensities at which these effects have been demonstrated (Adey, 1981; Polk and Postow, 1986) the difficulties of extrapolation of the experimental results to human exposure, this recommendation appears more to bring the problem into attention than to provide a well based health protection.

A few recent studies not discussed in the published reviews that have been considered in proposing the exposure standard are briefly discussed here.

Thermal responses of rhesus monkeys were recently investigated by Lotz (1985). Average rectal temperature increases were 0.4 and 1.7°C for 4-h exposures during the day at 1.7 and 3.4 Wkg^{-1}, respectively, close to resonant frequency (225 MHz). At a frequency much higher than the resonant frequency (1.29 GHz), higher SARs were necessary to cause the same temperature increases. For instance, for increases of 0.4, 0.7 and 1.3°C, SARs required were 2.9, 4.0 and 5.9 Wkg^{-1}. This study is very important for estimations of temperature increases in humans. The most important findings of the study are that exposures at the resonant frequency for a given species and in the E polarization are the most effective in inducing hyperthermia, and that an increase in the steady-state temperature is linearly dependant on the whole-body-average SAR for a given frequency, but the proportionality coefficients vary with frequency. High local SARs inside the body may be responsible for the higher efficiency of heating at the resonant frequency. While the rhesus monkey has many thermoregulatory characteristics similar to man's, the sweating responses differ. This must be considered when extrapolating results from the monkey study to human beings. The former indicates that the SAR at a resonant frequency approximately equal to that of the resting metabolic rate (RMR) increases rectal temperature of a rhesus monkey by 1°C (Lotz, 1985). An extrapolation of this result to humans leads to a prediction that at the resonant frequency SAR = 1 Wkg^{-1} may cause an increase of 1°C in the rectal temperature.

The effect of ambient conditions, and more specifically of ambient temperature, on thermal effects of RF exposure was clearly demonstrated in a recent study (Berman et al, 1985). An important finding of the study is that an increase in the ambient temperature from 20°C to 30°C lowers, by about half, the dose (SAR multiplied by exposure duration) sufficient to elicit a specific biological effect, in this case animal death from heat stress.

Exposures of monkeys' eyes to 100 Wm^{-2} pulsed fields at 2.45 GHz (SAR = 2.6 Wkg^{-1}) and 200 Wm^{-2} cw field (SAR = 5.3 Wkg^{-1}) for 4 h/day resulted in endothelial cell damage in the corneas. The changes appeared reversible, but some cell loss occurred (Kues et al, 1985).

Prenatally and perinatally exposed rats at 2 Wkg^{-1} showed an increased body weight and decreased swimming endurance at 30 days of age, but not at 100 days. No other reliable response was observed, although some differences in a startle response were noted (Gavin et al, 1986).

In a long-term low-level study male rats were exposed for 25 months to pulsed 2450 MHz RF radiation of 4.8 Wm^{-2}. The whole-body average SARs ranged from 0.4 Wkg^{-1} for a 200-g rat to 0.15 Wkg^{-1} for a 400-g rat (Guy et al, 1985). The primary objective of the study was to investigate possible cumulative effects on health and longevity. The following endpoints were investigated: behaviour (open field activity), corticosterone level, immunology, hematology, blood chemistry, protein electrophoresis, thyroxine, total-body analysis, organ weight, histopathology, and longevity. The exposed animals had a significantly increased response to ConA (Concanavallin A) and a decreased response to PPD (purified protein derivative of tuberculin) after 13 months of exposure. Other parameters of immune response were not significantly affected. No other endpoints were affected. There was statistical evidence that the number of primary malignancies was higher in the exposed animals, but the biological significance of this difference is reduced by several factors, as explained by the authors. The cumulative survival curves for both the exposed and sham-exposed animals indicated median survival times of 688 days for the exposed and 663 for the sham animals. The total number of all neoplastic lesions (benign and malignant) was 45 in the exposed and 40 in the control animals, found to be not statistically significant. While the incidence of malignant lesions was greater in the exposed group, the authors indicate that it was similar to the incidence reported in the literature for the animal species used. No single type of primary malignancy was enhanced in the exposed animals. In summary, the authors conclude "no defendable trends in altered longevity, cause of death or spontaneous aging lesions and neoplasia can be identified in the rats exposed to this long-term low-level radiofrequency radiation exposure". This position, though, has become a subject of controversy.

In the last few years there have been a few reports suggestive of increased rates of cancer in workers and some population groups exposed to electromagnetic fields, but in most cases the association stipulated was rather with exposure to fields at power-line frequencies than at radiofrequencies. This issue requires further investigation.

There has been a number of very important scientific studies, some of them discussed in the recent reviews (Elder and Cahill, 1984; Lary et al, 1985; NCRP, 1986; Polk and Postow, 1986), that show that RF fields modulated at extremely low frequencies (ELF) or ELF fields can affect various cellular systems. The responses are specific to the modulation frequency, intensity and system. For many of these effects interaction mechanisms and health implications are not clear. Overall, these otherwise very important findings cannot be effectively used in standard-setting at this time.

PROPOSED EXPOSURE LIMITS

Occupational Exposures

The limits given in Table 2 should not be exceeded, when averaged over a period of 0.1 h. The limits in Table 2 are shown graphically in Figures 1 and 2 for the electric and the magnetic field, respectively.

Exposure limits given in Table 2 can be exceeded for portable transmitters and other devices producing partial body or highly spatially-nonuniform exposures if it can be shown that the following conditions are satisfied:

(i)   the SAR averaged over any 0.2 of the body mass does not exceed 0.4 Wkg^{-1},
(ii) the local SAR in the eye does not exceed 0.4 Wkg^{-1},
(iii) the local SAR averaged over any gram of tissue does not exceed 8 Wkg^{-1}, except at the body surface or in the limbs (arms and legs) where the limit is 25 Wkg^{-1},
(iv) all above limit values are averaged over 0.1 h.

Simultaneously with the field strength limits specified in Table 2, the limits given in Table 3 for the current passing through the person making contact with any conducting object including grounded objects should not be exceeded.

Table 2. Occupational Exposure Limits - Field Strengths,
f is the frequency in MHz

Frequency (MHz)	Electric Field Strength; rms (V m^{-1})	Magnetic Field Strength; rms (A m^{-1})	Power Density (W m^{-2})
0.01 - 1.2	600	4.0	-
1.2 - 3	600	4.8/f	-
3 - 30	*(1800/f) or (3120/f$^{1.5}$)	4.8/f	-
30 - 100	*60 or 20	0.16	-
100 - 300	60 or 0.2 f	0.16	10
300 - 1500	3.45 f$^{0.5}$	0.0093 f$^{0.5}$	
1500 - 300 000	140	0.36	50

* The lower limits apply when the exposed person is separated less than 0.1 m from what can be considered as electrical ground; in all other cases the higher limits apply.

The General Population

The limits given in Table 4 should not be exceeded, when averaged over a period of 0.1 h. These limits are shown graphically in Figures 1 and 2.

Exposure limits given in Table 4 can be exceeded, for portable transmitters and other devices producing partial body or highly spatially-nonuniform exposures, if the following conditions are satisfied:

(i) the SAR averaged over any 0.2 of the body mass does not exceed 0.2 Wkg^{-1},

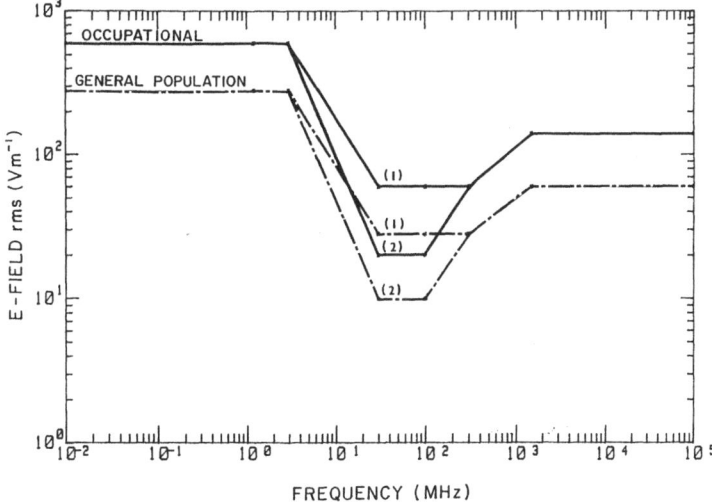

Figure 1. Proposed Canadian exposure limits for the electric field (E); option (1) refers to exposures 10 cm or more above electrical ground, option (2) refers to exposures of grounded persons.

(ii)   the local SAR in the eye does not exceed 0.4 Wkg^{-1},

(iii) the local SAR averaged over any gram of tissue does not exceed 4 Wkg^{-1}, except on the body surface or in the limbs (arms and legs), where the maximum SAR is 12 Wkg^{-1}.

(iv) all above limit values are averaged over 0.1 h.

Table 3. Contact current limits for occupational exposures
f - is the frequency in MHz

Frequency (MHz)	Current (mA)
0.01 - 0.1 0.1 - 100	$40(10f)^{0.7}$ 40

Table 4. Exposure Limits for the General Population
f is the frequency in MHz

Frequency (MHz)	Electric Field Strength; rms (V m⁻¹)	Magnetic Field Strength; rms (A m⁻¹)	Power Density (W m⁻²)
0.01 - 1.2	280	1.8	-
1.2 - 3	280	2.1/f	-
3 - 30	*(840/f) or $(1600/f^{1.5})$	2.1/f	-
30 - 100	*28 or 10	0.07	-
100 - 300	28 or 0.1 f	0.07	2
300 - 1500	$1.61\ f^{0.5}$	$0.004\ f^{0.5}$	
1500 - 300 000	60	0.16	10

* The lower limits apply when the exposed person is separated less than 0.1 m from electrical ground; in all other cases the higher limits apply.

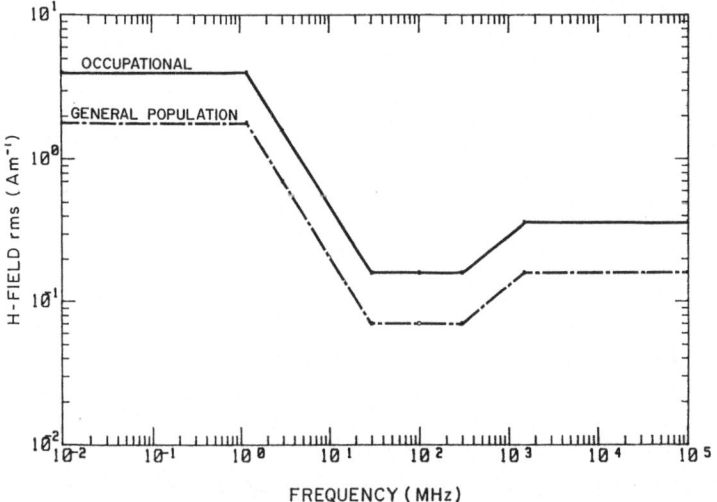

Figure 2. Proposed Canadian exposure limits for the magnetic field (H) (grounded or ungrounded).

Table 5. Contact-current limits for exposures of the general
population, f is the frequency in MHz

Frequency (MHz)	Current (mA)
0.01 - 0.1	$15(10f)^{0.5}$
0.1 - 100	15

Simultaneously with the field strength limits specified in Table 4,
the limits given in Table 5 for the current passing through the person
making contact with any conducting object including grounded objects
should not be exceeded.

Exemption

Portable transmitters operating below 1 GHz whose output power is
less than 7 W are exempt from the requirements specified in Tables 2-5.

RATIONALE

Different "measures" or "doses" are specified for protection against
effects of electromagnetic fields and for protection against effects of
currents passing through the human body contacting conducting objects.
The measures used in protection against the field effects are the
whole-body average SAR and the local SAR. Practical exposure limits
expressed in terms of the electric field strength, magnetic field strength
and power density are derived from the SAR values. The measure used for
protection against effects due to contact with conducting objects is the
total current passing through the person, and is applicable only to
frequencies below 100 MHz.

More restrictive exposure limits are recommended for the general
population than for occupationally exposed populations. The rationale for
a fivefold reduction in terms of SAR is similar to that given by
Scientific Committee 53 of the National Council on Radiation Protection
and Measurements (1986). Two main factors are considered, the duration of
exposure and potential for greater susceptibility to RF energy among some
parts of the general population such as aged, infants or chronically ill.
Furthermore, a worker is expected to be informed about potential hazards
related to RF exposure and can make a personal decision. Reduction
factors for contact currents are derived from the differences in
perception thresholds between adults and children and protection against a
possible startle due to perception rather than pain or burn (as in case of
occupational exposures).

From the existing scientific data it can be inferred that well
established (data corroborated by more than one laboratory) thresholds for
adverse effects are within 3-4 Wkg^{-1} range. Some analyses indicate that
this threshold may be as low as 1 Wkg^{-1} (EPA, 1986). The lower threshold
is derived from animal experiments on core-body temperature increases due
to RF exposure, and extrapolation to human beings. The threshold for the
whole-body SAR causing 0.5-1°C temperature increase is estimated to be
1 Wkg^{-1} (Elder and Cahill, 1984). This estimate is based on a
considerable number of assumptions, largely plausible but yet unproven.
Local SARs of about 2 Wkg^{-1} can cause adverse effects in some tissues, in
particular in the eye (Kues et al, 1985). However, in other tissues, e.g.

skin, much higher SARs are not likely to cause adverse effects, even though the exposure may be perceived.

Local SARs averaged over 1 g of tissue appear not to exceed 20 times the whole-body average SAR for the limits specified except at the body surface and in the extremities. Maximum values of local SAR in the ankle are expected to be about 20-30 Wkg^{-1} for occupational exposures and 10-15 Wkg^{-1} for the public (Gandhi et al, 1986).

At frequencies below 30 MHz different limits for electric and magnetic field strengths are recommended. This difference is due to much lower SARs deposited in the human body from exposures to magnetic fields than from exposures to electric fields (Durney et al, 1978; Stuchly et al, 1986).

Higher field limits than those specified in Tables 2 and 4 could be permitted, if only whole-body SAR was considered. Capping of the limits restricts the maximum induced current densities to approximately constant ratios between these current densities and threshold current densities for stimulation and perception (below hundreds of kHz) (Bernhardt, 1985).

Separation from ground of 0.1 m is considered as a reasonable approximation of "free-space" exposure (Gandhi et al, 1986).

For portable transmitters and similar devices that produce localized energy deposition in the human body, exposure limits cannot be easily specified in terms of strengths of external fields, but rather have to be specified in terms of SARs averaged over a limited body mass.

Since experiments indicate, as reviewed earlier, that for portable transmitters held close to the body and by inference for other sources producing similar exposure conditions, the energy is deposited within only about 20% of the body volume, instead of the SAR averaged over the whole-body mass, the SAR averaged over 0.2 of the body mass is considered as a more appropriate parameter. If the whole-body average of 0.4 Wkg^{-1} was used for a portable transmitter for a person weighing 70 kg, 28 W could be deposited in the head-neck region resulting in 2 Wkg^{-1} average SAR for this region, which is judged excessive.

The occupational and the general population limits for portable transmitters and other devices producing localized energy deposition differ only by a factor of two. This is sufficient in view of only intermittent exposure to such devices by the general public.

Although pulsed fields are more effective in some cases than cw fields in eliciting biological responses, the available scientific data are not sufficient for establishing thresholds for peak field strengths.

At frequencies below about 100 MHz hazardous RF current may be developed in a person contacting ungrounded conductive objects, e.g. vehicles, fences, metal roofs. It is not practical to limit perception, shock or burns through limiting the exposure-field strength, as the current among other parameters depends on the object size. On the other hand, engineering measures can be employed to effectively limit such currents or administrative measures may be employed to prevent access by people to the objects. Therefore, to protect workers current limits are specified in Table 3. These limits were derived from the data on pain thresholds for adults of both sexes (Chatterjee et al, 1986), and are set approximately equal to pain threshold for finger contact for an average female. A much higher threshold is associated with grasping contact. The limits specified in Table 3 can be considered as very conservative measures in preventing severe shock and burn hazard.

The current limits specified in Table 5 protect against perception for children and adults for finger contact, according to the data on the perception thresholds (Chatterjee et al, 1986).

The major unresolved issue remains the effect on human health of exposure to RF fields amplitude modulated at extremely low frequencies. Available scientific evidence indicates that such fields are capable of interactions with biological systems. Utilization of this evidence in standard-setting is not possible at this time.

COMPARISON WITH OTHER STANDARDS

There are several similarities between the Canadian proposal and the existing RF-exposure protection standards. The frequency dependence of the exposure limits is very similar for most current standards, and reflects the inverted "SAR curves". The electric field limits for ungrounded conditions are essentially the same as in the U.S. ANSI - 1982 recommendation for occupational exposures and as the NCRP (1986) and one of the proposed EPA options for the general population (EPA, 1986). The recommended levels are consistently lower than those in the FRG standard, and somewhat higher than the recent U.K. proposal. The lower limits recommended in the Canadian standard around the resonace frequencies are a new feature not present in other standards. It is interesting to note that these lower limits approach those in the USSR stardards, but are still two to four times higher.

Different limits for the general public from those for occupational exposures are recommended in many current standards and proposals, and appear to reflect the current thinking.

The Canadian proposal establishes higher limits for the magnetic field than the electric field below 3 MHz. Similar relaxation, even to a greater degree, is characteristic of the FRG and USSR standards, and the U.K. and U.S. ANSI (1987) proposals. Time averaging over 0.1 h is also typical for many standards.

One of the main different features of the Canadian proposal are the limits on contact currents. The only other standard-setting organization which is known to consider similar recommendations is the U.S. ANSI. The other new characteristic is the recommendations for portable transmitters, involving SAR averaging over a limited part of the body and limits on local SARs rather than field strength limits.

REFERENCES

Adey, W.R., 1981, "Tissue interaction with nonionizing electromagnetic radiation," Physiol. Rev. 61, 435-514.
Bernhardt, J.H., 1985, "Evaluation of Human Exposure to Low Frequency Fields", Advisory Group for Aerospace Research and Development (AGARD), Lecture Series No.138, North Atlantic Treaty Organization, 7 rue Ancelle, 92200 Neuilly sur Seine, France, 8.1-8.18.
Berman, E., Kinn, J.B., Ali, J., Carter, H.B., Rehnberg, B., and Stead, A.G., 1985, "Lethality in mice and rats exposed to 2450 MHz circularly polarized microwaves as a function of exposure duration and environmental factors," J. Appl. Toxic, 5, 23-31.

Chatterjee, I., Wu, D., and Gandhi, Ò.P., 1986, "Human body impedance and threshold currents for perception and pain for contact hazard analysis in the VLF-MF band," IEEE Trans. Biomed. Eng., 33, 486-494.

Chen, J.Y., and Gandhi, O.P., 1987, "Thermal implications of high SARs in the body extremities at the ANSI recommended MF-VHF safety levels," IEEE Trans. Biomed. Eng. (in press).

Dalziel, C.F., and Mansfield, T.H., 1950, "Effect of frequency on perception currents," AIEE Trans., 69, 1162-1168.

Durney, C.H., Johnson, C.C., Barber, P.W., Massoudi, H., Iskander, M.F., Lords, J.L., Ryser, D.K., Allen, S.J., and Mitchell, J.C., 1978, "Radiofrequency Radiation Dosimetry Handbook (Second Edition)," Report SAM-TR-78-22, USAF School of Aerospace Medicine, Brooks Air Force Base, TX 78235.

Elder, J.A., and Cahill, D.F., (Eds.), 1984, "Biological Effects of Radiofrequency Radiation," U.S. Environmental Protection Agency Document EPA-600/8-83-026F, Research Triangle Park, NC 27711.

Environmental Protection Agency, 1986, "Federal radiation protection guidance; proposed alternatives for controlling public exposure to radiofrequency radiation," Federal Register, 51(146) 27318-27339.

Gandhi, O.P., Chatterjee, I., Wu, D., and Gu, Y.G., 1985, "Likelihood of high rates of energy deposition in the human legs at the ANSI recommended 3-30 MHz RF safety levels," Proc. IEEE, 73, 1145-1147.

Gandhi, O.P., Chen, J.Y., and Riazi, A., 1986, "Currents induced in a human being for plane-wave exposure conditions 0-50 MHz and for RF sealers," IEEE Trans. Biomed. Eng., 33, 757-767.

Gavin, M.J., Tilson, H.A., Mitchell, C.L., Peterson, J., and McRee, D.I., 1986, "Influence of pre- and postnatal exposure of rats to 2.45-GHz microwave radiation on neurobehavioral function," Bioelectromagnetics, 7, 57-71.

Guy, A.W., and Chou, C.K., 1982, "Hazard Analysis: Very Low Frequency Through Medium Frequency Range," Report for USAF School of Aerospace Medicine, Brooks Air Force Base, TX 78235.

Guy, A.W., 1985, "Hazards of VLF Electromagnetic Fields," AGARD Lecture Series No.138, North Atlantic Treaty Organization, 7 rue Ancelle, 92200 Neuilly sur Seine, France.

Guy, A.W., Chou, C.K., Kunz, L.L., Crowley, J., and Krupp, J., 1985, "Effects of Long-Term Low-Level Radiofrequency Radiation Exposure of Rats," Vol.9, Summary, Report No. USAFSAM-TR-85-64, USAF School of Aerospace Medicine, Brooks AFB, Tx 78235-5301.

Health and Welfare Canada, 1979, Safety Code-6, "Recommended Safety Procedures for Installation and Use of Radiofrequency and Microwave Devices in the Frequency Range 10 MHz - 300 GHz", Publication 79-EHD-30, HWC, Brooke Claxton Bldg., Ottawa, Ont. K1A 0K9.

Hill, D.A., and Walsh, J.A., 1985, "Radiofrequency current through the feet of a grounded man," IEEE Trans. Electromagn. Compat., EMC-27, 18-23.

Kraszewski, A., Stuchly, S.S., Stuchly, M.A., and Hartsgrove, G., 1985, "Energy deposition in a model of man: frequency effects in the near field," Paper presented at the 7th Annual Meeting of Bioelectromagnetics Soc., June 16-20, San Francisco, CA.

Kues, H.A., Hirst, L.W., Lutty, G.A., D'Anna, S.A., and Dunkelberger, G.R., 1985, "Effects of 2.45 GHz microwaves on primate corneal endothelium," Bioelectromagnetics, 6, 177-188.

Lary, J.M., Conover, D.L., and Murray, W.E., 1985, "Assessment of the biological effects of radiofrequency radiation," Presented at the 7th Annual Meeting of Bioelectromagnetics Soc., June 16-20, San Francisco, CA.

Lotz, W.G., 1985, "Hyperthermia in radiofrequency-exposed rhesus monkeys: a comparison of frequency and orientation effects," Rad. Res., 102, 59-70.

National Council on Radiation Protection, 1981, "Radiofrequency Electromagnetic Fields. Properties, Quantities and Units, Biophysical Interactions and Measurements," Report No.67, Washington, DC 20014, National Council on Radiation Protection and Measurements.

National Council on Radiation Protection, 1986, "Biological Effects and Exposure Criteria for Radiofrequency Electromagnetic Fields," Report No.86. Washington, D.C., 20014, National Council on Radiation Protection and Measurements.

Polk, C., and Postow, E. (Eds.), 1986, "Handbook of Biological Effects of Electromagnetic Radiation," Boca Raton, FL 33431, (City Publisher) CRC Press, Inc.

Stuchly, S.S., Kraszewski, A., Stuchly, M.A., Hartsgrove, G., and Adamski, D., 1985, "Energy deposition in a model of man in the near-field," Bioelectromagnetics, $\underline{6}$, 115-129.

Stuchly, M.A., Kraszewski, A., and Stuchly, S.S., 1985, "Exposure of human models in the near- and far-field. A comparison," IEEE Trans. Biomed. Eng., $\underline{BME-32}$, 609-616.

Stuchly, M.A., Spiegel, R.J., Stuchly, S.S., and Kraszewski, A., 1986, "Exposure of man in the near-field of a resonant dipole, comparison between theory and measurements," IEEE Trans. Microwave Theory Tech., $\underline{MTT-34}$, 26-31.

Stuchly, S.S., Stuchly, M.A., Kraszewski, A., and Hartsgrove, G., 1986, "Energy deposition in a model of man; frequency effects," IEEE Trans. Biomed. Eng., $\underline{BME-33}$, 702-711.

Tell, R.A., Multiply, E.D., Durney, C.H., and Massoudi, H., 1987, "Electric and magnetic field intensities and associated induced body currents in man in close proximity to a 50 kW AM standard broadcast station", IEEE Trans. Broadcast (in press).

INTERNATIONAL HEALTH CRITERIA DOCUMENTS

AND GUIDELINES FOR ELECTROMAGNETIC FIELDS

P. Czerski[1,2] and J.H. Bernhardt[1,3]

(presented by M.L. Swicord[2])

1. Member International Non-Ionizing Radiation
    Committee of the International Radiation
    Protection Association (IRPA/INIRC)
2. Center for Devices and Radiological Health
    Food and Drug Administration, PHS, DHHS
    Rockville, MD, USA
3. Institute for Radiation Hygiene, Neuherberg
    Federal Republic of Germany

INTRODUCTION

Due to the growing concerns about safety of NIR exposure and the expanding uses of NIR in medicine, science and technology, the International Radiation Protection Association (IRPA) organized a session on non-ionizing radiation (NIR) during the 3rd International IRPA Congress in Washington, D.C., in 1973. As a follow-up to the discussions during this meeting, the Executive Council of IRPA instituted a study group to review the needs and the international situation in the field of NIR protection. Based on the recommendations of this group, the IRPA constitution was amended during the 4th Congress in Paris, France, in 1977, to include NIR protection, and the International Non-Ionizing Radiation Committee (IRPA/INIRC) was established. IRPA charged the Committee with the tasks of assessing the present knowledge on biological and health effects of NIR, developing criteria documents, health protection recommendations, and exploring ways and means for furthering NIR protection activities in cooperation with other interested bodies. During the past ten years an international framework was established for this purpose, and is presented schematically in Fig.1.

Under the auspices of the United Nations Environment Programme (UNEP), the Headquarters of the World Health Organization (WHO) in Geneva and IRPA/INIRC develop environmental health criteria (EHC) documents for particular parts of the NIR spectrum. These documents serve then as a rationale for IRPA/INIRC guidelines on NIR exposure limits. IRPA/INIRC cooperates also with the International Labour Office (ILO) in Geneva, developing technical reports and codes of practice on NIR safety. Support for IRPA/INIRC activities is provided by the Commission of European Communities. An important part of the international cooperation is the co-sponsorship of scientific symposia

organized by URSI, the present one being an instance of this
collaboration. The international base of IRPA/INIRC activities is
expanded by the input provided by national member societies of IRPA.

ENVIRONMENTAL HEALTH CRITERIA

These documents present the physical characteristics, dosimetry,
natural and man-made sources of NIR, mechanisms of interaction,
biological and health effects, health hazard assessment, and a review
of protection standards and their rationales. Each EHC contains a
section called summary and recommendations. Levels for health effects
are identified, however, EHCs do not establish protection standards.
These documents are intended as advice to national health authorities
or other bodies involved in NIR protection, as well as for personnel
active in this field. To ensure that EHC present accurate information
and a balanced assessment of health implications, consecutive drafts
are subjected to an extensive review process, the final text being the
result of deliberations of an international task group.

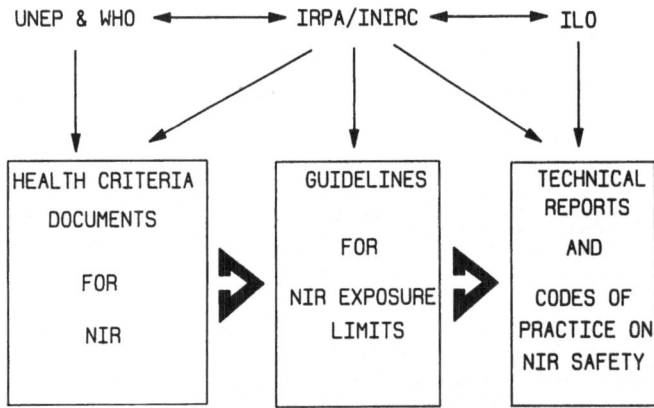

Fig. 1. Schematic representation of international cooperation in the
development of radiofrequency protection documents.

IRPA/INIRC develops a preliminary draft of EHCs of the NIR series.
The drafting group consists of IRPA/INIRC members and invited experts,
who are willing to devote their time and energies to this task. The
preliminary draft is reviewed by invited external experts and all
IRPA/INIRC members. Following necessary corrections, the text is
submitted to the whole IRPA/INIRC membership for approval. Once
approved, the text is transmitted as a first draft to the Environmental
Health Division at WHO Headquarters in Geneva. WHO distributes the
first draft for review to Member States, Focal Points, Collaborating
Centers, and selected individual experts. A joint WHO and IRPA/INIRC
editorial group collates the comments, and prepares a revised second
draft. Members of a task group are identified, care being exercised to

provide balance in the representation of scientifc disciplines involved
in this field, different points of view and schools of thought, and
geographical distribution of WHO Member States. Members of the
WHO/IRPA task group are provided the text well in advance of a meeting,
during which a final text is developed. Following such a meeting,
additional editorial work is usually needed, however, this does not

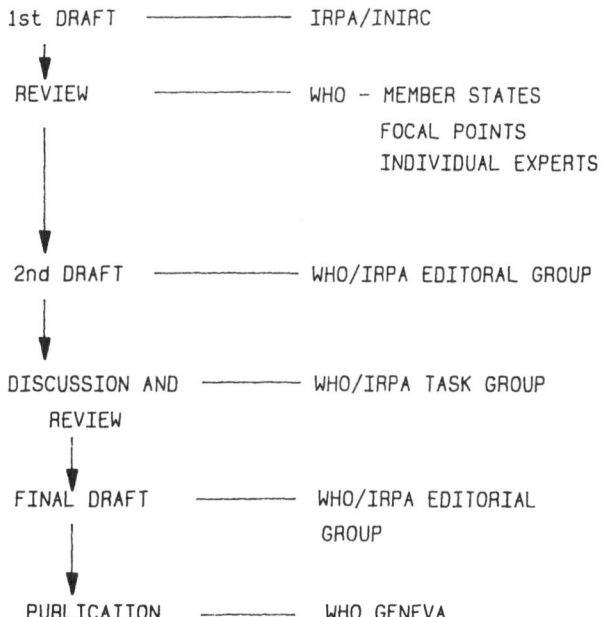

Fig. 2. Schematic representation of the steps in the development of
environmental health criteria (EHC) documents of the NIR series.

Involve any changes in the substance of the body of the text, and no
changes in the wording of the summary and recommendations section.
Thus EHCs represent an international consensus on the state of art at
the time of the task group meeting. Fig. 2 presents a flow-chart of
the development of EHCs of the NIR series.

To date the UNEP/WHO/IRPA cooperation resulted in the preparation
of three Environmental Health Criteria of interest to the present
Symposium:

    EHC 16. Radiofrequency and Microwaves. 1981.
    EHC 35. Extremely Low Frequency (ELF) Fields. 1984.
    EHC 69. Magnetic Fields. 1987.

All of these were published by WHO, Geneva, and are obtainable from national distributors of WHO publications. Questions on orders may be also addressed to WHO, Distribution and Sales Service, 1211 Geneva 27, Switzerland.

EHC 16 is now outdated, and a new revised version on "Electromagnetic Fields in the Frequency Range from 300 Hz to 300 GHz" is in preparation. It is hard to predict a publication date, the present author's best estimate is 1990.

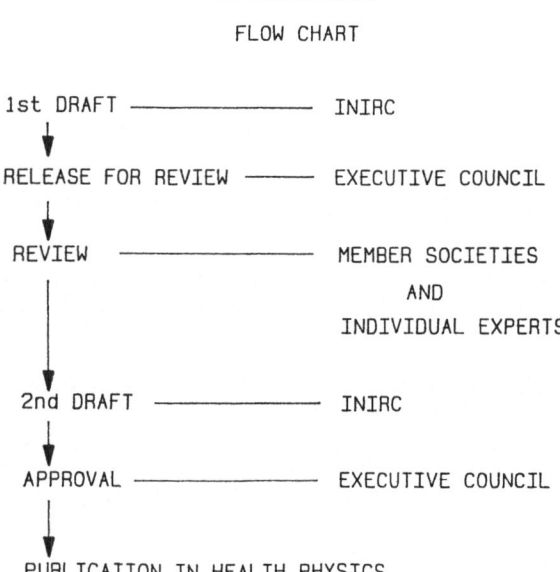

Fig. 3. Schematic representation of the procedure for the development of IRPA/INIC guidelines.

RADIOFREQUENCY GUIDELINES

IRPA/INIRC uses the relevant EHC, supplemented by reports published after the finalization of its text, as a scientific data base for the development of guidelines on exposure limits (EL). The Committee recognizes that when standards on ELs are established various value judgements are necessary, including cost-versus-benefit analyses, economic impact of controls, and national priorities. It is not possible to take into account all such judgements applicable world-wide, therefore the limits in IRPA/INIRC guidelines are based solely on scientific data from currently available knowledge. The IRPA/INIRC limits should provide a safe from NIR, healthy working and living environment under all normal conditions. However, some modifications may be necessary to suit particular local needs, conditions, and priorities. A brief rationale, and a reference list are appended to each guideline. IRPA/INIRC guidelines are also subjected to an extensive review, as presented on Fig.3.

In 1984 IRPA/INIRC published "Interim guidelines on limits of exposure to radiofrequency electromagnetic fields in the frequency range from 100 kHz to 300 GHz" (IRPA/INIRC, 1984). These guidelines were recently revised (IRPA/INIRC, 1988), and are briefly summarized below.

The guidelines distinguish between basic and derived ELs. For frequencies above 10 MHz the specific absorption rate (SAR, W/kg) was chosen as a quantity for establishing basic ELs. Derived limits in this range are expressed in terms of equivalent plane wave power density (P_{eq}, W/m^2) or energy flux density, and unperturbed RMS electric (E, V/m) or magnetic (H, A/m) field strengths. These quantities may be considered as "radiometric" ones, as opposed to the SAR, which is a dosimetric quantity (Czerski, 1986). Below 10 MHz the ELs are expressed in terms of RMS electric and magnetic field strengths. The derived limits for whole body exposure, as averaged over any 6 min. period are presented in Table 1.

The derived limits for frequencies above 10 MHz were obtained primarily from the frequency dependence of whole body average (WBA) SAR, and modified by considerations of the non-uniformity of energy deposition in various body parts, or the spatial peak SAR. For occupational exposure the basic WBA-SAR limit is 0.4 W/kg. Based on thermal diffusion and heat exchange characteristics of the human body IRPA/INIRC selected 0.1 kg or approximately 100 cm^3 as the mass or volume over which spatial peak energy deposition should be averaged. The spatial peak SAR limit for the wrists and ankles is 2 W per 0.1 kg, and 1 W per 0.1 kg in any other part of the body. In the near field, exposure may be predominatly to the H or to the E field, or even to one of these fields alone. To account for such situations the field strengths indicated in Table 1 may be modified according to a formula, derived from the analysis of the contribution of the H and E field to the WBA-SAR:

$$5/6(E^2/120\pi) + 1/6(120\pi H^2) < P_{eq}$$

where E is the electric field strength, H is the magnetic field strength, and P_{eq} is power density limit indicated in Table 1.

For public exposure the WBA-SAR of 0.08 W/kg was selected as the basic limit. At this level, a local thermal injury due to an unacceptably high spatial peak SAR is extremely unlikely.

The 0.4 W/kg value was originally selected (IRPA/INIRC, 1984) on the assumption that a SAR of 4 W/kg corresponds to a threshold for untoward effects, the most sensitive endpoints being based on effects on behavior (ANSI, 1982, see also Chou and Guy, 1985). Extensive reviews of literature (EPA, 1984, NCRP, 1986) indicate that untoward effects relatable to temperature increases may occur under certain exposure and environmental conditions at a WBA-SAR of 1 W/kg. Such situations are not common, and the 0.4 W/kg value and the corresponding derived limits were still considered adequate for occupational ELs in the 1988 IRPA/INIRC revision. The 0.08 W/kg basic limit for public exposure is 5 times lower than the occupational one. It was selected to afford additional protection to the public keeping in mind that there are approximately 2000 working hours during a year, which is 8766 hours long. Rounding the latter value up to 10,000 results in a ratio of 1:5 between the duration of occupational versus public exposure per year. Several recent reports indicate that exposure to microwaves at levels between 0.3 and 0.7 W/kg alters the effects of some psycho-active drugs and substances (see Hjeresen et al., 1988, Lai et al.,

TABLE 1. IRPA/INIRC EXPOSURE LIMITS TO
RADIOFREQUENCY ELECTROMAGNETIC FIELDS
(FROM IRPA/INIRC, 1988).

Frequency (f) MHz	Unperturbed RMS field strength electric (E) V/m	magnetic (H) A/m	Equivalent plane wave power density (W/m^2)
Occupational			
0.1 - 1	614	1.6/f	–
> 1 - 10	614/f	1.6/f	–
> 10 - 400	61	0.16	10
> 400 - 2x10^3	3f$^{1/2}$	0.008f$^{1/2}$	f/40
> 2x10^3 - 3x10^5	137	0.36	50
Public			
0.1 - 1	87	0.23/f$^{1/2}$	–
> 1 - 10	87/f$^{1/2}$	0.23/f$^{1/2}$	–
> 10 - 400	27.5	0.073	2
> 400 - 2x10^3	1.375f$^{1/2}$	0.0037f$^{1/2}$	f/200
> 2x10^3 - 3x10^5	61	0.16	10

Note to occupational exposure limits:

Hazards of radiofrequency burns should be eliminated by limiting
currents from contact with metal objects (see text). In most
situations this can be achieved by reducing the E values from 614 to
194 V/m in the range from 0.1 to 1 MHz, and from 614/f to 194/f$^{1/2}$ in
the range from >1 to 10 MHz.

1987), and that administration of certain ophthalmologic drugs (Monahan
et al., 1988) lowers the threshold for microwave (2450 MHz, pulsed)
ocular injury, damage to corneal endothelium occurring at a SAR of 0.26
W/kg. It follows that sensitive subgroups exist within the general
population, and provides an additional rationale for recommending lower
ELs for the public, as compared to the occupationally exposed subpopulation. As far the latter is concerned, a warning about combined
effects of radiofrequency exposure and neurotropic and ophthalmologic
drugs seems to be advisable.

For frequencies below and up to 10 MHz occupational exposure should
not exceed the limits in Table 1 , provided that the body-to-ground
current does not exceed 200 mA. The latter provision was intended to
eliminate excessive heating in ankles and wrists in persons grounded

over the limbs. The limits for public exposure are also indicated in
Table 1. Both for occupational and public exposure hazards of radiofrequency burns and shocks should be eliminated. Burns and shocks can
result from touching ungrounded metal objects that have been charged up
by the field, or from contact of a charged up body with a grounded
metal object. If the current at point contact exceeds 50 mA, there is
a risk of burns.

For the case of pulse modulation the guidelines state: "Although
very little information is presently available on the relation of
biological effects with peak values of pulsed fields, it is suggested
that the equivalent plane wave power density as averaged over the pulse
width not exceed 1000 times the P_{eq} limits or the field strength not
exceed 32 times the field strength limits for the frequency concerned",
provided that the limits for occupational (or public) exposure
"averaged over 6 minutes are not exceeded, and hazards of radiofrequency burns are eliminated."

The guidelines, as presented above, provide protection against
systemic thermal overload and excessive local heating. The ELs are
also below levels, which would result in stimulation of excitable cell
membranes in nerve and muscle cells (Bernhardt, 1979, 1988). Another
endpoint taken into account are radiofrequency shocks and burns. An
important point is the distinction between occupational and public
exposure limits. Apart from the differences in the duration of exposure
during a given year and over the life-time, and the presence of
sensitive subpopulations within the general public, other considerations were taken into account. Occupational exposure concerns adults,
who can be made aware of, and trained to avoid risks. Moreover,
exposure conditions can be controlled. Members of the public may be
unaware that exposure takes place, cannot be expected to take
precautions to avoid radiofrequency shocks and burns, and may be
unwilling to take risks, however slight.

In a final section of the rationale IRPA/INIRC states: "The
committee considered the recent data linking electric and magnetic
field exposure to increased cancer risks or or congenital anomalies in
various exposed human populations. Available data are inconclusive and
cannot be used for establishing exposure limits".

A special appendix to the guidelines lists briefly protective
measures. More data on this subject are presented in two documents
prepared by IRPA/INIRC in cooperation with ILO in the Occupational
Safety and Health Series:

No 53. Occupational Hazards from Non-Ionising
       Electromagnetic Radiation. 1985

No 57. Protection of Workers against Radio-Frequency
       and Microwave Radiation: A Technical Review. 1986

both published by ILO, Geneva, Switzerland.

NOTE: At the time of this presentation the IRPA/INIRC was chaired
by H.Jammet (France), A.Duchene (France) was the scientific secretary,
and J.H. Bernhardt (Federal Republic of Germany), B.F.M. Bosnjakovic
(The Netherlands), P.Czerski (U.S.A.), M.Grandolfo (Italy), D.Harder
(Federal Republic of Germany), B.Knave (Sweden), J.Marshall (United
Kingdom), M.H.Repacholi (Australia), D.H.Sliney (U.S.A.), and
J.A.J.Stolwijk (U.S.A.) were members. During the recent (April, 1988)

IRPA Congress in Sydney (Australia) H.Jammet was appointed chairman emeritus, and M.H.Repacholi (Australia) was appointed chairman. A.Duchene is the scientific secretary (Departement de Protection Sanitaire, B.P. N° 6, 92265 Fontenay-aux-Roses Cedex, France) and the membership consists of J.H.Bernhardt (Federal Republic of Germany), B.F.M.Bosnjakovic (The Netherlands), L.A.Court (France), P.Czerski (U.S.A.), M.Grandolfo (Italy), B.Knave (Sweden), A.F.McKinley (United Kingdom), M.G.Shandala (U.S.S.R.), D.H.Sliney (U.S.A.), J.A.J.Stolwijk (U.S.A.), M.A.Stuchly (Canada), and L.D.Szabo (Hungary).

REFERENCES

American National Standards Institute (ANSI), 1982, American national standards safety levels with respect to human exposure to radiofrequency electro-magnetic fields, 300 kHz to 100 GHz. IEEE, New York, N.Y., ANSI C95.1.1982.

Bernhardt, J.H., 1979, The direct influence of electromagnetic fields on nerve and muscle cells in man within the frequency range of 1 Hz and 30 MHz. Radiat. Environ. Biophys. 16: 309-329.

Bernhardt, J.H., 1988, The establishment of frequency dependent limits for electric and magnetic fields and evaluation of indirect effects. Radiat.Environ. Biophys. 27: 1-27.

Chou, C.-K., and Guy, A.W., 1985, Research on nonionizing radiation: Physical aspects in extrapolating infrahuman data to man. In: Monahan, J.C., and D'Andrea, J.A. (eds), Behavioral Effects of Microwave Radiation Absorption. HHS Publ. FDA 85-8238, U.S. D.H.E.W., PHS, FDA, Center for Devices and Radiological Health, Rockville, MD.

Czerski, P. 1986, The developent of biomedical approaches andconcepts of radiofrequency radiation protection. J.Microwave Power 21: 9-23.

Environmental Protection Agency (EPA), 1984, Biological Effects of Radiofrequency Radiation. Elder, J.A., and Cahill, D.F. (eds). EPA-600/8-83/026F. U.S.EPA, Research Triangle Park, N.C.

Hjeresen, D.L., Francendese, A., and O'Donnell, J.M., 1988, Microwave attenuation of ethanol-induced hypothermia: Ethanol tolerance, time course, exposure duration, and dose response studies. Bioelectromagnetics 9: 63-78.

IRPA/INIRC, 1984, Interim guidelines on limits of exposure to radiofrequency electromagnetic fields in the frequency range from 100 kHz to 300 GHz. Health Physics 46: 975-984.

IRPA/INIRC, 1988, Guidelines on limits of exposure to radio-frequency electromagnetic fields in the frequency range from 100 kHz to 300 GHz. Health Physics 54: 115-123.

Lai, H., Horita A., Chou C.-K., and Guy, A.W. 1987, Microwave irradiation and actions of psychoactive drugs: A review. IEEE Eng.Med.Biol. 6: 31-36.

Monahan, J.C., Kues, H.A., McLeod, D.S., D'Anna, S.A., and Lutty, G.A.,1988, Lowering of microwave exposure threshold for induction of primate ocular effects by timolol maleate. 10th Ann. Mtg of the Bioelectromagnetics Society, Stamford, CT, abstr. p. 48.

National Council on Radiation Protection and Measurements (NCRP), 1986, Biological Effects and Exposure Criteria for Radiofrequency Electromagnetic Fields. NCRP Report No 86. NCRP, Bethesda, MD.

PANEL DISCUSSION ON STANDARDS

Moderator: John M. Osepchuk

Raytheon Research Division
Lexington, MA   02173

INTRODUCTION   (Moderator)

The art and science of standards-making relative to safe exposure to electromagnetic fields is not trivial and includes some subtle issues. The thresholds for effects (hazardous or not) in animals at a given frequency are most simply expressed in terms of some exposure parameter, like power density, and an exposure duration.

This determines a curve A as depicted on the "exposure diagram" shown in Figure 1. A similar curve results after extrapolation to man.

The threshold exhibits a characteristic time τ above which the threshold closely follows a line of constant SAR (specific absorption rate) and below which the threshold closely follows a line of constant SA (specific absorption). This characteristic time may be related to a thermal time constant in the animal or some other process. If the latter is true it may be aptly denoted as recovery time. It is the normal business of standards-setting to determine a curve B (see Figure 1) which is sufficiently below the curve A for man. If the curve B has the same characteristic time τ as for curve A then the safety limit B has the same safety factor for $t < \tau$ and $t > \tau$. For the curve B represented by the solid line in Figure 1 there is a smaller characteristic time than for curve A. This means that there is a higher safety factor for short-duration exposures than for long-duration exposures ($t > \tau$).

The characteristic time for the safety limit (curve B) corresponds to what is called averaging time.

Historically, ANSI C95 chose a 0.1 hour averaging period because it corresponded to the thermal time constant (approximately) 10 minutes) in localized diathermy and localized heating of critical organs, e.g., the eye of a rabbit. Since the thermal time constant in heating a whole human being with RF is more like an hour[1], this means ANSI C95 is more conservative (e.g., safety factor of 100) for short-duration (<1 hour) exposure than for long-term exposure (>1 hour; safety factor approximately 10-20).

The meaning of the averaging time τ in RF exposure standards

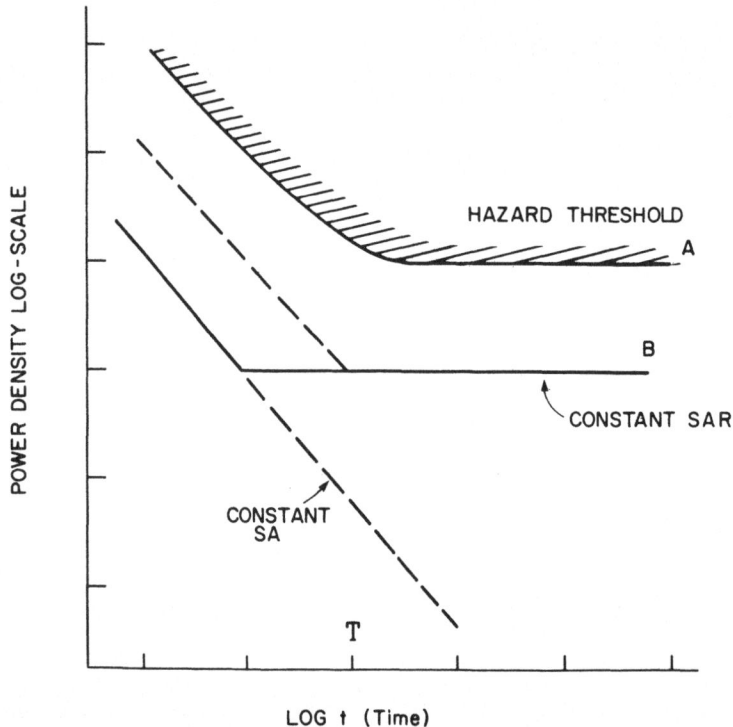

Figure 1. Depiction of Hypothetical Hazard Threshold (Curve A)
and derived safe-exposure limit (Curve B) in an
exposure diagram. The coordinates are exposure
power density and time each on an arbitrary
logarithmic scale.

is not generally well-understood. To clarify, consider Figure 2.
An exposure standard specifies the maximum power density p_0 (or fields) to
which a person should be exposed continuously (for long intervals), as well
as an energy density flux limit, $p_0 \times \tau$, for isolated periods of exposure
shorter than τ. As we see in Figure 2, the "energy" limit permits an
exposure flux greater than p_0 for short periods of time:

$$p = p_0 \times \frac{\tau}{t} \qquad t < \tau$$

Note, though, that in any period τ, once such an exposure occurs, there
must be no further exposure for the rest of that period. After that,
further isolated or even continuing exposure can occur. Several cases are
depicted in Figure 2.

In ANSI C95.1-1982, $\tau = 0.1$ hour. In extending C95 for protection of
the general public (the environmental limit), both Massachusetts and NCRP
extend τ to 0.5 hours while dropping p_0 by a factor of 5 -- keeping
$p_0 \times \tau$ constant. The EPA proposal [2] for averaging time seems confused.
It permits exposure levels up to 10 p_0 for $t < \tau$ where $\tau = 0.1$ hour (averaging time for all options), but if a single exposure of 10 p_0 for a period
of 0.1 hour is incurred, then exposure must be zero for 0.9 hours. Clearly,
the Massachusetts/NCRP procedure is simpler and more sensible.

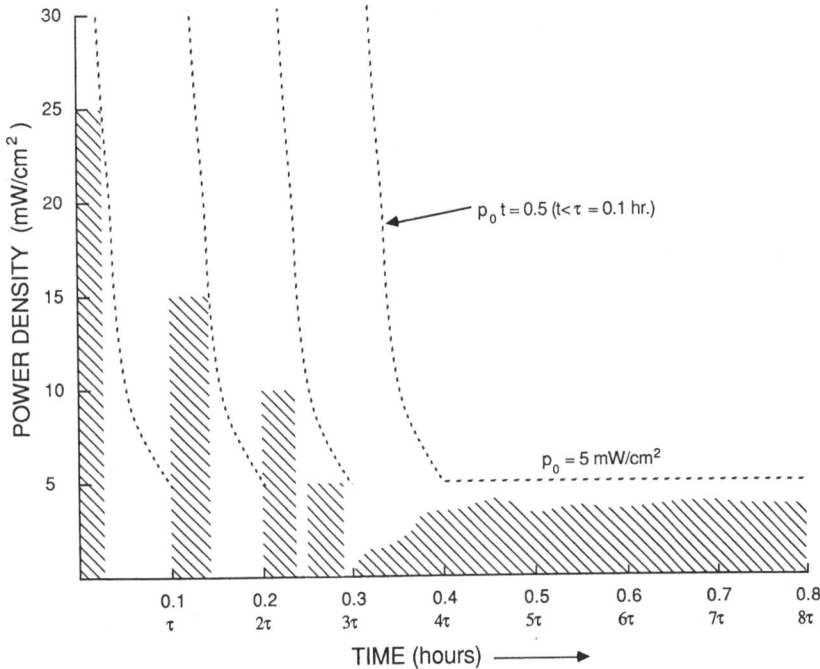

Figure 2. Depiction Of Several Isolated Exposures (Shaded Areas)
as permitted by time averaging under the C95 microwave
standard (τ = 0.1 hour) (f>1.5 GHz).

Groups outside the U.S. deviate markedly in how they treat time averaging; most simply adding to the confusion, as depicted in Table 1. We
show not only the disparate range of values for τ , but also the long-term
limit p_0 in the microwave range. Note that short-term limits in Australia
and Canada are even more conservative than in the USSR.

TABLE 1

Average Times In Some Microwave Standards
(Microwave Range Only)

Country	τ (hour)	p_0 (mW/cm^2)	$p_0 \times \tau$ (mWhr/cm^2)
Australia* (1985)	1/60	1.0 Occ. 0.2 GP	1/60 1/300
Canada (1979)	0.1	1.0 Occ. 0.2 GP	0.1 0.02
IRPA	0.1	5.0 Occ. 1.0 GP	0.5 0.1
USSR** (1984)	8.0 24.0	0.025 Occ. 0.010 GP	0.2 0.24
Poland***	8.0 24.0	0.20 Occ. 0.01 GP	1.6 0.24

Comments

*Occupational exposures up to 5 p_o ceiling allowed for t<30 minutes but with 40 minute recovery required.

**These are true averaging times. The new USSR limits are on dose or energy.

***No true averaging time. Instead a formula $p^2xt=0.32$ is used for t<8.0 hours up to 10 mW/cm^2.

A review of this table not only illustrates the confusion among the world's standards-setters, but also some arbitrary rules unsupported by scientific rationale in the microwave range -- e.g., τ = 1/60 in Australia or the Polish formula of $p^2x\tau$ =0.32 (the latter resembles those used in small area optical exposure situations). It is hoped that standard-setters will follow a rationale based on human response to RF in establishing reasonable time averaging policy. In my opinion, that followed by Massachusetts or NCRP is superior to the rest, scientifically.

We have considered only the parameter, time. Safe exposure limits also are a function of frequency, degree of body exposed, and possibly many other factors like polarization parameters and instantaneous peak field levels. It is clear that complete detailed safety standards could be quite complicated. Since it is desirable to achieve as much agreement as possible throughout the world, what hope is there of this when there is so much disagreement even when focusing on one parameter, that of averaging time? I hope this presentation will evoke the different views held throughout the world and their rationales.

DISCUSSION   John M. Osepchuk, Reporter

Dr. Ross Adey pointed out observations in various fields that indicate intermittent exposures may be more effective than continuous exposure. For example, in the field of cancer chemotherapy and cancer modeling, with continuous exposure of an animal to cancer-promoting agents, cancer will not result. The limiting time cycle is approximately once a week and not months. Also commonly overlooked is the work by the Navy, 10 years ago, which reviewed clearly abnormal peripheral lymphoblasts in Navy personnel which resulted from intermittent exposures. If exposures were more than once every three days the effect was not seen, but with intermittent exposure every 72 hours, it was possible to see lymphoblasts in a significant number of these people.

"Therefore, the question of time constants cannot be discussed simply on the comfortable notions of Schwan and others that we're going to take a look at the question of where and how quickly the body heating occurs and how quickly it cools off. That is frivolous, trivial and irrelevant."

The moderator invited Dr. Adey to become involved in standards setting. He pointed out that after debate by biological scientists, whatever the resulting data base and models, scientists and engineers must translate this information into workable rules for limiting human exposure as a function of space, time, frequency, etc.

Dr. Korniewicz presented an overview on standards. He stated the original objective of standards as the protection of workers during 45 years of work from injury to the workers or their progeny. This is basic and must be enforced during a lifetime and not just for a few years. The threshold of highest permissible dose derives from ideas on maximum levels of exposure to chemicals. In the field of electromagnetics the analogous level is that of thermal effects. Time averaging is in many cases inappropriate - e.g., 6 minutes is too long for pulse fields.

The next idea is to consider the "dose" not just for 6 months but for a long period. In Poland there is a trend to use dose limits in RF exposure standards. It is the opinion in Poland that the approach with dose rate limits by IRPA and western countries is unwise.

Another concept accepted in Poland is the time for restitution. After 8 hours of exposure it is desirable to have at least 10 hours of no exposure before the next exposure. This is consistent with normal daily work schedules.

The resonance theory inherent in western standards is oversimplified according to Dr. Korniewicz. Though average SAR is higher at resonance the distribution of absorbed power is fairly uniform inside the body. At much higher frequencies, however, the SAR is concentrated in the skin and body averages are somewhat meaningless. Clearly microwave standards should match those of infrared at the highest microwave frequencies.

Finally Dr. Korniewicz made a plea for more international cooperation. He cited similar work in Canada, Poland and the U.S. on measuring to prevent RF shock and burns at low RF frequencies. Similar concepts of calculation and measurement of induced currents are involved.

The moderator expressed a belief that the URSI meeting is a useful step in developing international cooperation. He added that modification of averaging time is an important task when developing microwave standards to match those of infrared at 300 GHz.

Dr. Gandhi was then asked to report on work in the U.S. to develop more realistic H-field limits at low frequency and rules for prevention of RF shock and burns. Dr. Gandhi referred to a chart describing various options under consideration by the EPA in the U.S. This was presented by Ric Tell at the 1987 meeting of Subcommittee and of the Associated Standards Committee C95 and is reproduced in Figure 3.

The options considered by EPA basically refer to either that of C95, one fifth of C95, or one tenth of C95. Ric Tell had used data from Gandhi to deduce the lower limits for E shown in Figure 3. Those are required to insure a limit of 50mA for induced currents in the general population subject to possible contact with large conducting objects present in environmental fields.

Dr. Gandhi then reviewed his own proposals for revised occupational limits, also presented at the 1987 C95 meeting and reproduced in Figure 4. One can see that SAR theory permits large relaxation in H-field limits at low frequency. The E-field limit retains a somewhat arbitrary cap at low frequencies but this is desirable because of the relevance of nerve stimulation rather than just heating as the basis for exposure limits at low frequency.

The proposed E-field limit is not adequate to insure against RF shock and burns. Instead for occupational standard separate requirements for limiting induced currents to 50mA are included.

In the public setting the 50mA requirement dictates the lower E-field limit of 10 V/m for the range of 30 to 300MHz. The elevated H-field limit seems perfectly reasonable for both occupational or general public based on conservative ideas of induced current limits. It is not clear what H-field limits will evolve under EPA options, however.

In response to questions, Dr. Gandhi pointed out that E-field and induced current limits must be more stringent at very low frequencies where humans can detect lower current densities. While one question posed an apparent widening disparity between the U.S. and Eastern Europe in RF standards, Dr. Gandhi countered by pointing out the closeness of microwave oven emission limits and short duration (<10 minutes) exposure limits at microwave frequencies throughout the world.

Dr. Korniewicz agreed in part with Dr. Gandhi. He referred to the historical differences in the U.S. and the USSR. In the USSR work started in 1936 emphasizing acceptable exposure levels for an 8 hour work day. In the U.S. work started with Dr. Schwan and his 6 minute average of heating by microwaves. Thus the supposed 1000:1 difference is misleading because the Soviets applied a level for 8 hours and the U.S. for 6 minutes. In Poland the ideas of Schwan were accepted for 6 minutes and the ideas of the USSR for 8 hours. This has led all COMECON countries toward acceptance of dose limits per daily exposure.

Dr. Marian Stuchly of Canada expressed some confusion over Dr. Korniewicz's minimizing of differences between the U.S. and Eastern Europe. She asked for answers to the following:

1) Is there little difference in practice between the U.S. and the USSR?

2) How are the conservative Polish RF limits compatible with the high fields near Polish RF sealers reported in the literature.

Dr. Korniewicz acknowledged high H-fields near induction coils working at 0.1 to 1.0 MHz but these are below Polish limits. High E-fields can exist near sealers operating at 27 MHz but screening is used to reduce these fields at operator positions. With very large machines used to heat furniture this screening is difficult. In this case operator exposure time is limited.

An anonymous speaker (audience) then spoke up to insist that there remain large differences between the U.S. and USSR because U.S. limits apply not only for 6 minutes but for unlimited duration of exposure. He asked if standards-setters account for environmental temperatures and physical stress.

Dr. Mild offered the opinion that consideration of environmental temperature and humidity is generally unnecessary. It is a significant factor only in the case of exposure of "tower climbers".

Dr. Justesen then opined that environmental temperature and humidity undoubtedly are more important at the equator then in Sweden. In defense of the C95 concept of SAR limits for long duration exposure, he referred to animal (rats) data on lethal thresholds. These show that for the first one half hour that SA is the important criteria, whereas for times greater than one half hour SAR is the determining criterion. This agrees with the one half hour averaging times in the NCRP standard for exposure of the general public. He agreed that much more remains to be

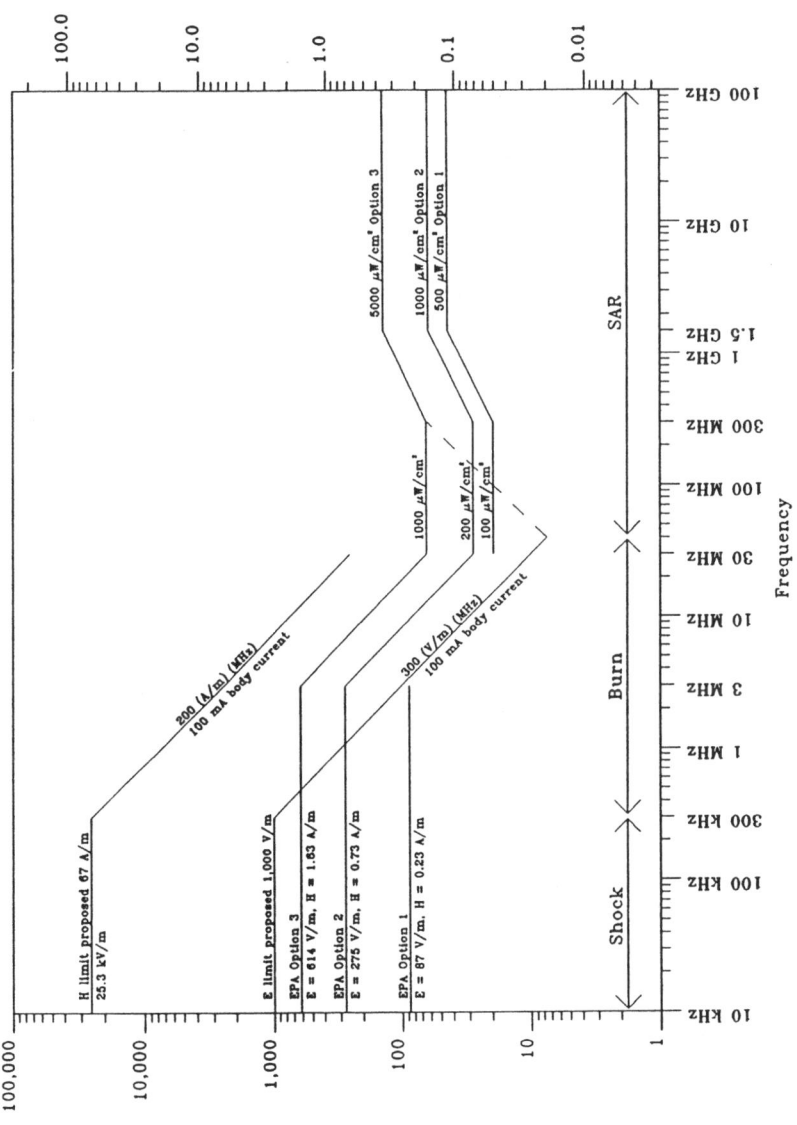

Figure 3. NPR Options and Proposed RF Shock/Burn Limits (US–EPA) 1987.

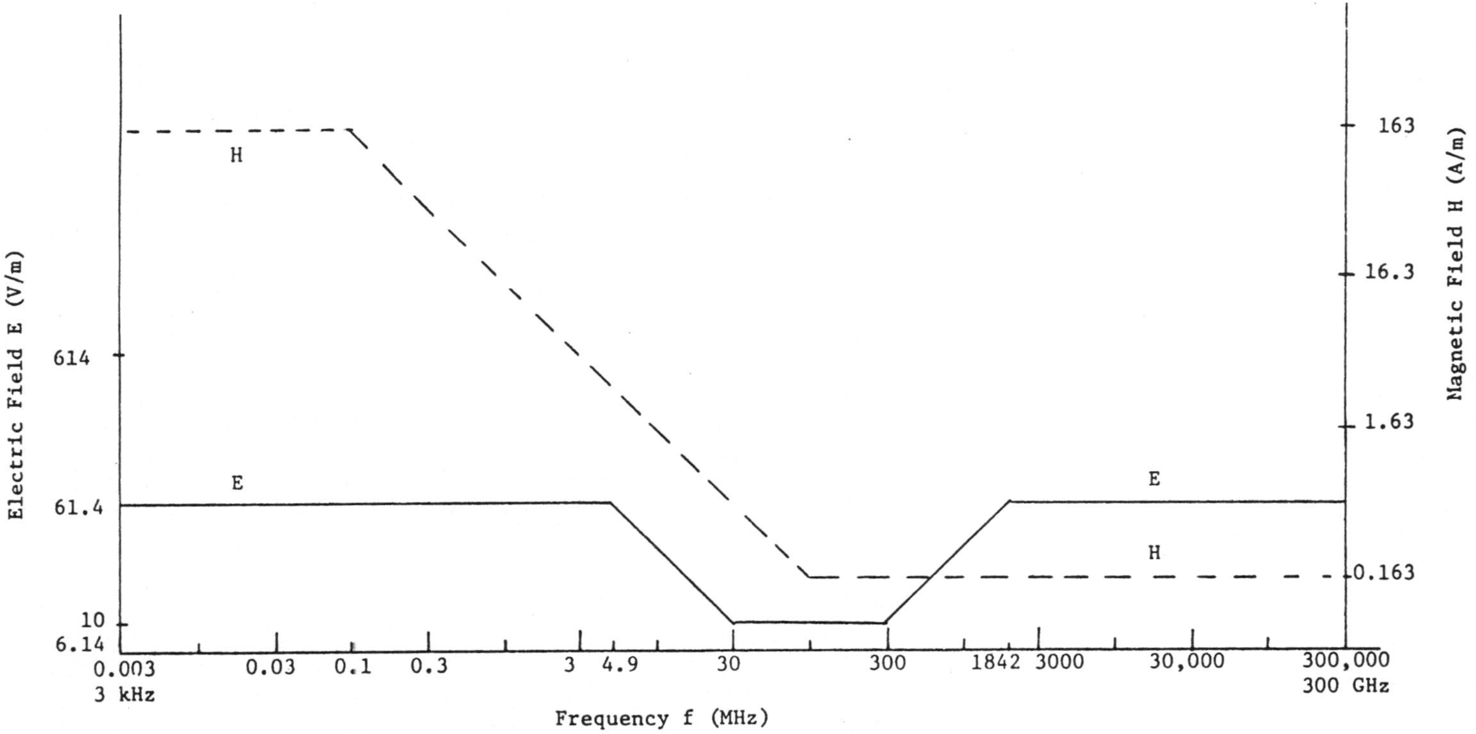

Figure 4. Radio-Frequency Protection Guide For Occupational Exposures. (Proposed by Gandhi, 1987).

288

done in support of improved standards-e.g., study of possible synergistic effects of RF exposure with other physical or chemical agents.

An anonymous speaker (audience) then opined that SA and SAR concepts may be irrelevant if recent indications on existence of "windows" in biological effects are established.

Dr. Lin commented that standards are developed for normal environments. It is understood that they should be modified for peculiar environments. One should not overreact to the criticism aired by Dr. Adey. Many factors go into standards-setting of which practicality is a leading example. A standard for man is not a standard for the rat. The published standards generally do not extensively describe the details of conservatism resulting when animal data are extrapolated to man.

Dr. Mild added that because of uncertainty in measurements (say $\pm$ 3dB), standards must incorporate some conservatism. The moderator added that this point has long been recognized. For example in the Soviet literature[3] allowance for even greater uncertainty ($\pm$ 6dB) is said to underlie the need for conservatism in standards.

In conclusion, the moderator recalled the review of many different factors during the discussion. This complexity perhaps partially explains the differences in various countries. Standards making is an imperfect process in a very imperfect world. Discussions at meetings such as those of URSI will continue the trend toward international cooperation if not unanimity.

REFERENCES

1. Tell, R.A. and Harlen, F., "A Review of Selected Biological Effects and Dosimetric Data Useful for Development of Radio Frequency Safety Standards for Human Exposure", J. Mic. Power 14 (1), pp.404-424, 1979.

2. U.S. Environmental Protection Agency, "Federal Radiation Protection Guidance; Proposed Alternatives for Controlling Public Exposure to Radio Frequency Radiation", Federal Register Vol. 51, No. 146 pp.27318-27339, July 30, 1986.

3. Minin, B.A. Microwaves and Human Safety; Moscow; U.S. Translation JPRS-65506-2, 1974.

CONTRIBUTORS

W. Ross Adey
Veterans Administration Medical
Center and Loma Linda University
School of Medicine
Loma Linda, CA 92357 USA

Paolo Bernardi
Department of Electronics
University of Rome "La Sapienza"
Via Eudossiana 18, 00184 Rome
Italy

J.H. Bernhardt
Institute for Radiation Hygiene
Neuherberg, Federal Republic of
Germany

B. Bocquet
Centre Hyperfrequences et
Semiconducteurs
Universite des Sciences et
Techniques de Lille
Flandres Artois
59655 Villeneuve D'Ascq Cedex
France

Huai Chang
Microwave Research Laboratory
Zhejiang Medical University
Hangzhou, 310006, China

P. Czerski
Center for Devices and
Radiological Health
Food and Drug Administration
PHS, DHHS
Rockville, MD USA

Guglielmo D'Inzeo
Department of Electronics
University of Rome "La Sapienza"
Via Eudossiana 18, 00184 Rome
Italy

J.W. Hand
Medical Research Council Cyclotron
Unit
Hammersmith Hospital
Ducane Raod
London W12 U.K.

Don Justesen
Medical Research Laboratories (151)
USVA Medical Center
Kansas City, MO 64128 USA

Y. Leroy
Centre Hyperfrequences et
Semiconducteurs
Universite des Sciences et
Techniques de Lille
Flandres Artois
59655 Villeneuve D'Ascq Cedex
France

James C. Lin
Department of Bioengineering
University of Illinois
Chicago, IL 60680-4348 USA

A. Mamouni
Centre Hyperfrequences et
Semiconducteurs
Universite des Sciences et Techniques
de Lille Flandres Artois
59655 Villeneuve D'Ascq Cedex
France

Kjell Hansson Mild
National Institute of
Occupational Health
Box 6104, S-900 06 UMEA, Sweden

John M. Osepchuk
Raytheon Research Division
Lexington, MA 02173 USA

Binjie Shao
Microwave Research Laboratory
Zhejiang Medical University
Hangzhou, 310006, China

Maria A. Stuchly
Bureau of Radiation and Medical
Devices
Health and Welfare Canada
Ottawa, Ontario K1A OL2, Canada

Stanislaw Szmigielski
Department of Biological Effects
of Non-ionizing Radiations
Center for Radiobiology and
Radiation Safety
128 Szaserow, 00-909 Warsaw
Poland

T.S. Tenforde
Research Medicine and Radiation
Biophysics Division
Lawrence Berkeley Laboratory
Univeristy of California
Berkeley, CA 94720 USA

Present Address:

Life Sciences Center (K4-14)
Battelle Pacific Northwest
Laboratories
P.O. Box 999
Richland, WA 99352

J.C. Van De Velde
Centre Hyperfrequences et
Semiconducteurs
Universite des Sciences et
Techniques de Lille
Flandres Artois
59655 Villeneuve D'Ascq Cedex
France

Hermann Weiss
Philips GmbH Forschungslaboratorium
Hamburg
Vogt Koellnstrasse 30
D 2000 Hamburg 54, West Germany

INDEX